WILLIAM F. MAAG LIBRARY
YOUNGSTOWN STATE UNIVERSITY

ELECTROANALYTICAL CHEMISTRY

A SERIES OF ADVANCES

edited by

Allen J. Bard
Department of Chemistry
University of Texas
Austin, Texas

Israel Rubinstein
Department of Materials
and Interfaces
Weizmann Institute of Science
Rehovot, Israel

VOLUME 20

Marcel Dekker, Inc.   New York • Basel • Hong Kong

The Library of Congress Cataloged the First Issue of This Title as Follows:

Electroanalytical chemistry: a series of advances, v. 1
  New York, M. Dekker, 1996–
    v. 23 cm
    Editors: 1966–1995   A.J. Bard
            1996–   A.J. Bard and I. Rubinstein
    1. Electromechanical analysis—Addresses, essays, lectures
    1. Bard, Allen J., ed.
QD115E499   545.3   66-11287
Library of Congress
0-8247-9996-8 (v. 20)

The publisher offers discounts on this book when ordered in bulk quantities. For more information, write to Special Sales/Professional Marketing at the address below.

This book is printed on acid-free paper.

Copyright © 1998 by MARCEL DEKKER, INC. All Rights Reserved.

Neither this book nor any part may be reproduced or transmitted in any form or by any means, electronic or mechanical, including photocopying, microfilming, and recording, or by any information storage and retrieval system, without permission in writing from the publisher.

MARCEL DEKKER, INC.
270 Madison Avenue, New York, New York 10016
http://www.dekker.com

Current printing (last digit):
10 9 8 7 6 5 4 3 2 1

PRINTED IN THE UNITED STATES OF AMERICA

INTRODUCTION TO THE SERIES

This series is designed to provide authoritative reviews in the field of modern electroanalytical chemistry defined in its broadest sense. Coverage is comprehensive and critical. Enough space is devoted to each chapter of each volume so that derivations of fundamental equations, detailed descriptions of apparatus and techniques, and complete discussions of important articles can be provided, so that the chapters may be useful without repeated reference to the periodical literature. Chapters vary in length and subject area. Some are reviews of recent developments and applications of well-established techniques, whereas others contain discussion of the background and problems in areas still being investigated extensively and in which many statements may still be tentative. Finally, chapters on techniques generally outside the scope of electroanalytical chemistry, but which can be applied fruitfully to electrochemical problems, are included.

Electroanalytical chemists and others are concerned not only with the application of new and classical techniques to analytical problems but also with fundamental theoretical principles upon which these techniques are based. Electroanalytical techniques are proving useful in such diverse fields as electro-organic synthesis, fuel, and radical ion formation, as well as with such problems as the kinetics and mechanisms of electrode reactions, and the effects of electrode surface phenomena, adsorption, and the electrical double layer on electrode reactions.

It is hoped that the series is proving useful to the specialist and non-specialist alike—that it provides a background and a starting point for graduate students undertaking research in the areas mentioned, and that it also proves valuable to practicing analytical chemists interested in learning about and applying electroanalytical techniques. Furthermore, electrochemists and industrial chemists with problems of electrosynthesis, electroplating, corrosion, and fuel cells, as well as other chemists wishing to apply electrochemical problems, may find useful material in these volumes.

A. J. B.
I. R.

CONTRIBUTORS TO VOLUME 20

ROSE A. CLARK   Pennsylvania State University, University Park, Pennsylvania

ROBERT M. CORN   University of Wisconsin–Madison, Madison, Wisconsin

ANDREW G. EWING   Pennsylvania State University, University Park, Pennsylvania

BRIAN L. FREY   University of Wisconsin–Madison, Madison, Wisconsin

DENNIS G. HANKEN   University of Wisconsin–Madison, Madison, Wisconsin

CLAIRE E. JORDAN   University of Wisconsin–Madison, Madison, Wisconsin

STUART LICHT   Technion Israel Institute of Technology, Haifa, Israel

BIRGIT MEYER   Institute of Chemistry, Humboldt University, Berlin, Germany

FRITZ SCHOLZ   Institute of Chemistry, Humboldt University, Berlin, Germany

SUSAN E. ZERBY   Pennsylvania State University, University Park, Pennsylvania

CONTENTS OF VOLUME 20

Introduction to the Series	iii
Contributors to Volume 20	v
Contents of Other Volumes	ix

VOLTAMMETRY OF SOLID MICROPARTICLES IMMOBILIZED ON ELECTRODE SURFACES
Fritz Scholz and Birgit Meyer   1

I.	Introduction	2
II.	A Survey of Methods Used for Electrochemical Studies of Solid Materials	4
III.	Voltammetry of Microparticles of Solid Compounds Immobilized on an Electrode Surface	9
IV.	Conclusions	80
	References	82

ANALYSIS IN HIGHLY CONCENTRATED SOLUTIONS: POTENTIOMETRIC, CONDUCTANCE, EVANESCENT, DENSOMETRIC, AND SPECTROSCOPIC METHODOLOGIES
Stuart Licht   87

I.	Introduction	88
II.	Potentiometric Analysis	89
III.	Evanescent (Solvent) Activity Analysis	103
IV.	Conductometric Analysis of Concentrated Solutions	113
V.	Differential Densometric Analysis	121
VI.	Submicron Path UV/VIS/IR Spectroscopy	127
VII.	Final Comments	137
	References	137

SURFACE PLASMON RESONANCE MEASUREMENTS
OF ULTRATHIN ORGANIC FILMS AT ELECTRODE
SURFACES
Dennis G. Hanken, Claire E. Jordan, Brian L. Frey, and
Robert M. Corn   141

I.	Introduction	142
II.	Background	143
III.	SPR Measurements of Monolayer Thickness on Gold Surfaces	148
IV.	SPR Imaging Experiments	179
V.	SPR Electric Field Measurements	195
VI.	Future Directions	217
	References	218

ELECTROCHEMISTRY IN NEURONAL
MICROENVIRONMENTS
Rose A. Clark, Susan E. Zerby, and Andrew G. Ewing   227

I.	Introduction	228
II.	Electrochemistry: Methods and Electrodes	234
III.	Intracellular Voltammetry	252
IV.	Extracellular Voltammetry	259
V.	Concluding Remarks and Other Directions	286
	References	287

Author Index   295

Subject Index   313

CONTENTS OF OTHER VOLUMES

VOLUME 1

AC Polarograph and Related Techniques: Theory and Practice, Donald E. Smith
Applications of Chronopotentiometry to Problems in Analytical Chemistry, Donald G. Davis
Photoelectrochemistry and Electroluminescence, Theodore Kuwana
The Electrical Double Layer, Part I: Elements of Double-Layer Theory, David M. Mohilner

VOLUME 2

Electrochemistry of Aromatic Hydrocarbons and Related Substances, Michael E. Peover
Stripping Voltammetry, Embrecht Barendrecht
The Anodic Film on Platinum Electrodes, S. Gilaman
Oscillographic Polarography at Controlled Alternating Current, Michael Heyrovksy and Karel Micka

VOLUME 3

Application of Controlled-Current Coulometry to Reaction Kinetics, Jiri Janata and Harry B. Mark, Jr.
Nonaqueous Solvents for Electrochemical Use, Charles K. Mann
Use of the Radioactive-Tracer Method for the Investigation of the Electric Double-Layer Structure, N. A. Balashova and V.E. Kazarinov
Digital Simulation: A General Method for Solving Electrochemical Diffusion-Kinetic Problems, Stephen W. Feldberg

VOLUME 4

Sine Wave Methods in the Study of Electrode Processes, Margaretha Sluyters-Rehbach and Jan H. Sluyters

The Theory and Practice of Electrochemistry with Thin Layer Cells, A. T. Hubbard and F. C. Anson

Application of Controlled Potential Coulometry to the Study of Electrode Reactions, Allen J. Bard and K. S. V. Santhanam

VOLUME 5

Hydrated Electrons and Electrochemistry, Geraldine A. Kenney and David C. Walker

The Fundamentals of Metal Deposition, J. A. Harrison and H. R. Thirsk

Chemical Reactions in Polarography, Rolando Guidelli

VOLUME 6

Electrochemistry of Biological Compounds, A. L. Underwood and Robert W. Burnett

Electrode Processes in Solid Electrolyte Systems, Douglas O. Raleigh

The Fundamental Principles of Current Distribution and Mass Transport in Electrochemical Cells, John Newman

VOLUME 7

Spectroelectrochemistry at Optically Transparent Electrodes; I. Electrodes Under Semi-infinite Diffusion Conditions, Theodore Kuwana and Nicholas Winograd

Organometallic Electrochemistry, Michael D. Morris

Faradaic Rectification Method and Its Applications in the Study of Electrode Processes, H. P. Agarwal

VOLUME 8

Techniques, Apparatus, and Analytical Applications of Controlled-Potential Coulometry, Jackson E. Harrar

Streaming Maxima in Polarography, Henry H. Bauer

Solute Behavior in Solvents and Melts, A Study by Use of Transfer Activity Coefficients, Denise Bauer and Mylene Breant

Contents of Other Volumes

VOLUME 9

Chemisorption at Electrodes: Hydrogen and Oxygen on Noble Metals and their Alloys, Ronald Woods

Pulse Radiolysis and Polarography: Electrode Reactions of Short-lived Free Radicals, Armin Henglein

VOLUME 10

Techniques of Electrogenerated Chemiluminescence, Larry R. Faulkner and Allen J. Bard

Electron Spin Resonance and Electrochemistry, Ted M. McKinney

VOLUME 11

Charge Transfer Processes at Semiconductor Electrodes, R. Memming

Methods for Electroanalysis In Vivo, Jiří Koryta, Miroslav Brezina, Jiří Pradáč, and Jarmila Pradáčová

Polarography and Related Electroanalytical Techniques in Pharmacy and Pharmacology, G. J. Patriarche, M. Chateau-Gosselin, J. L. Vandenbalck, and Petr Zuman

Polarography of Antibiotics and Antibacterial Agents, Howard Siegerman

VOLUME 12

Flow Electrolysis with Extended-Surface Electrodes, Roman E. Sioda and Kenneth B. Keating

Voltammetric Methods for the Study of Adsorbed Species, Etienne Laviron

Coulostatic Pulse Techniques, Herman P. van Leeuwen

VOLUME 13

Spectroelectrochemistry at Optically Transparent Electrodes, II. Electrodes Under Thin-Layer and Semi-infinite Diffusion Conditions and Indirect Coulometric Iterations, William H. Heineman, Fred M. Hawkridge, and Henry N. Blount

Polynomial Approximation Techniques for Differential Equations in Electrochemical Problems, Stanley Pons

Chemically Modified Electrodes, Royce W. Murray

VOLUME 14

Precision in Linear Sweep and Cyclic Voltammetry, Vernon D. Parker

Conformational Change and Isomerization Associated with Electrode Reactions, Dennis H. Evans and Kathleen M. O' Connell

Square-Wave Voltammetry, Janet Osteryoung and John J O'Dea

Infrared Vibrational Spectroscopy of the Electron-Solution Interface, John K. Foley, Carol Korzeniewski, John L. Dashbach, and Stanley Pons

VOLUME 15

Electrochemistry of Liquid-Liquid Interfaces, H. H. J. Girault and D. J. Schiffrin

Ellipsometry: Principles and Recent Applications in Electrochemistry, Shimson Gottesfeld

Voltammetry at Ultramicroelectrodes, R. Mark Wightman and David O. Wipf

VOLUME 16

Voltammetry Following Nonelectrolytic Preconcentration, Joseph Wang

Hydrodynamic Voltammetry in Continuous-Flow Analysis, Hari Gunasingham and Bernard Fleet

Electrochemical Aspects of Low-Dimensional Molecular Solids, Michael D. Ward

VOLUME 17

Applications of the Quartz Crystal Microbalance to Electrochemistry, Daniel A. Buttry

Optical Second Harmonic Generation as an In Situ Probe of Electrochemical Interfaces, Geraldine L. Richmond

New Developments in Electrochemical Mass Spectroscopy, Barbara Bittins-Cattaneo, Eduardo Cattaneo, Peter Königshoven, and Wolf Vielstich

Carbon Electrodes: Structural Effects on Electron Transfer Kinetics, Richard L. McCreery

VOLUME 18

Electrochemistry in Micelles, Microemulsions, and Related Microheterogeneous Fluids, James F. Rusling

Mechanism of Charge Transport in Polymer-Modified Electrodes, György Inzelt

Scanning Electrochemical Microscopy, Allen J. Bard, Fu-Ren F. Fan, and Michael V. Mirkin

VOLUME 19

Numerical Simulation of Electroanalytical Experiments: Recent Advances in Methodology, Bernd Speiser

Electrochemistry of Organized Monolayers of Thiols and Related Molecules on Electrodes, Harry O. Finklea

Electrochemistry of High-T_c Superconductors, John T. McDevitt, Steven G. Haupt, and Chris E. Jones

VOLTAMMETRY OF SOLID MICROPARTICLES IMMOBILIZED ON ELECTRODE SURFACES

Fritz Scholz and Birgit Meyer

Institute of Chemistry, Humboldt University,
Berlin, Germany

I. Introduction   1
II. A Survey of Methods Used for Electrochemical Studies of Solid Materials   4
   A. Compact electrodes made of solid material   4
   B. Carbon paste electrodes with an organic binder   5
   C. Carbon paste electrodes with an electrolytic solution serving as a binder   6
   D. Suspensions of solid particles   7
   E. Voltammetry of solid particles immobilized on an electrode in a polymer film   8
   F. Voltammetry of solid compounds sandwiched between solid electrodes   8
III. Voltammetry of Microparticles of Solid Compounds Immobilized on an Electrode Surface   9
   A. Immobilization of solid particles on an electrode surface   10
   B. Mechanisms of electrochemical reactions of immobilized solid particles   14
   C. Quantitative analysis of the composition of solids   34
   D. Phase analysis   44
   E. Structure analysis   52
   F. Determination of thermodynamic data of solid compounds   64
   G. In situ and ex situ combinations of the voltammetry of solid microparticles with other analytical techniques   72
   H. Problems of a theoretical description of the voltammetry of microparticles   74
Conclusions   80
References   82

I. INTRODUCTION

This chapter is devoted to the discussion of a novel technique serving to study the electrochemistry of solid materials. But what is solid state electrochemistry (SSE) and what kind of information can it provide? The term "solid state electrochemistry" has been mainly used in conjunction with solid electrolytes [1,2]. However, in a strict sense, this term should be applied to all electrochemical reactions in which at least one solid phase is involved. A more pragmatic approach is to consider solid state electrochemistry as the science that investigates the role and the fate of solid phases in the course of electrochemical reactions. This more comprehensive understanding of solid state electrochemistry has now become widely accepted [3]. Aspects of solid state electrochemistry are involved in almost all fields where electrochemistry is applied. Examples are simple galvanic cells like the Leclanché element, accumulators such as lead batteries, electrochromic devices, galvanic plating and surface treatments, electrochemical machining, and many stripping voltammetric analyses, only to name a few. Indeed, such cases are rather rare where there are no solid phases involved in electrochemical reactions. This is the case in polarography of dissolved compounds using mercury electrodes as well as when ions are transferred between immiscible electrolyte solutions with the help of an externally applied potential difference.

A remarkable increase in the number of publications, including books and reviews [3–8], and even a new journal* bears witness to the growing importance of solid state electrochemistry in contemporary science.

In the past the electrochemistry of solid phases was extensively studied mainly in the following areas: primary and secondary batteries, electrochemical synthesis, and corrosion. Despite corrosion studies [9], it was only in a comparably small number of publications that methods were described using electrochemical measurements to obtain *analytical information* about a solid.

The relatively rare use of electrochemical methods for the *analysis* of solid compounds can be attributed to the fact that it is difficult to carry out electrochemical measurements on solid phases. Developments in the field of solid state electrochemistry have certainly been discouraged by the great number of studies that indicate how much the results of electrochem-

---

Journal of Solid State Electrochemistry, Springer, 1997.

Voltammetry of Microparticles

ical measurements depend on the state of the surface studied, its crystallographic orientation, its chemical or electrochemical pretreatment, the adsorption of foreign compounds, and other aspects. This led to the popular belief that information on solids can only be obtained from extremely carefully prepared surfaces requiring procedures which, in most cases, can only be followed in specialized laboratories.

This is a regrettable situation in that electrochemical reactions of solid phases have a high potential for providing important information on solids. Thus, it is possible, in principle, to analyze the elemental composition of a solid phase, study its behavior in electrochemical and chemical reactions, determine thermodynamic and kinetic data of these reactions, and, finally, obtain information on the bonding of elements in the solid phase. *Spectroscopy* mirrors the changes that take place in a compound upon energetic excitation that is usually insufficient to cause a chemical transformation. Therefore, the compound remains unchanged after relaxation and spectroscopy will provide information on the ground and excited states of the compound. *Electrochemistry* uses energies that initiate reactions, i.e., one compound is transformed into another. The thermodynamics of these reactions depends on Gibbs free energies of reactants and products, and the rates of these reactions depend on reaction pathways and activation barriers between the reactants and products. Hence, electrochemical measurements provide information on *reactions* of compounds. Since the reactivity of a compound always depends on its composition and structure, the respective information can be deduced from electrochemical measurements.

Having been developed much later than solution electroanalysis, solid state electroanalysis has only recently become an analytical tool. This is because of its more complicated nature, from both an experimental and a theoretical point of view. In this respect, electrochemistry follows a similar pattern as spectroscopy where solid state techniques developed later than solution techniques, i.e., in NMR, IR, and UV-Vis spectroscopies.

During the past 20–30 years, a number of different electrochemical techniques have been published which can be applied to studies of solid materials. These techniques will be briefly reviewed in the following section. The main part of this chapter, however, is devoted to a detailed description of a novel technique used to study the electrochemistry of solid microparticles immobilized on the surface of an electrode. This technique was introduced by F. Scholz in 1989 [10,11]. With the respective reviews

[6,7,12] being either not up-to-date or not comprehensive, it seems to be the proper time for a detailed description. This novel technique is a very worthy addition to the established techniques as it allows electrochemical studies to be performed with extremely small sample amounts (down to 10^{-12} mol), with single microparticles, and, finally, even with insulator particles. The use of minimized sample amounts makes it possible to explore much higher scan rates than in the case of compact, large electrodes of the sample material.

II. A SURVEY OF METHODS USED FOR ELECTROCHEMICAL STUDIES OF SOLID MATERIALS

This section gives a short survey of the different techniques that have been developed to study the electrochemistry of solid compounds. It is the aim of this chapter to explain how to bring a solid compound into an electric field where its electrochemical reactions can be studied.

A. Compact Electrodes Made of Solid Material

The easiest way to analyze solid compounds using electrochemical techniques is to use the solid as a compact electrode. A sample with sufficient electrical conductivity is needed, hence, this method is restricted to metals, alloys, and semiconductors. Compact electrodes have been extensively studied to assess their electrochemical corrosion properties [9].

For a review of the electroanalysis of solid materials used as compact electrodes, see Refs. 8 and 13 to 15. In addition to laborious electrode preparation preventing broader application, this technique presents several problems. The currents obtained with compact electrodes are very high due to high sample amounts. This leads to iR-distorted voltammograms with a poor signal resolution, making a multicomponent analysis difficult or even impossible to attain. Additionally, electrochemical reactions often cause a coverage of the electrode surface.

To avoid some of the previously mentioned problems, especially with regard to time-consuming electrode preparation, so-called *pressed cells* have been developed [16,17]. These are electrochemical cells without a base. A rubber ring is placed around the missing base. This cell is pressed onto the substance to be analyzed and filled with electrolyte solution.

B. Carbon Paste Electrodes with an Organic Binder

One way to minimize the disadvantages of solid electrodes is the use of carbon paste electrodes. Carbon paste electrodes are made of high purity graphite powder mixed with organic binders such as Nujol or silicone oil to form a paste. In *modified* carbon paste electrodes, this paste also contains small amounts of the solid substance to be analyzed. Therefore, the sample should be finely ground and thoroughly mixed with the carbon paste. The advantages of this technique as compared to compact electrodes are as follows:

1. There is no need for the sample to have a high electrical conductivity because the conduction is mainly accomplished by the graphite.
2. The currents obtained are within a suitable range due to the small sample amounts. This means that signal resolution is improved. Additionally, surface coverage during electrode reactions is rarely observed.
3. This technique helps in investigating substances in such cases where only small sample amounts are available and no compact electrodes can be manufactured. The electrochemical reactions proceed at the surface of the paste electrode. At this location, an ion transfer between the solid sample and the electrolyte solution is possible.

The disadvantages of the technique mainly concern the use of organic binders. The respective voltammograms often show a higher irreversibility than in cases where no organic binders are used.

Kuwana and French [18] introduced this technique to perform voltammetric measurements on water-insoluble organic compounds, such as ferrocene or anthraquinone. The technique of modified paste electrodes was extensively developed in the former Soviet Union by W. G. Barikov, O. A. Songina, N. F. Zakharchuk, and K. Z. Brainina. A number of applications have been described for the quantitative analysis of powder mixtures, e.g., Ag_2O-AgO, cubic and hexagonal In_2O_3, and iron magnetite-wüstite mixtures [8]. A very convincing application of modified paste electrodes has been the determination of the composition of thin oxide films (2–100 nm), which are formed on GaAs by anodic or thermal oxidation [19]. The thin film was mechanically removed with the help of diamond powder, and the resulting mixture of film material and diamond powder was attached to the surface of a graphite paste electrode before starting voltammetric measurements. The following phases have been detected: α-, β-, γ-, δ-, ε-Ga_2O_3, amorphous $Ga(OH)_3$, claudetite and arsenolite (both As_2O_3), α- and β-Sb_2O_3, As_2O_4, As_2O_5, Sb_2O_5, $GaAsO_4$, As and Sb (amorphous and crystalline), and Ga. At the time of publication, it has not been possible to make similar analyses by using spectroscopic techniques such as ESCA (Electron Spectroscopy for Chemical Analysis) or Auger electron spectroscopy. In a recent paper, the fundamentals and limitations of modified graphite paste electrodes were critically discussed with respect to the influence of particle sizes on the reproducibility of results [20]. The influence of the organic binder on the electrochemistry of solid compounds has not yet been fully elucidated; however, it was shown [21] that the organic compound is able to at least partly obscure the faradaic reactions of the solid compound since the hydrophobic binders spread themselves on the surface of the graphite and sample particles, and, in this way, they can inhibit the electrochemical reaction.

The use of modified carbon paste electrodes as tailored electrodes for solution analyses [22] became very popular due to the fact that they can be made more easily and more reliably than surface modified electrodes.

C. Carbon Paste Electrodes with an Electrolyte Solution Serving as a Binder

This kind of carbon paste electrode is prepared using an electrolytic binder instead of an organic binder. Therefore, the electrochemical reaction can, in principle, proceed throughout the whole paste and not only at the paste-electrolyte interface. However, the iR drop within the paste will restrict the reaction to a certain paste layer adjacent to the electrolyte solution. In order to obtain stable electrodes, the electrode paste must be poured into small cups, with the upper surface being exposed to the electrolyte solution, which has the same composition as the electrolytic paste binder. Bauer and Gaillochet [23] studied the behavior of these electrodes with respect to such parameters as sweep rate, paste volume, and concentration of sample in the paste. They proved that the solid itself can be involved in the electrochemical processes. Lamache and Bauer [24] considered the oxidation of Cu_2S in 1 M H_2SO_4 at very low scan rates and found that the oxidation proceeds in steps forming copper sulfides of different stoichiometry. Some of these sulfides are minerals occurring in nature. Chouaib et al. [25] studied the behavior of different manganese oxides, enabling them to distinguish different modifications resulting from different electrochemical reactions. Eguren et al. [26] proposed a method for the determination of the β-SnO_2 contents in commercial tin dioxide. A comprehensive review on carbon paste electrodes with electrolytic binders was published by Batanero et al. [27]. These paste electrodes do not suffer from an undesirable influence of organic compounds on the electrochemical reactions. However, the electrolytic binder has to be chosen very carefully as aggressive binders like mineral acids or bases are able to dissolve the solid compound chemically before the electrochemistry starts, thus making erratic results possible.

D. Suspensions of Solid Particles

Kolthoff and Stock were the first to perform voltammetry of suspended silver bromide [28]. Micka later observed specific signals in the polarography of suspended charcoal. Thereafter, he and Kalvoda systematically studied the polarography of various solid compounds as suspensions in electrolyte solution [29,30]. To avoid a precipitation of particles, their suspensions had to be stirred. But this stirring led to very noisy polarograms, especially when a dropping mercury electrode was applied; however, well-defined voltammograms were obtained using a rotating disc electrode, as this type of electrode provides very stable hydrodynamic conditions. Micka's early results point to the fact that the voltammetric response of suspensions is connected with the point of zero charge (p.z.c.) insofar as the suspensions gave peak-shaped current signals near to the p.z.c. Micka's interpretation was that the peak-shaped curves are due to an adsorption-like behavior of the suspended particles on the electrode surface and that the adsorption of particles reaches its maximum at the p.z.c. Dausheva and Songina [31] showed that suspended particles of mercury iodide are strongly bound to the surface of a hanging mercury drop around the p.z.c. Applying a theory of Frumkin on the potential dependence of the stability of surface films, they derived the following conclusion: At the p.z.c., the electrolyte film between the electrode surface and the suspended particles has such an instability that the particles can come into closest contact with the electrode and are really deposited on the surface by electrocapillary forces. They call this phenomenon electrocapillary deposition. This deposition ensures that the particles come into such close contact that electrons can be transferred between them and the electrode. At potentials far away from the p.z.c., the stability of the electrolyte film is so high that even a forced transport of the particles to the electrode surface by stirring the suspension cannot bring the particles into close cntact with the electrode.

Decreasing the particle size of the suspended material will eventually lead to colloidal solutions. The electrochemistry of colloids possesses features of both solutions and solids. This has been shown recently by M. Heyrovský et al. [32–35], who studied the electrochemistry, especially polarography, of colloidal tin(IV), titanium(IV), and mixed iron(III)-titanium(IV) oxides. Since the electrochemistry of colloids is a research area in its own right [36], and since the preparation of colloids is not well suited to give access to the electrochemistry of the parent solids, we shall not pay any further attention to colloids in this chapter.

E. Voltammetry of Solid Particles Immobilized on an Electrode in a Polymer Film

The electrochemical behavior of polymer films on electrodes has been studied on a broad scale [37,38], and the term solid-state voltammetry has been used for those systems in which redox active sites are incorporated in the polymer film [39].

Particles of solid compounds have been immobilized on the surface of a solid electrode using a polymeric binder [15]. Franklin et al. [40] described the voltammetry of suspended solid compounds in cationic surfactant–styrene–aqueous sodium hydroxide emulsions using platinum electrodes. Under these conditions a hydrophobic polymeric film is formed on the electrode surface, which has been made responsible for an enlarged potential window for measurements. Solid particles are held on the electrode surface by adsorption of surfactants embedding these particles. From the literature, it is not entirely clear what the influence of the polymer film on the electrochemistry of solid particles is. It is possible that the polymer film can have a negative effect on the voltammograms, but this has not yet been proved.

F. Voltammetry of Solid Compounds Sandwiched Between Solid Electrodes

Kulesza et al. [41] reported that solid compounds that possess ion conductivity and mixed valence sites give voltammetric signals when sandwiched between two solid electrodes. In these systems no deliberately added electrolyte solution is present. The degree of hydration strongly influences the voltammetric response because a certain water content is essential for ion mobility. This kind of voltammetry has been reported only for metal hexacyanoferrates and solid heteropoly compounds [42–44] since the described technique is limited to compounds that possess both sufficient electronic and ionic conductivity.

III. VOLTAMMETRY OF MICROPARTICLES OF SOLID COMPOUNDS IMMOBILIZED ON AN ELECTRODE SURFACE

Studying the electrochemistry of a solid compound requires solving two major experimental problems: The first is that a solid compound may provide or consume large amounts of electric charge because the electrochemically active constituents are concentrated in a small volume. This is contrary to the case of diluted or highly diluted solutions, as they are studied in solution electrochemistry. The resulting large currents can lead to undesirable iR distortions and bad signal resolution. The second problem may be a very low electrical conductivity. In this case it is not easy to apply a voltage across the interface of the insulator with an electrolyte solution. A technique solving both of these problems must employ very small amounts of a solid compound and bring these small amounts as closely into an electric field as possible.

The technique where microparticles of a solid compound are embedded into the surface of a suitable solid electrode meets these two requirements to a considerable extent. An important feature of this technique is that reproducible measurements can be performed with such immobilized microparticles, which reflect the electrochemistry of the *bulk* of the microparticles. Only in such cases where macroelectrodes of the studied material exhibit surface phenomena (e.g., electrochemical passivation) can this also be seen on microparticles. In these studies the particles have sizes in the μm range or below.

In the first publications regarding this new technique [10,11,45,46], immobilized particles were dissolved in many cases by electrochemical reactions. Therefore, the term "abrasive stripping voltammetry" was introduced to refer to an abrasive transfer step and the electrochemical stripping step. Many electrochemical reactions of immobilized microparticles were later studied, which are not associated with a dissolution process. Hence, it is more reasonable to use the general term *voltammetry of microparticles* (VMP).

A. Immobilization of Solid Particles on an Electrode Surface

A simple way to immobilize solid particles on the surface of an electrode is by embedding them mechanically in a soft electrode surface. This has the great advantage that no chemical bonding or alteration of the nature of the solid particle or its surface is required. Abrasion from solids was the technique chosen when the method was introduced. Later, a number of other immobilization techniques was developed.

1. Electrodes

Paraffin-Impregnated Graphite Electrodes

For the immobilization of solid particles, different electrodes were tested in our and other laboratories, but the initial choice of paraffin-impregnated graphite rods still seems to be the most successful [10,11]. These *paraffin-impregnated graphite electrodes* (PIGEs) are fabricated from graphite rods, which are used in spark spectroanalysis. These rods may have different sizes, but those 50 mm long and 5 mm in diameter have the best handling properties. As graphite rods are microporous, the pores would lead to a high background current due to an ingress of electrolyte solution into the graphite rods. Of course, such a penetration is also undesired as it can lead to a contamination of the electrodes with solution constituents. Therefore, impregnation of these graphite rods with solid paraffin is suggested.

To prepare PIGEs, paraffin with a melting point between 56 and 70°C is melted in a closed vessel in a water bath. The graphite rods are given to the melt and the vessel is evacuated. The impregnation is performed until no more gas bubbles evolve from the rods. Usually, this is the case within a time span of 2–4 hours. After ambient pressure has been established, the rods are removed from the melt before the paraffin solidifies. The warm electrodes are put onto filter paper to allow their cooling and drying. The lower end of the electrode is carefully polished on smooth paper. This surface will later accommodate the solid particles to be studied. The electrodes are now ready for use. For easy cleaning, the PIGE is used without an insulation of its cylindrical surface, as will be described below. The upper end of the electrode is connected with the electrochemical instrument by a crocodile clip.

Basal Plane Pyrolytic Graphite Electrode

These commercially available graphite electrodes are also very well suited for VMP as they are not penetrated by the electrolyte solution [47–49].

These electrodes can be cleaned after measurements by removing the upper surface layer with a razor blade. A comparison between the electrochemical behavior of solid compounds immobilized on this type of electrode and of those immobilized on PIGEs did not reveal any differences.

Glassy Carbon Electrodes

Glassy carbon electrodes are less suited because many solid substances do not adhere to their polished surface. A number of organic compounds do adhere and can therefore be studied using these electrodes. For a special investigation, a surface roughened glassy carbon electrode was used for the immobilization of inorganic compounds, which otherwise do not adhere at the electrode [50]. In this case an electrolyte film separated the glassy carbon surface from the immobilized particles. This gave rise to special phenomena associated with the electron transfer from the carbon to the sample particles.

Pencil-Lead Electrodes

Recently, pencil-lead electrodes were proposed for the voltammetry of microparticles [51]. It seems they are equally suited in that they are also impervious to solutions and basically consisting of graphite.

Metal Electrodes

Metal electrodes were also applied in VMP studies, although they are less convenient to clean and microparticles can be immobilized on their surface only with some difficulty. For these reasons platinum electrodes were used to immobilize metal samples for measurements in nonaqueous solutions [45], while gold electrodes were used for in situ quartz microbalance measurements [48,49,52,53].

2. *Transfer and Immobilization of Solid Particles on the Electrode Surface*

Usually, most of the solid samples are powders or small particles. These samples may be additionally ground, if necessary. Different procedures have been reported for immobilizing samples on electrodes:

Transfer via a smooth plate [46,54,55]: Some µg to mg amounts of sample are placed on a carefully cleaned and smooth glass plate or glazed tile. The polished graphite electrode is pressed onto the sample spot and can be slightly moved over the plate in such a way that the polished electrode surface remains parallel to the plate surface. The result is an embedding of the sample particles in the soft surface of the graphite electrode. Of course, this procedure can lead to an additional crushing of crystals and, depending on the sample properties, also to a certain "smearing" and film formation on the electrode surface.

Transfer via filter paper [56–59]: When the sample powder is placed on filter paper and the electrode is gently rubbed and pressed on the sample spot as already described, the result is a very even distribution of the particles on the electrode surface. Applying this procedure, a possible immobilization of large particles is avoided.

Transfer from smooth sample surfaces [10,60,61]: For compact samples possessing a smooth surface, e.g., pieces of metals and alloys, the transfer can simply be accomplished by rubbing the polished electrode surface on the smooth side of the sample. This procedure is applicable even to materials harder than the graphite electrode because even then traces of the sample will be transferred. In case of very hard materials, e.g., steel and hard metal alloys, an addition of small amounts of an abrasive powder (corundum, diamond) is of assistance in the transfer process.

Transfer from microsamples: Especially in mineralogy, it is desirable to obtain a sample from single crystals under a microscope. For this purpose, a crystal is scratched with a steel needle and the practically invisible sample amount is then transferred from the needle to a glass plate and finally onto the graphite electrode. A better approach is to use a glassy carbon or graphite rod with a very thin tip, which helps in taking a sample and later serves as a working electrode.

Transfer of thin films: Sometimes it is necessary to sample very thin films from a support. If this support has a smooth surface, a direct transfer can be achieved by rubbing the polished electrode surface on it. Of course, one has to be aware of the fact that some support material may be transferred simultaneously.

Sonochemical transfer: Madigan et al. [62] proposed a method in which metal particles dispersed in a solvent are melted onto the surface of a solid electrode after being accelerated by microjets formed by acoustic cavitation. This method allows exact control of the amount of deposited metal, however, its application requires preliminary preparation of a suitable metal slurry suspension. The authors recommend their technique for the analysis of metal particles dispersed in used motor oil.

3. *The Measurement*

After the immobilization of the sample particles on the surface of the electrode, the electrode is ready for measurement. Figure 1 depicts a most ap-

Voltammetry of Microparticles

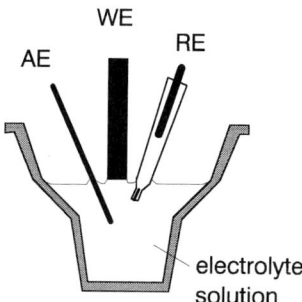

FIG. 1. Configuration of electrodes in an electrochemical cell for the voltammetry of microparticles.

propriate configuration of the working electrode in the electrochemical cell. The working electrode is dipped into the electrolyte solution and then slightly raised to make the solution adhere to the lower circular electrode surface. This ensures that the active electrode surface area is reproducible from experiment to experiment and no insulation of the electrode shaft is necessary. But even if the electrode is dipped some millimeters into the solution, this will only affect the capacitive background current as the cylindrical surface is not contaminated with sample particles.

The described electrode configuration in the cell does not cause any limitations with respect to the application of measuring techniques. Normally, one electrolyte solution can be used throughout a large number of measurements.

4. Cleaning of the Electrode Surface After Measurement

After measurement, the electrodes are polished on filter paper to clean their surface. It is important to ensure that the electrodes are not rubbed twice on the same spot; otherwise, the electrode might be polluted by the preceding sample. The cleaning has to be done in a way that the graphite traces on the filter paper are frequently interrupted. Only then can one be sure that the old sample material does not interfere. The success of cleaning has to be controlled by a blank voltammogram. This is mandatory to ensure reliable results. Since some solid compounds are electrochemically dissolved during measurements, this can lead to a pollution of the electrolyte solution with electroactive or interfering species. In such cases it is necessary to exchange the electrolyte solution after measurement, although experience shows that interferences occur only after a large number of measurements.

The renewal of the basal plane pyrolytic graphite electrodes is better accomplished by scraping off the surface layer with a razor blade. Metal and glassy carbon electrodes can also be cleaned chemically by dissolving residual samples of solid reaction products in appropriate solvents or chemicals.

B. Mechanisms of Electrochemical Reactions of Immobilized Solid Particles

Once the solid particles are immobilized on the electrode surface, one can perform all known electrochemical measurements. In this section we explain some strategies on investigating the electrochemical properties of immobilized solid particles and discuss their reaction mechanisms. The latter is very diverse in the case of solid substances. Generally, it is important to know the physical and chemical properties of a compound, for this allows the design of appropriate experiments.

1. Oxidative Dissolution of Metals and Alloys

The oxidation of a metal can proceed according to the following simple equation:*

$$\{Me\} \rightarrow Me^{n+} + ne^- \qquad (I)$$

Moreover, it is well known that the anodic dissolution can also proceed along complex reaction pathways. Most of the information concerning this matter was gathered in corrosion studies and concerns compact pieces of metal. VMP studies of metals and alloys per se concern small particles. This may cause one to wonder what kind of information can be obtained. In Sec. III.C, we suggest how it is possible to perform a quantitative analysis of alloys. Some general considerations regarding the anodic dissolution of metals will be given here. From the published literature it can be ascertained that one distinct advantage of VMP is the signal resolution [10,61]. The resolution of peaks is improved due to the small amounts of sample, which allow a complete successive dissolution of all constituents of an alloy during one voltammetric scan. This could never have been achieved

---

*Braces denote solid phases.

with compact electrodes since there are practically infinite amounts of solid sample and electrolyte solution available. Figure 2 offers some voltammograms of metals, and Fig. 3 shows voltammograms of alloys to illustrate their behavior in VMP. The "different constituents" which are anodically dissolved can be either the elements themselves or alloy phases. This depends on both the alloys and the electrolyte composition. For a strong interaction of two or more metals in the alloy phase, i.e., the existence of an intermetallic compound, a reaction pathway will be favored in which all constituents of this phase are oxidized together, whereas for a weak interaction of the metals in the alloy, a separate dissolution will be favored. The latter can also be achieved by using an electrolyte solution which strongly differentiates between the dissolutions of the alloy metals,

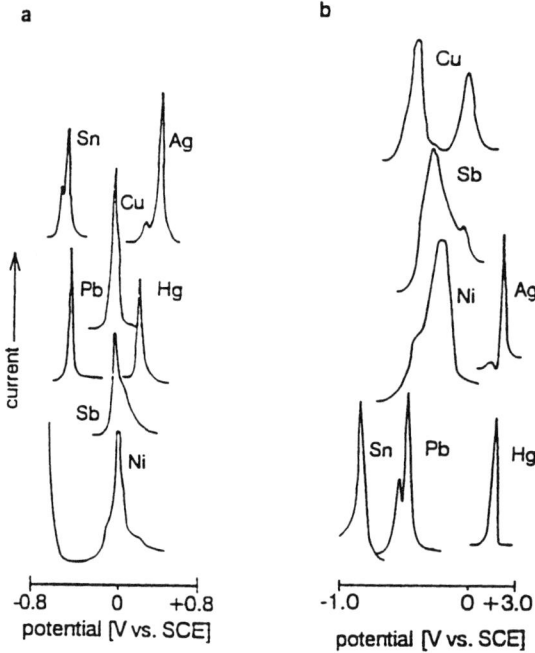

FIG. 2. Differential pulse voltammograms of microparticles of several metals in two electrolyte solutions. (a) 0.1 mol/l oxalic acid. (b) 1 mol/l NH_3/NH_4Cl buffer. Current in arbitrary units. (From Ref. 79.)

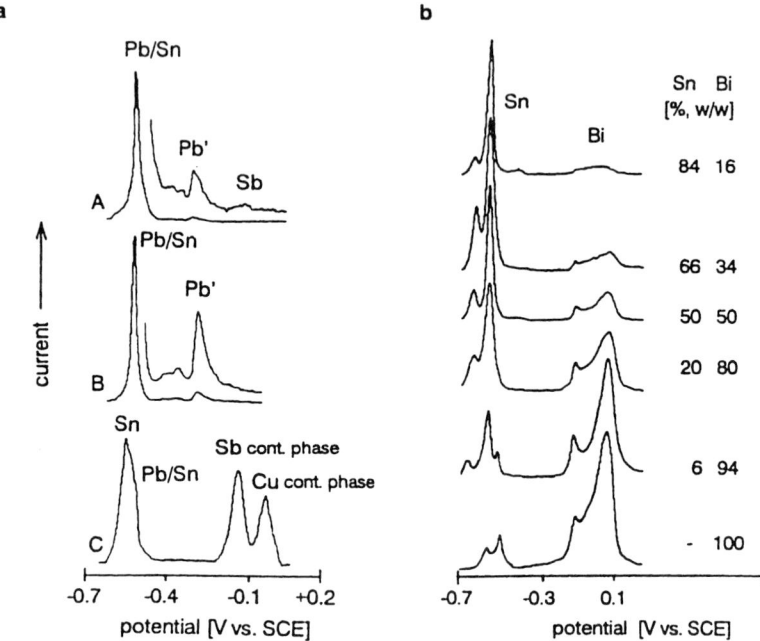

FIG. 3. (a) Voltammograms of microparticles of solder samples of different composition. Differential pulse voltammetry, supporting electrolyte: 0.1 M oxalic acid. A: Sn = 60.8% Pb = 37.8% Sb = 1.3%; B: Sn = 29.3% Pb = 70.5%; C: Sn = 80.0% Pb = 3.0% Cu = 5.5% Sb = 11.4. (From Ref. 88.) (b) Voltammograms of microparticles of bismuth and tin bismuth alloys with differential pulse voltammetry. ΔE_{pulse} = 10 mV; ΔE (staircase ramp) = 2 mV; ΔT_{pulse} = 10 msec; $E_{initial}$ = −0.7 V. (From Ref. 60.)

e.g., by the formation of strong complexes with one metal and weak complexes with the other one. This was observed in the case of VMP studies of dental alloys [61]. Komorsky-Lovrić et al. [63] compared the anodic dissolution of lead and mercury after mechanical immobilization with that of electrochemically deposited lead and mercury. From the results they concluded that the mechanical transfer leads to the deposition of well-separated patches of polycrystalline lead on the electrode surface. The overall behavior of both metals was not very different in the square wave voltammetric dissolution experiments which had been performed.

Voltammetry of Microparticles

Unfortunately, this method has not yet been fully evaluated with respect to the available information on metal powders. It is expected that chronoamperometric experiments will produce data on particle sizes and particle size distributions, as they were derived from similar measurements of the cathodic dissolution of iron oxides [56].

2. Reductive Dissolution of Oxides and Oxide Hydrates

A number of metal oxides and oxide hydrates can be reductively dissolved to form soluble species with a lower oxidation state rather than being reduced to the elementary metal. Examples of these are the iron oxides. Grygar [56–58,64] studied these systems in detail. The cathodic dissolution of hematite (α-Fe$_2$O$_3$) and goethite (α-FeOOH) can be described with the following equations:

$$\tfrac{1}{2}\{Fe_2O_3\} + 3H^+ + ne^- \rightarrow (1-n)Fe^{3+} + nFe^{2+} + \tfrac{3}{2}H_2O \qquad (II)$$

$$\{FeOOH\} + 3H^+ + ne^- \rightarrow (1-n)Fe^{3+} + nFe^{2+} + 2H_2O \qquad (III)$$

Obviously, the dissolution is a proton-promoted process. When $n = 0$, a pure chemical dissolution occurs and for $n = 1$, an electrochemically assisted dissolution with a complete reduction of iron (III) to iron (II) takes place. Grygar showed that the chronoamperometric curves can be described by the following equation [56]:

$$\frac{I(t)}{Q_0} = k\left[\frac{Q(t)}{Q_0}\right]^\gamma \qquad (1)$$

$I(t)$ is the current at a time t, Q_0 is the overall amount of charge necessary for reduction, $Q(t)$ is the charge consumed during a time t, and k is the electrochemical rate constant defined according to the following basic equation:

$$k = k_0 \exp\left[-\frac{\alpha n F E}{RT}\right] \qquad (2)$$

The exponent γ in Eq. (1) characterizes the particle size distribution of the powder sample. In this way, it was possible to determine the rate constant of the reductive dissolution and to show that the process is limited by charge transfer as the logarithm of the rate constant is linearly dependent on the electrode potential [64] (see Fig. 4). Further, it was shown that the presence of ions that chemisorb on goethite considerably increases the dis-

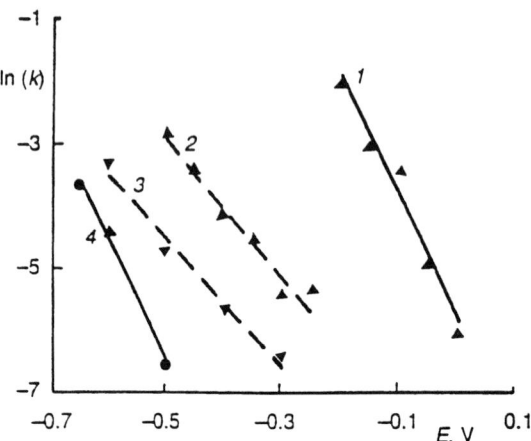

FIG. 4. Dependence of the logarithm of the electrochemical dissolution rate constant k on the working electrode potential E for ferrihydrite, goethite, and hematite. Curves: 1: ferrihydrite, 0.05 M chloroacetate; 2: goethite, 0.2 M chloroacetate + 0.5 M KCl; 3: goethite, 0.05 M chloroacetate; 4: hematite, 0.2 M chloroacetate. (From Ref. 64.)

solution rate. Table 1 shows the rate constant of the reductive dissolution of goethite in different electrolyte solutions. In Table 2, the rate constants and γ-values for a number of iron compounds together with the voltammetric peak potentials are summarized. Grygar also studied the dependence of the differential pulse peak potential of the reduction of iron oxides on the specific surface area σ of the starting material [56]. For the crystalline compounds, barium hexaferrite, hematite, goethite, maghemite, and magnetite, the following equation holds

$$E_{p,dp} = const. + \left(\frac{RT}{\alpha n F}\right) \ln \sigma \qquad (3)$$

Equation (3) is not maintained in the case of lepidokrokite and ferrihydrite, as these phases exhibit a high rate of chemical dissolution. Recently, Grygar showed by VMP of Al-, Cr-, and Mn-substituted goethite that these compounds are reductively dissolved in a shape-preserving reaction [65].

Unlike iron compounds, where a reduction always produces Fe^{2+} ions, copper ions are usually reduced to metallic copper. Only in the case

TABLE 1
Effect of Supporting Electrolyte Composition on the Rate of Dissolution of Goethite at –0.4 V

Supporting electrolyte	$k\ 10^{-3}$, s^{-1}			
	0.01 M	0.1 M	0.2 M	0.5 M
MCAc[a]		3.5	3.2	2.3
0.2 M MCAc				
0.2 M MCAc + KNO$_3$	3.1	3.6		
0.2 M MCAc + K$_2$SO$_4$	13	20		
0.2 M MCAc + KCl		3.6	10	17
0.2 M MCAc + KH$_2$PO$_4$	15	42	37[b]	
0.2 M MCAc + oxalate	11	15[b]		

[a]Monochloracetate buffer.
[b]Pure buffer (pH approaching that of the chloroacetate buffer).
Source: Ref. 64.

of the YBaCu high-temperature superconductor was it observed that the reduction of copper ions leads to a copper(I) compound and not to metallic copper [66,67]. This property has been used to probe the purity of the high-T_c phase.

3. *Reductive Conversion of Oxides and Salts to Metal Deposits and Dissolved Anions*

Many oxides and salts can be reduced to a metal deposit. This process is accompanied by the liberation of the anion, which will be protonated depending on its basicity:

$$\{MeA\} + ne^- \rightarrow \{Me\} + A^{n-} \quad \text{(IV)}$$

The following reactions are typical examples:

$$\{PbO\} + 2e^- + 2H^+ \rightarrow \{Pb\} + H_2O \quad \text{(V)}$$

$$\{Pb_3O_4\} + 8e^- + 8H^+ \rightarrow 3\{Pb\} + 4H_2O \quad \text{(VI)}$$

$$\{PbCO_3\} + 2e^- + 2H^+ \rightarrow \{Pb\} + CO_2 + H_2O \quad \text{(VII)}$$

$$\{Tl_2S\} + 2e^- + 2H^+ \rightarrow 2\{Tl\} + H_2S \quad \text{(VIII)}$$

The mechanisms of reactions of this type are certainly most diverse. They share one common facet, namely, that one solid phase is destroyed and an-

TABLE 2

Values of the k and γ Parameters in the Kinetic Eq. (1) for the Dissolution of Phases in the 0.1 M Oxalate Buffer at Working Electrode Potentials of –0.2 and –0.4 V vs. SCE and the Voltammetric Half-Wave and Peak Potentials in 0.1 M Chloroacetate Buffer[a]

Phase (sample)		$k \cdot 10^{-3}$, sec^{-1} (–0.2 V)	γ (–0.2 V)	$k \cdot 10^{-3}$, sec^{-1} (–0.4 V)	γ (–0.4 V)	$E_{1/2}$ (V)
Ferrihydrite	(A)	45	1.32	>100	—	0.00
	(B)	33	1.38	>100	—	–0.10
α-FeOOH	(A)	26	1.15	22	1.36	–0.34
	(B)	8.0	1.18	15	1.13	–0.49
	(D)	3.8	1.18	6.9	1.28	–0.53
	(E)	7.1	1.17	5.9	0.93	–0.55
	(F)	4.0	1.01	5.6	1.38	–0.50
	(G)	3.6	1.38	—	—	–0.47
β FeOOH		4.1	0.88	[b]		–0.16[c]
γ FeOOH		55	1.35	>100	—	–0.01
δ FeOOH		43	1.31	>100	—	0.00
Fe_3O_4		65	1.98	>100	—	–0.02
α-Fe_2O_3		0.6	1.12	2.7	1.04	–0.56
γ-Fe_2O_3		63	2.62	>100	—	–0.08
Sediments:	BRS			15	3.31	[d]
	SLI	16	1.73	16	2.81	[d]
Fe^{3+} (aq)		Reduction process obeying Cottrell's equation				–0.08 +0.30

[a]Scan rate 10 mV sec^{-1}.
[b]Chronoamperometric curve is not monotonous.
[c]Broad peak, slow current decrease after the maximum.
[d]No clear maximum, slowly rising wave.
[e]Clean electrode surface.
[f]Electrode surface with deposited FeOOH.
[g]c_{Fe} = 1 mmol liter^{-1}.
Source: Ref. 64.

other solid phase is formed. All of them require a transfer of anions from the starting phase into the adjacent solution. In principle the reaction rate can be controlled (a) by the electron transfer from the graphite (or any other electron conductor) to the reactant phase, (b) by the electron transfer from the product phase to the reactant phase, (c) by the diffusion of anions into the solution, (d) by the rate of diffusion of ions in the reactant phase, (e) by preceding chemical reactions, e.g., a protonation or complex formation on the surface of the reactant, or (f) by chemical follow-up reactions. Most of these reactions are chemically irreversible since the soluble products of the electrode reaction are lost by diffusion into the solution. Thus, it is easy to imagine that the reaction rate control will be very different in each case, depending on the chemical and physical properties of each participating phase and species.

A very interesting aspect of these reactions is which type of product phase will be formed. This aspect was studied by Meyer et al. [68] by combining VMP and in situ x-ray diffraction for the following two reactions:

$$\{PbO\} + 2e^- + 2H^+ \rightarrow \{Pb\} + H_2O \qquad (IX)$$

$$\{Pb(OH)Cl\} + 2e^- + 2H^+ \rightarrow \{Pb\} + H_2O + Cl^- \qquad (X)$$

Bearing in mind that these electrochemical reactions are performed at room temperature, one could expect that the lead thus formed may not have the chance to crystallize, as the activation energy for the crystallization may not suffice under these conditions. The in situ study was performed with a cell as shown in Fig. 5. It was observed that reaction (IX) formed cubic lead simultaneously with the reduction of the PbO, whereas reaction (X) resulted in the formation of an x-ray amorphous phase, which only slowly crystallized to form cubic lead. Figure 6 depicts linear scan voltammograms together with the simultaneously recorded x ray reflections for the reduction of PbO and Pb(OH)Cl, respectively. The explanation for these results is found in the fact that the PbO phase has a tetragonal structure while the Pb(OH)Cl phase has an orthorhombic structure. Therefore, it is much easier, i.e., less activation energy is needed, to transform the PbO into cubic Pb than to transform Pb(OH)Cl into Pb. If the reactant phase is reduced, the metal atoms will initially fall into positions not too different from those in the reactant phase. If these positions are very different from those of the thermodynamic stable product phase, additional activation will be needed. It is, of course, only possible to say that the crystallization of cubic lead occurs simultaneously with PbO reduction with respect to the time scale of the experiment. Since the experiment was performed on a long-time scale (115–135 sec for one scan of the main reflection), it would be interesting to study the process in a range of seconds or even shorter. Another observation made in this study was that the immobilization of PbO on the electrode surface led to a preferential

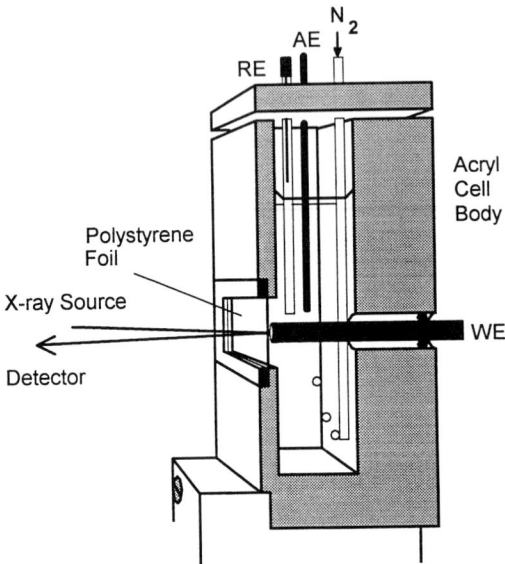

FIG. 5. Electrochemical cell for in situ x-ray diffraction analysis. AE: Auxiliary electrode; RE: reference electrode; WE: working electrode. (From Ref. 68.)

orientation of the crystals. This preferential orientation was maintained after their electrochemical conversion to Pb. The resulting lead crystals were then allowed to react with dissolved oxygen and chloride ions to form Pb(OH)Cl, which was later reduced back to metallic lead in a second electrochemical reaction. Even after these numerous conversions, the described preferential orientation was still detectable. Further, it is interesting to learn that this orientation was distinctly different from that of lead, which was deposited from a Pb^{2+} solution. All these results strongly support the hypothesis of topotactic electrochemical solid state reactions and discourage the assumption of reactions proceeding via dissolved lead ions.

4. *Solid Compounds That Possess Redox Sites and Are Able to Exchange Ions with an Adjacent Solution*

The compounds considered in this section exhibit a very interesting electrochemistry, lending themselves much more to solid state electrochemical

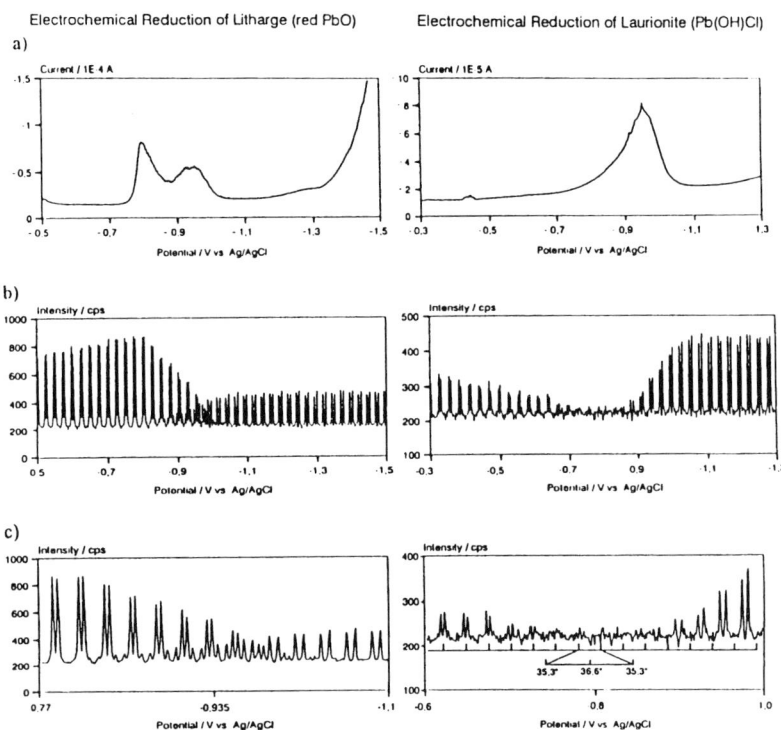

FIG. 6. Electrochemical reduction of red tetragonal PbO (litharge) and Pb(OH)Cl (laurionite) in situ coupled with x-ray diffraction. PbO was mechanically immobilized on the surface of a paraffin-impregnated graphite electrode, electrolyte 1 M KCl. Left: (a) linear sweep voltammogram with scan rate 0.1 mV sec^{-1}; (b) corresponding diffraction pattern, signal pairs between −0.5 and −0.8 V are PbO reflections at $d = 2.51$ Å, signal pairs between −1.1 and −1.5 V are Pb reflections at $d = 2.48$ Å. The 2Θ range from 26.00° to 45.00° was cyclically scanned and recorded simultaneously with the voltammetric scan. This causes one reflection to occur twice in each cycle: (c) enlarged part of (b) showing the simultaneous decrease in the PbO reflections and increase in the Pb reflections. Right: (a) linear sweep voltammogram with scan rate 0.1 mV sec^{-1}; (b) corresponding diffraction pattern, signal pairs between −0.65 V are Pb(OH)Cl reflections at $d = 2.52$ Å, signal pairs between −0.9 and −1.3 are Pb reflections at $d = 2.48$ Å (for further details see text); (c) enlarged part of (b) showing a potential (time) range with an amorphous phase. (From Ref. 68.)

experiments than any other group of compounds. Generally, the electrochemistry of these compounds can be completely reversible because the electron transfer is accompanied by a charge compensating ion transfer. Solid compounds with redox active sites inside the lattice are anything but rare. In principle, the constituents of every solid compound can be reduced or oxidized. Of course, this may be rather difficult from a thermodynamic point of view, as in the case of fluoride F^- and oxide O^{2-}, potassium K^+ and sodium Na^+ ions. For many other ions, it holds that a reduction can occur at more positive potentials or the oxidation at more negative potentials. Here, we discuss only such solid compounds, which have redox systems with formal potentials within the usual measuring range for aqueous solutions, i.e., roughly between +1.7 and −1.3 V vs. SHE. Many solid compounds accept foreign ions from a solution because of their appropriate structure [69]. Solid compounds can be classified according to the host lattice: (1) the first class comprises compounds with a three dimensional network of channels; (2) the second class shows two dimensional channels, i.e., interstitial layers; (3) the third class has one dimensional channels. If the solid compound possesses cavities and channels, the incorporation of ions is called *insertion*, while it is called *intercalation* when the solid has a layered structure and the ions move into the interlayer space. Most of the known host lattices belong to the two- or three-dimensional class. Solid compounds possessing such a host lattice as well as redox sites provide the possibility for electrochemical activities to be exhibited, i.e., they undergo electron transfer reactions which are accompanied by an ion transfer to maintain the charge balance inside the solid. This is the prerequisite that solids can preserve their structure when being oxidized or reduced, i.e., they undergo reversible topotactic redox reactions [69]. For the sake of simplicity, we discuss here the behavior of Prussian blue. Upon reduction of a C- or N-coordinated iron ion of the backbone lattice, an alkali metal ion enters the solid compound from the adjacent solution and goes to an interstitial position. Upon oxidation, it leaves the hole position and goes into the solution (see Fig. 7).

The electrochemical reactions of Prussian blue can be described by the following equations:

$$\{K_2Fe_N^{(2+)}[Fe_C^{(2+)}(CN)_6]\} \qquad \text{Everitt's salt}$$

$$\downarrow\uparrow E_{f1} \qquad\qquad\qquad\qquad\qquad\qquad\qquad\qquad (XI)$$

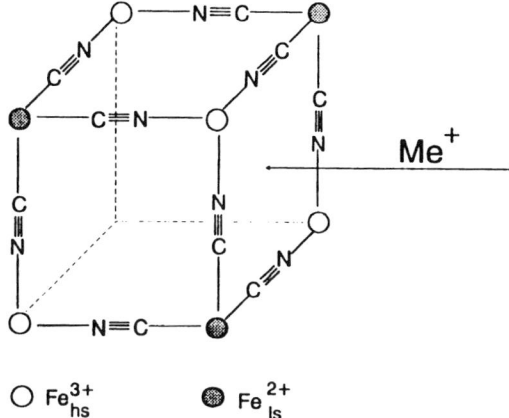

FIG. 7. Structure of solid PB and the insertion/expulsion of cations M⁺ during reduction/oxidation of iron ions. The full circles are low-spin iron(II) ions and the empty circles represent high-spin iron(III) ions. The depicted cube is 1/8 of the elementary cell of PB.

$$\{KFe_N^{(3+)}[Fe_C^{(2+)}(CN)_6]\} + K^+ + e^- \qquad \text{Prussian blue}$$

$$\downarrow\uparrow E_{f2} \qquad\qquad\qquad\qquad\qquad\qquad\qquad\qquad\qquad (XII)$$

$$\{Fe_N^{(3+)}[Fe_C^{(3+)}(CN)_6]\} + K^+ + e^- \qquad \text{Prussian yellow}$$

where Fe_N and Fe_C denote iron ions coordinated to N or C of the cyanide ions, respectively. Other hexacyanoferrates such as $\{K_2Cd^{(2+)}{}_N [Fe^{(2+)}{}_C (CN)_6]\}$ and $\{K_2Cu^{(2+)}{}_N [Fe^{(2+)}{}_C (CN)_6]\}$ show only a redox activity of the low-spin iron ions. Figure 8 shows multiple cyclic voltammograms of Prussian blue.

The electrochemistry of most of the metal hexacyanoferrates is reversible and stable over almost infinite cycles. The voltammetric peak currents exhibit a behavior typical for diffusion limitation at scan rates above approximately 50 mV/sec. This diffusion limitation is caused by the diffusion of alkali metal ions through the hexacyanoferrate particles. The limitation of the diffusion space in microparticles can lead to severe deviations

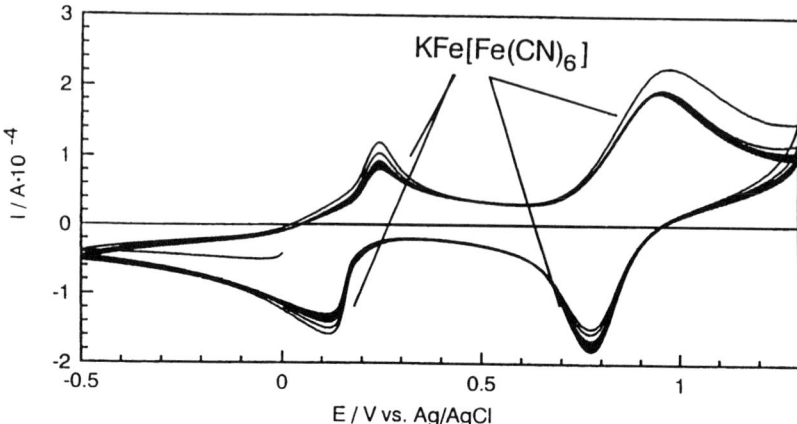

FIG. 8. The first 10 cyclic voltammograms of Prussian blue immobilized on a PIGE. Electrolyte solution: 0.1 M KCl, scan rate: 0.1 V sec^{-1}. (From Ref. 97.)

from a true diffusion control, especially at low scan rates. From electron-microscopic studies, it can be seen that the size of the particles is in the μm range. The "thick layer behavior" of hexacyanoferrates at elevated scan rates is lost at very low scan rates where the voltammograms show the "thin layer behavior." If the scan rates are very small, the electrochemical reactions can proceed through the entire amount of immobilized substance, and, therefore, cyclic voltammograms with symmetric peaks will be obtained, which in some cases hardly show any separation between the anodic and cathodic peaks. Whether compounds exhibit a thin or thick layer behavior will depend upon particle sizes and particle size distribution, diffusion coefficients, and the scan rate of the voltammetric experiment. Many hexacyanoferrates and also octacyanomolybdates and octacyanotungstates exhibit a pronounced electrochromism [70]. This can be easily observed with a recently developed cell for in situ light microscopy [71]. This cell was developed to observe the colors of the sample particles, which are immobilized on the surface of a graphite electrode.

Of course, there are many other compounds that undergo this kind of solid state electrochemical reactions. Some, such as uranium oxides, have been studied by VMP [72]. This study supports a mechanism in which the reduction of U(VI) to U(V) is accompanied by the insertion of cations in a topochemical reaction.

Bond et al. [73] studied the solid state electrochemistry of Dawson molybdate anion salts. These solid compounds show a fairly weak response due to the reaction:

$$\{[H_nS_2Mo_{18}O_{62}]^{x-}\} + 2e^- + 2H^+ \leftrightarrow \{[H_{n+2}S_2Mo_{18}O_{62}]^{x-}\} \qquad (XIII)$$

At more negative potentials and in the presence of appropriate cations (Ca^{2+}, Ba^{2+}, Cs^+), additional strong reduction signals are observed, which are due to the following reaction:

$$\{(NR_4)_4[S_2Mo_{18}O_{62}]\} + xCs^+ + yH^+ + (x+y)e^- \leftrightarrow$$
$$\{(NR_4)_4Cs_xH_y[S_2Mo_{18}O_{62}]\} \qquad (XIV)$$

Bond et al. [73] developed a very straightforward approach to the task of determining the redox state of solid compounds after an electrochemical reaction has been performed. First, they electrolyzed the immobilized compound at the electrode surface during a sufficient period of time to ensure a complete reaction at a certain potential. Then they took the electrode off the cell and added a small drop of organic solvent onto the surface of the electrode in order to dissolve the reaction product. In this small droplet, they recorded a steady-state voltammogram with a microelectrode, using the graphite electrode as a combined reference and auxiliary electrode. This methodology seems applicable to a great number of other compounds.

Very recently, Bond et al. [74] discovered a case in which oxidation is not accompanied by the expulsion of a cation but by incorporation of an anion. They studied the behavior of immobilized microcrystals of $[(C_4H_9)_4N][Cr(CO)_5I]$ and observed the incorporation of perchlorate ions.

Further examples of this type of solid state electrochemical reactions are discussed in Section E.

5. Electrochemistry of Molecular Solids

Molecular solids are solid materials composed of discrete molecular compounds. Their electrochemistry has a lot in common with the compounds considered above. Electrochemical synthesis and the electrochemical properties of molecular solids have attracted considerable interest [75], especially with respect to possible applications in electronic devices. VMP expanded these studies because of its facility to attach these compounds to electrodes. The molecular solids studied thus far indicate that their electrochemistry is surface-confined on a short-time scale, but that this may be followed by their proceeding more slowly through the bulk of crystals on a longer time scale. The ion insertion is obviously accompanied by a swallowing of crystals. The counterions force the molecules to make space for their diffusion in the solid.

Bond et al. [47,48,59] performed a detailed study of the electrochemistry of the chromium organic compounds cis- and trans-$Cr(CO)_2(dpe)_2$ with dpe = $Ph_2PCH_2CH_2PPh_2$:

Structure I
cis-$Cr(CO)_2(dpe)_2$

Structure II
trans-$Cr(CO)_2(dpe)_2$

Figure 9 shows a comparison of the solution voltammogram of a mixture of cis and trans compounds with the solid state voltammogram of the trans complex. The overall electrochemistry of these two compounds in solution and in the solid state can be described by the following square scheme:

$$\begin{array}{ccccc}
& \text{process I} & & \text{process III} & \\
\text{Trans} & \underset{+e^-}{\overset{-e^-}{\rightleftarrows}} & \text{Trans}^+ & \underset{+e^-}{\overset{-e^-}{\rightleftarrows}} & \text{Trans}^{2+} \\
\updownarrow & & \updownarrow & & \updownarrow \\
& \text{process II} & & \text{process IV} & \\
\text{cis} & \underset{+e^-}{\overset{-e^-}{\rightleftarrows}} & \text{cis}^+ & \underset{+e^-}{\overset{-e^-}{\rightleftarrows}} & \text{cis}^{2+}
\end{array}$$

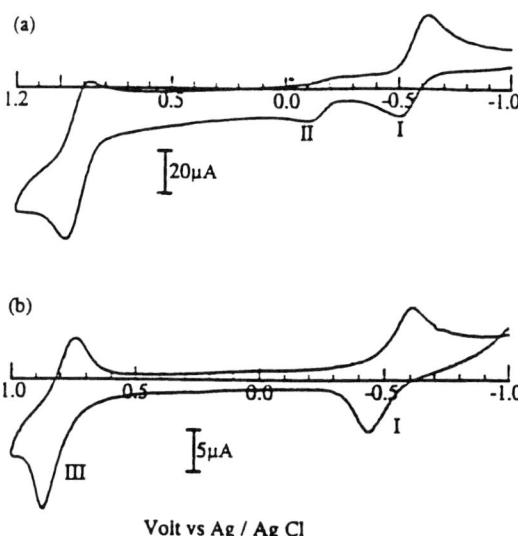

FIG. 9. Cyclic voltammogram obtained at 20°C at (a) a platinum disk electrode in dichlormethane (0.1 M Bu_4NClO_4) for a mixture of cis-$Cr(CO)_2(dpe)_2$ and trans-$[Cr(CO)_2(dpe)_2]^+$ showing processes I, II, and III (scan rate = 200 mV sec^{-1}) and (b) in aqueous (0.1 M $NaClO_4$) media at 2°C for solid trans-$Cr(CO)_2(dpe)_2$ mechanically attached to a polished basal plane pyrolyte graphite electrode (scan rate = 50 mV sec^{-1}). Initial and final potential = -1.0 V; switching potential = $+1.0$ V vs. Ag/AgCl. (From Ref. 59.)

All the species shown in the scheme are insoluble in water but soluble in less polar solvents. All electron transfer processes are coupled with isomerizations. In solution, the conversion of cis$^+$ to trans$^+$ is fast and the equilibrium is such as to favor trans$^+$. It is most remarkable that this isomerization does not proceed when the solid compound cis^0 is oxidized to cis$^+$. Obviously, the solid matrix hinders the isomerization which occurs in solution as a twist mechanism. The electrode product cis$^+$ A$^-$ (A$^-$ is an anion) has been identified on the electrode surface with the help of ex situ specular reflectance infrared spectroscopy.

7,7,8,8-Tetracyanoquinodimethane (TCNQ) is a molecular solid built up of the following molecule:

Structural Formula of TCNQ
(7,7,8,8-tetracyanoquinodimethane)

The electrochemistry of its solution in organic solvents is very well studied and understood.

Figure 10 shows a steady-state voltammogram of TCNQ in acetonitrile solution at a platinum microelectrode and a non–steady-state voltammogram at a conventionally sized platinum electrode. Figure 11 shows the cyclic voltammogram of solid microcrystals immobilized on the surface of a basal-plane pyrolytic graphite electrode with an aqueous KCl solution as the electrolyte [49]. This figure includes steady-state voltammograms obtained after dissolution of the electrochemically generated TCNQ⁻ and TCNQ° in a droplet of acetonitrile on the electrode surface (method as described in Sec. III.B.4). To prevent a slow dissolution of the TCNQ, the electrode with the immobilized TCNQ particles was surface-covered with a thin Nafion layer. The system denoted I in Fig. 11 was shown to occur according to the following equilibrium:

$$x\{TCNQ\} + ye^- + yM^+ \Leftrightarrow \{(M^+)_y(TCNQ^-)_y(TCNQ)_{x-y}\} \qquad (XV)$$

The second system denoted II in Fig. 11 results from a further reduction, which is, however, irreversible to a certain extent. System I exhibits an inertness in a certain potential range, as is better seen in Fig. 12. The authors give the following interpretation: The reduction of solid TCNQ leads to the formation of a new solid phase.

Figure 13 shows a model of the constellation with TCNQ being phase 1 and the reduction product being phase 2. Phase 3 is the electrolyte solution and phase 4 the solid support, i.e., the graphite electrode. After having performed a simple thermodynamic treatment of the phase formation using a previously developed nucleation theory [76], they obtained the following expression for the dependence of the critical overpotential for

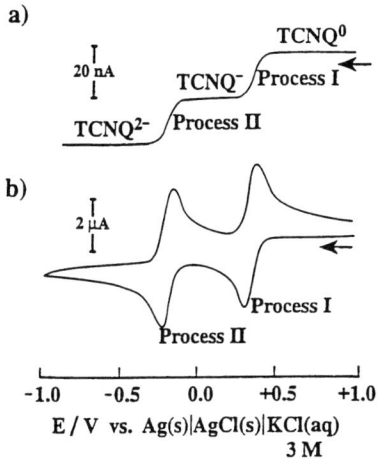

FIG. 10. Voltammograms for the reduction of a solution of 1 mM TCNQ in acetonitrile (0.1 M NBu$_4$PF$_6$). (a) Steady-state voltammogram at a 10 μm diameter Pt microdisk electrode (scan rate 10 mV sec^{-1}). (b) Non steady-state voltammogram at a 1 mm diameter Pt macrodisk electrode (scan rate 200 mV sec^{-1}). (From Ref. 49.)

phase formation on the interfacial tension γ_{12}, the interfacial tension between phases 1 and 2:

$$\eta_{crit} = \frac{\gamma_{12}^{3/2}}{nF\rho_m} \left[\frac{4\pi}{3kT \ln(\alpha_0/\chi)} \right]^{\frac{1}{2}} \quad (4)$$

where χ is the minimum detectable nucleation rate, α_0 the maximum possible nucleation rate, and ρ_m the molar density of TCNQ. It can be seen that for $\alpha_0 = 10^{20}$ and $\chi = 10$, 100, and 1000, the term enclosed in brackets is almost constant within a range of 6%. For large values of γ_{12}, a large range of the overpotential η exists where the nucleation rate is negligible, i.e., the system is inert. Additionally, the authors showed that the extent of the inert zone in the voltammogram depends on the nature of the ingressing cation, which is obviously due to the different structures of the reduction products formed. In this excellent study, the authors could for the first time present proofs of a solid state electro-

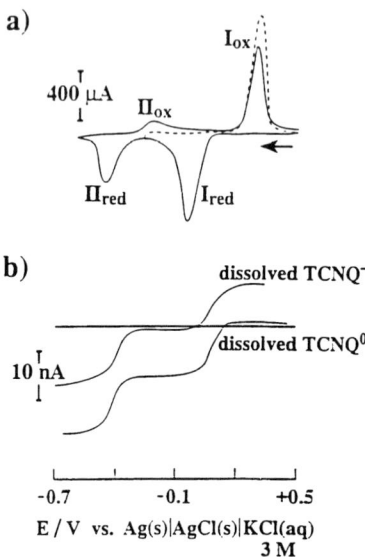

FIG. 11. (a) Cyclic voltammogram (scan rate 100 mV sec^{-1}) for the reduction and re-oxidation of nanocrystals of TCNQ immobilized on a basal-plane pyrolytic graphite electrode in 0.1 M KCl (aq.) (dashed line potential scanned only over the range of Process I). (b) Steady-state voltammogram at a 10 µm diameter Pt macrodisk electrode of material dissolved into minimal volumes of acetonitrile from the basal-plane pyrolytic graphite electrode after holding for 5 minutes at different potentials in 0.1 M KCl (aq.) (acetonitrile electrolyte 0.1 M NBu$_4$PF$_6$, scan rate 10 mV sec^{-1}). Top curve: TCNQ generated in the solid state at –0.10 V. Bottom curve: TCNQ generated in the solid state at +0.50 V. (From Ref. 49.)

chemical reaction which is under rate control by nucleation and growth of the solid product phase.

Tetrathiafulvalene is another molecular solid that undergoes a solid state electrochemical oxidation reaction accompanied by the ingress of anions [53]. Also in this case, there are indications that nucleation processes of the product phase are involved.

In the case of solid decamethylferrocene, Bond and Marken [77] demonstrated that the oxidized form remains solid and combines with anions from the electrolyte solution in a surface-confined reaction. This has been revealed with the help of electron probe x-ray measurements after

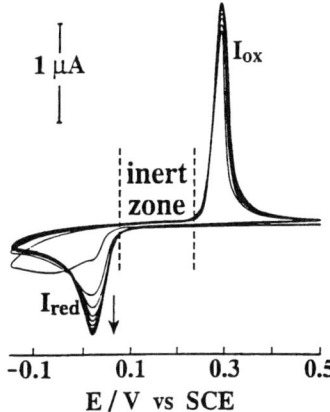

FIG. 12. Development of the voltammetric response of nanocrystals of TCNQ beneath a freshly prepared layer of Nafion® on a RAM™ electrode immersed in 1 M KCl(aq) (scan rate 100 mV sec^{-1}, solution flow rate 1 ml min^{-1}). The slow approach to the steady state has been attributed to the slow ingress of solution into initially dry Nafion® film. Note also the inert zone between the reduction and reoxidation peaks: something not seen in the redox cycling of solution species. (From Ref. 49.)

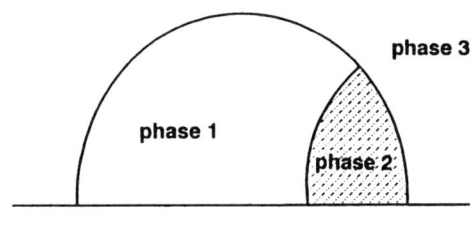

FIG. 13. Schematic representation of nucleation at the three-phase boundary electrode (phase 4)/ solid (phase 1 and phase 2)/ solution (phase 3). (From Ref. 49.)

performing the electrochemical reactions. Hydrophobic anions exhibit a preferential interaction with the oxidized decamethylferrocene and are able to ingress into the solid.

Cobalt and manganese phthalocyanines can also be regarded as molecular solids. Their solid state electrochemistry was studied by Komorsky Lovrić [78], who showed that the reduction of metal ions is accompanied by the transfer of cations from the electrolyte solution to the solid compound.

C. Quantitative Analysis of the Composition of Solids

1. Quantitative Analysis of Metal Alloys and Assessment of their Electrochemical Corrosion Stability

The voltammetry of microparticles (or abrasive stripping voltammetry, as it was called at the time) was introduced with the publication of a method for the quantitative analysis of lead-antimony alloys [10]. The authors demonstrated a mechanical rubbing of a graphite electrode on the surface of Pb-Sb alloys leads to the transfer of representative sample traces. The differential pulse anodic stripping voltammograms showed distinct dissolution peaks (Fig. 14). Since the absolute amounts of transferred sample could not be controlled, the evaluation was made by calculating the percentage of peak current for each peak, considering the sum of both peak currents to be 100%. This led to a calibration plot as shown in Fig. 15. The absolute standard deviation of a determined percentage was in the range from 0.3 to 5.0%. The calibration plot is necessarily bent as the slopes of the absolute calibration plots of both components differ from each other. (This is explained in detail in Ref. 46.) The possibility of quantitatively analyzing binary alloys was further demonstrated for tin-bismuth alloys [60], mercury-tin, and tin-silver alloys [61]. The latter systems were studied in connection with dental amalgams. The electrochemical oxidation of dental amalgams was used to assess their corrosion stability. With the help of abrasive stripping voltammetry, it was possible to detect the so-called γ_2 phase $Sn_{7.8}Hg$ in the amalgams. It is this phase that is responsible for the high corrosiveness of some dental amalgams. The presence of this phase led to the appearance of a distinct peak in VMP, whereas it was only visible as a badly developed shoulder in current voltage curves, usually being measured with compact samples of the amalgams. The voltammograms of microparticles of alloys allowed determining the peak potentials of the an-

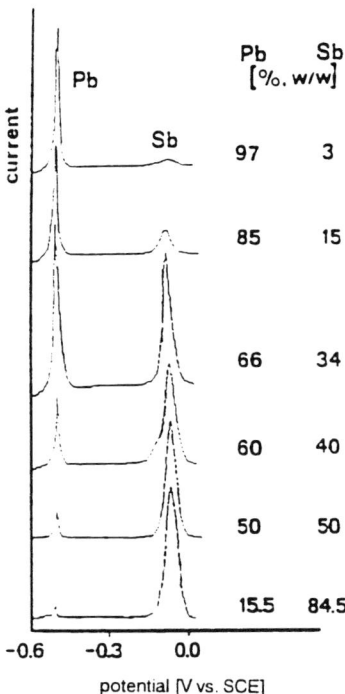

FIG. 14. Differential pulse stripping voltammograms obtained after abrasive transfer of trace amounts of different lead-antimony alloys onto a paraffin-impregnated graphite electrode. Voltammograms are offset for clarity. Electrolyte solution: 0.1 M oxalic acid. (From Ref. 10.)

odic dissolution of the constituents, being either pure metal components or alloy phases. These peak potentials (e.g., in DP and SW voltammetry) of metals and alloy phase were in some cases practically identical with the formal potentials. Thus, they provide important thermodynamic information. But even in cases of kinetic control, they allowed assessing the anodic stability of the alloys. For dental amalgams [61] and a number of nickel alloys [79], it was shown that VMP provides the same general information on corrosiveness as conventional corrosion studies. It especially helps in comparing different alloys very quickly. Figure 16 offers voltammograms of different nickel alloys. It can be seen that the anodic peak po-

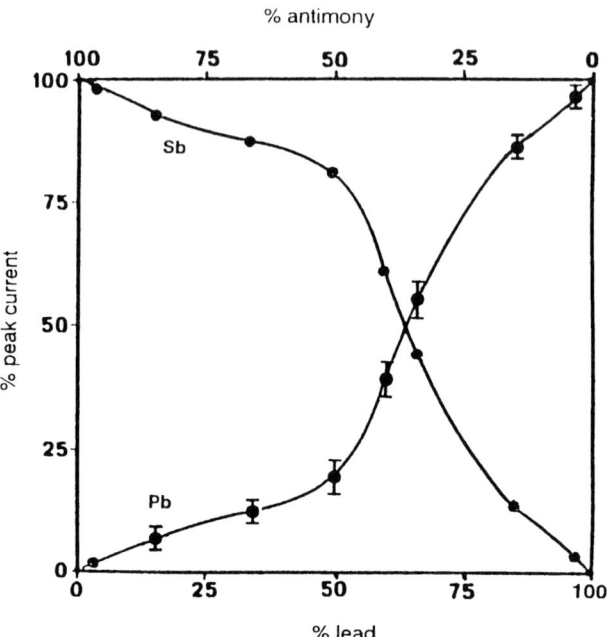

FIG. 15. Dependence of the percentage of the peak current of lead and antimony on the composition of the lead-antimony alloy. Standard deviations are indicated by vertical bars. (From Ref. 10.)

tentials show the same order as for the corrosiveness of these alloys. For quantitative measurements of the rate of corrosion, however, classical corrosion techniques are indispensable, requiring an exact knowledge of the surface area of the specimen. It can be expected that VMP will be developed in future, especially for corrosion studies of alloy powders being that these materials are much more difficult to study with other techniques.

Although it has been shown that a quantitative analysis of the composition of binary alloys is possible, one should critically discuss the analytical limitations of the voltammetry of microparticles: the calibration method (measurement of intensity ratios) and the very small sample amounts (usually 10^{-6}–10^{-8} mol) lead to relative standard deviations between 5 and 15%. This is not as effective as typical solution analyses, but it compares well with other solid state techniques by which very small

Voltammetry of Microparticles 37

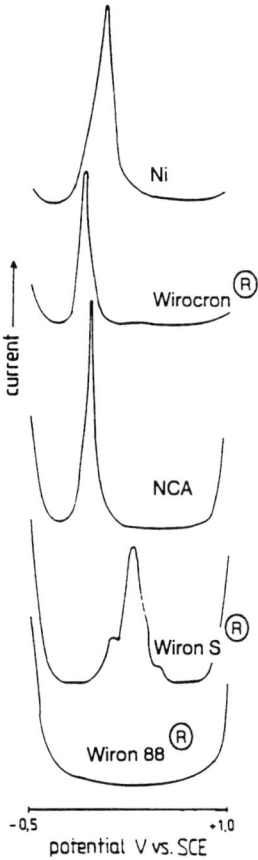

FIG. 16. Differential pulse voltammograms of microparticles of nickel and nickel alloys in 1 mol/liter sulfuric acid. (From Ref. 79.)

samples can be analyzed. A further limitation is that this technique can only be used for the relative quantification of the electroactive components. Nonactive components do not appear in the voltammogram. When the alloy has more than two components, the quantification becomes even less accurate because of the use of signal ratios for quantification. AbrSV should be applied carefully in such cases where a solution analysis is not possible and/or where the highest precision is not necessary.

2. Metal Compounds

In several cases the quantitative composition of solid metal compounds can be determined with the *coulometry* of microparticles. The strategy is the following: A component that contains the metals $M_{(1)}, M_{(2)}, \ldots M_{(n)}$ and the anions $A_{(1)}, A_{(2)}, \ldots A_{(n)}$ is transferred onto the electrode and reduced to the metal deposits $M_{(1)}, M_{(2)}, \ldots M_{(n)}$. These metals are then oxidized and the charge for their oxidation is measured, either by integration of voltammograms or in potential step experiments with the

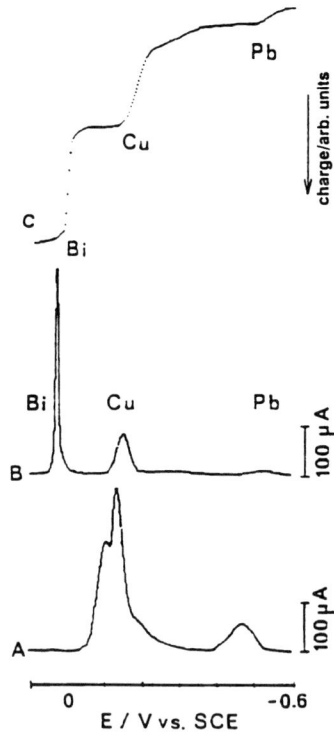

FIG. 17. Voltammograms of microparticles of the high-T_c superconductor $Bi_{1.8}Pb_{0.39}Sr_{1.99}Ca_{2.06}Cu_{3.15}O_{10.5}$ after preliminary reduction for 30 sec at −1.5 V vs. SCE in 1 mol/liter HCl. Differential stripping voltammetry. (A) Without calomel; (B) With calomel; (C) Integrated curve of B. (From Ref. 80.)

help of chronocoulometry. From these charges $Q_{(1)}, Q_{(2)}, \ldots Q_{(n)}$, the ratios of the metals, e.g., $M_{(1)} : M_{(2)}$; $M_{(1)} : M_{(3)}$, etc. can be calculated. This has been done for the quantitative analysis of high temperature superconductors of the composition $Bi_{1.8}Pb_{0.39}Sr_{1.99}Ca_{2.06}Cu_{3.15}O_{10.5}$ [80]. Of course, only bismuth, lead, and copper were measurable, as the other elements do not provide electrochemical signals. For the VMP analysis of this material, a special technique was developed that allowed determining the copper content relative to lead with a standard deviation of 1% and the bismuth content relative to lead, even with an 0.5% relative standard deviation. This was remarkable as such precise determinations are otherwise only accessible by solution analysis. This success was

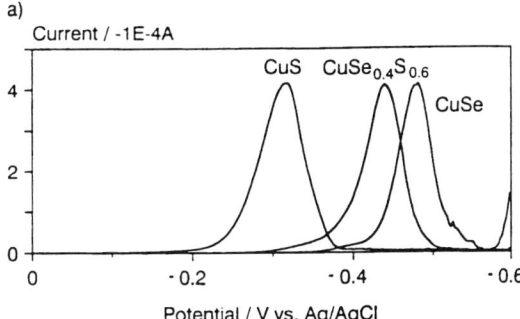

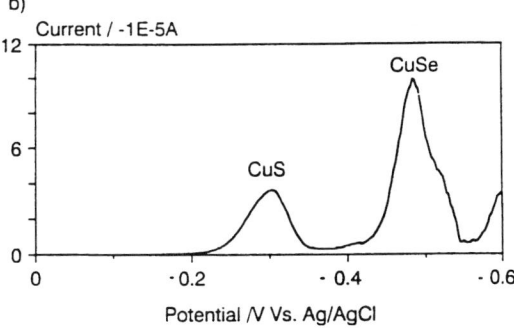

FIG. 18. Cathodic voltammograms of (a) CuS, CuSe, and $CuSe_{0.4}S_{0.6}$ and (b) a mechanical mixture of CuS and CuSe (1:2). Electrolyte: 1 mol/liter H_2SO_4; scan rate: 0.011 V/sec. (From Ref. 83.)

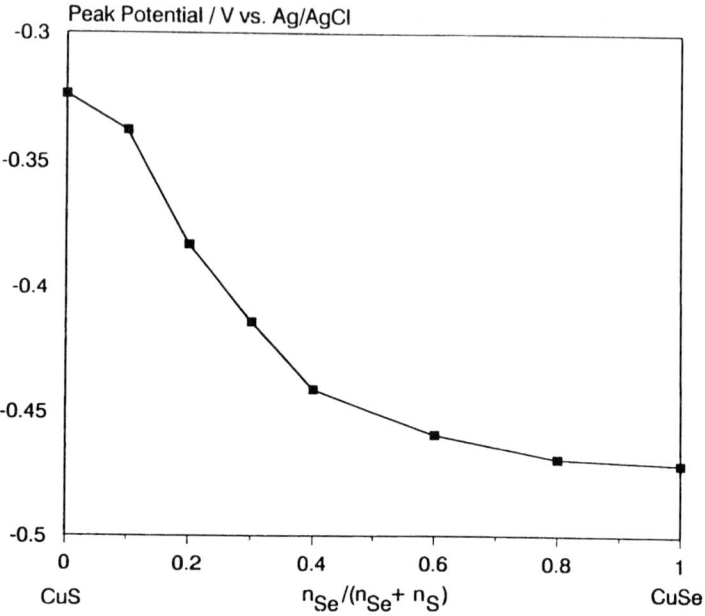

FIG. 19. Plot of standardized peak potentials (extrapolated to values for $Q = 0$) of different mixed crystals of $CuSe_{1-x}S_x$ versus molar ratio $n_{Se}/(n_{Se} + n_S)$. (From Ref. 83.)

achieved by an in situ deposition of mercury droplets during the reduction of the superconductor material. Therefore, the sample was thoroughly mixed in a mortar with calomel to have a source for mercury. This in situ plating of mercury droplets led to the dissolution of such metals as copper, lead, and bismuth in the mercury, which ensured well-resolved anodic voltammograms and a precise measurement of the oxidation charges. Figure 17 depicts the anodic dissolution voltammograms in the absence and presence of calomel and also the integrated curve for the latter case. In later experiments it was observed that the mercury can also be added to the electrolyte solution in the form of soluble salts. Six different thallium-tin sulfides were analyzed in the same way with only small deviations from the theoretical values [55]. Further examples of the analysis of metal compounds are discussed in Sec. III.D where phase analysis is emphasized.

Voltammetry of Microparticles

The voltammetry of microparticles is very well suited for a quantitative analysis of mixed crystals. The thermodynamics of such mixed phases as $MeA_xB_{(1-x)}$, which were synthesized from two different metal salts, MeA and MeB, is the reason why the reduction of the mixed phase does not occur at the two different potentials where that of the parent salts is affected. The mixed phase is reduced at a potential that depends on the molar ratio of the two salts forming the mixed crystal. This was shown first for silver chloride–bromide mixed crystals [81], later for nickel-iron hexacyanoferrates [82] and recently for copper sulfide–selenide mixed crystals [83]. Figure 18 shows cathodic single sweep voltammograms of the reduction of the pure phases CuS and CuSe, the mixed-phase $CuSe_{0.4}S_{0.6}$, and of a powder mixture of CuS and CuSe. Figure 19 shows a plot of the peak potentials of the mixed phases versus the molar ratio $x = n_{Se}/(n_{Se} + n_s)$. To obtain this plot it is necessary to take into account that the peak potential depends to some extent on the amount of immobilized compound. The

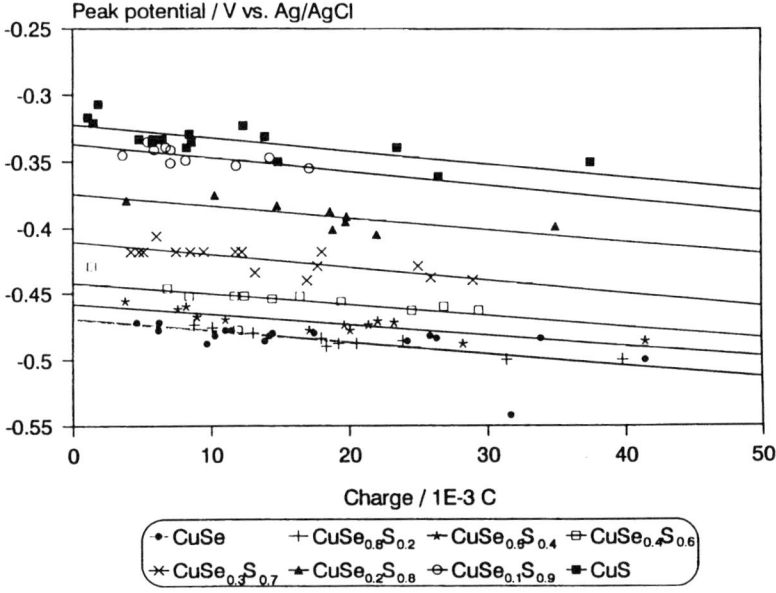

FIG. 20. Plot of the peak potential E_p of different mixed crystals of $CuSe_{1-x}S_x$ versus the charge Q consumed in the electrochemical reaction. (From Ref. 83.)

task was approached as follows: Every solid sample was measured 12 times with arbitrarily chosen, different amounts of sample. The resulting voltammograms were integrated to obtain the charge consumed for the reduction in each experiment. A plot of peak potentials versus the charge consumed, i.e., versus the amount of compound undergoing the reaction, is linear (Fig. 20) and allows one to compare the different samples easily, either at an arbitrarily chosen charge or with the help of the peak potentials extrapolated to zero charge. For all possible compositions, the span of results was found to be $x = \pm 0.03$.

3. Powder Mixtures

Mechanical mixtures of different compounds can also be analyzed by VMP, provided that each compound has at least one electroactive com-

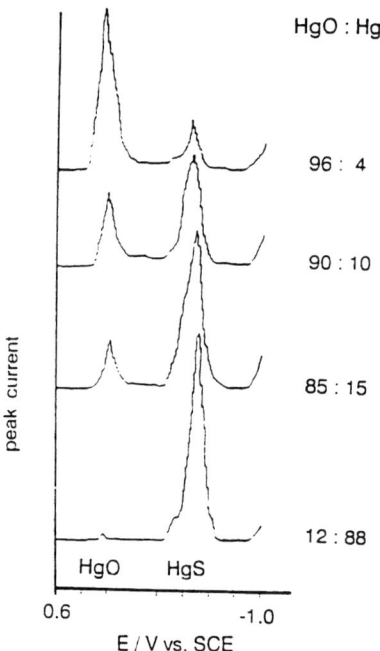

FIG. 21. Voltammograms of powder mixtures of HgO:HgS and the corresponding calibration plot. Electrolyte: 0.1 M oxalic acid, differential pulse voltammetry. (From Ref. 46.)

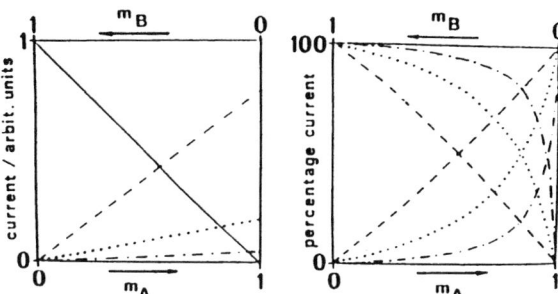

FIG. 22. (Left) Theoretical calibration plots for the absolute peak current when the amount of mixture is constant for each measurement. (Right) Corresponding theoretical calibration plots for the percentage peak current. (From Ref. 46.)

ponent. As in previous cases, it is necessary to transfer a representative amount of sample onto the electrode surface. This is only possible with well-ground samples, otherwise, a preferential deposition of one component would occur. Even this can be tolerated as long as the calibration mixtures have the same size distribution as the samples. The following powder mixtures were subjected to a calibration analysis: MnO_2-FeOOH, HgS-HgO, HgO-PbO [46]. Figure 21 shows the differential pulse voltammograms for the reduction of HgO-HgS mixtures and the corresponding calibration plots. The origin of the bent structure of the calibration plots is the mathematical result of different slopes of the calibration graphs of peak current versus absolute amounts of sample. Figure 22 compares the absolute with the relative calibration plots based on a simple theoretical calculation. Since the components of a mechanical mixture do not influence each other with respect to their thermodynamic activities in electrochemical reactions, this phenomenon is the only reason for the bent structure of the calibration plots. In the case of alloys, a bending of the calibration plot can also be caused by a variation of the activity coefficients of the constituents. For the case of AgCl-AgI mixtures, a calibration-free determination of the ratio of components is possible using coulometry [46]. In this study, it is also shown that the detection limit of electroactive solid compounds is of the order of 10^{-12} mol.

4. Quantitative Assessment of Organic Residues on Surfaces

With the help of a mechanical transfer of pesticide residues from plant leaves onto a PIGE, it is possible to screen the surface for pesticides and even determine their surface concentration [84]. The pesticide chlorothion could be detected and determined down to a surface concentration of 0.1 ng/cm^2.

D. Phase Analysis

The term *phase analysis* is used here to describe an analytical procedure used to identify a pure solid compound to be a certain solid phase, i.e., a compound with a certain chemical composition and a certain structure. Hence, a phase analysis is a *qualitative* analysis. A qualitative analysis is occasionally regarded as something very simple, which is of no special interest to contemporary science. This is unjustified, indeed, as the qualitative analysis is one aim of practically all spectroscopic techniques. Moreover, it is the same with VMP, as with all spectroscopic techniques, that a qualitative analysis also involves quantitative features, the utilization of which provides a wealth of important information on the compound studied.

Mineralogists especially need analytical techniques for phase analyses. Dominating techniques are powder x-ray diffraction and microprobe techniques with laser irradiation, electron, ion, or atom bombardment. X-ray diffraction is usually performed with powders and requires sufficient sample amounts. The generation of a powder diffraction pattern using a single microcrystal is, in principle, possible (Gandolfi camera [85]), but this demands so much time that it is rarely done. Microprobe techniques provide information on elemental composition but they do not provide information on structure. Thus one cannot distinguish structural modifications having the same chemical composition. A third tool for phase identification is classical light microscopy, which is very powerful but needs experience gained over many years of extensive practice. The voltammetry of microparticles contributes to mineralogical phase identification in two ways: it provides information on *quantitative and qualitative elemental composition* and allows distinguishing *structural modifications*. With the help of VMP, voltammograms showing specific signals for different elements can be obtained and the ratio of these elements can be measured using coulometric techniques. Furthermore, the electrochemical oxidation or reduction of a mineral also depends on the structural modification of the sample allowing us to distinguish them.

Voltammetry of Microparticles

The following scheme shows the different possibilities of performing electrochemical measurements with mechanically immobilized mineral particles:

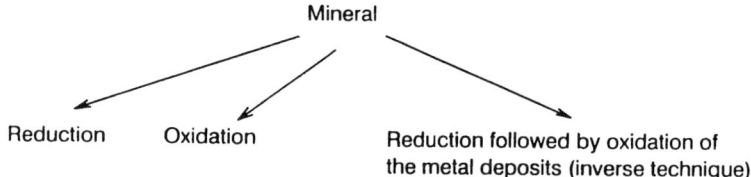

In many cases, the reduction of a mineral will lead to the deposition of metal constituents while, in other cases, it will lead to the liberation of soluble metal ions of lower oxidation states. Electrochemical oxidations of minerals can induce a dissolution accompanied by the formation of metal ions in a higher oxidation state or by the oxidation of anions, especially sulfide anions. If a mineral is preliminarily reduced to form metal deposits, and if this is followed by an anodic dissolution of these metals, the resulting voltammograms can often be interpreted very easily. Henceforth, this approach is referred to as the "inverse mode" of VMP. The metals give dissolution signals, which are specific to each metal. The ratio of the peak currents of these dissolution peaks depends on the composition of the mineral. Thus, these inverse voltammograms are reliable fingerprints of each mineral. Figure 23 shows differential pulse voltammograms of the reduction, the oxidation, and the inverse voltammogram of a lead antimony sulfide mineral. Figure 24 shows a selection of inverse voltammograms of a number of different minerals. These voltammograms are such specific fingerprints that a compilation of spectra of minerals and inorganic compounds is already available [86]. Naturally, VMP can be used for pigment analysis as well [87]. It can also be applied to the fingerprint identification of metal alloys [88].

If the voltammograms do not suffice for an unambiguous identification of certain phases, this can in some cases be elegantly done with the help of coulometric measurements. In Sec. III.C.2, the coulometric determination of the ratio of elements in solid compounds has been considered. However, this approach can also be chosen for phase identifications. Reductive and also inverse voltammograms of thallium sulfides (Tl_2S, TlS, Tl_4S_3, and Tl_2S_5) showed the same signal for their reduction and, of course,

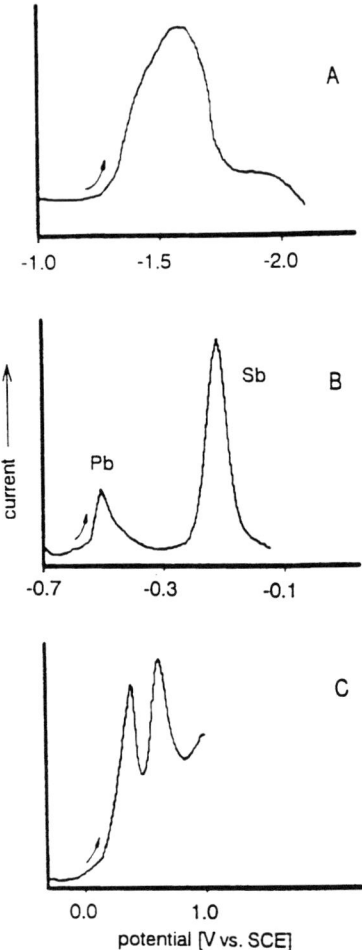

FIG. 23. Differential pulse voltammograms of boulangerite ($Pb_5Sb_4S_{11}$) obtained after mechanical transfer of traces of the mineral onto the surface of a paraffin impregnated graphite electrode. (A) Reduction; electrolyte: 0.1 M sodium oxalate, pH 5. (B) Inverse voltammogram after reduction at −1.5 V; electrolyte; 0.1 M sodium oxalate, pH 2. (C) Oxidation; electrolyte: 0.1 M oxalic acid. (From Ref. 11.)

Voltammetry of Microparticles

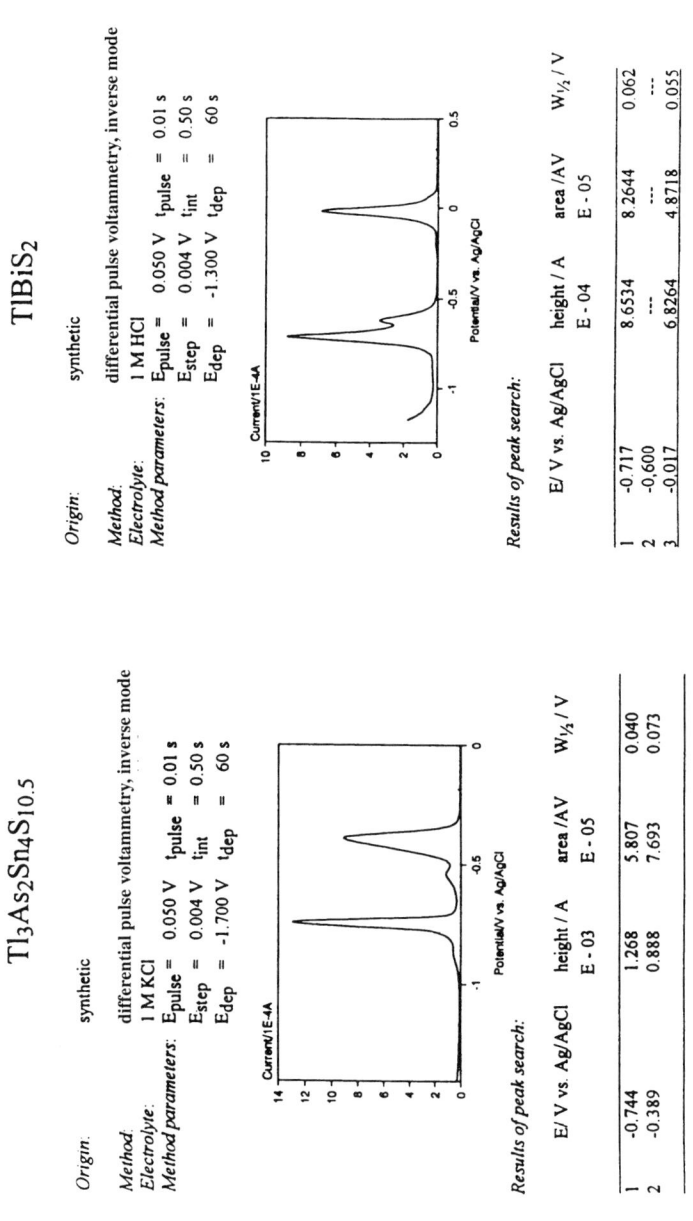

FIG. 24. Voltammograms of microparticles of minerals. Inverse mode, i.e., anodic stripping after a preliminary reduction. (Selection of examples from Ref. 86.)

Miargyrite (AgSbS$_2$)

Origin: San Genaro (Peru)
Method: differential pulse voltammetry, inverse mode
Electrolyte: 1 M KCl
Method parameters: E_{pulse} = 0.010 V t_{pulse} = 0.07 s
E_{step} = 0.005 V t_{int} = 0.50 s
E_{dep} = −1.200 V t_{dep} = 60 s

Results of peak search:

	E / V vs. Ag/AgCl	height / A E - 05	area / C E - 07	W$_{1/2}$ / V
1	−0.073	0.1800	3.0255	0.171
2	0.113	2.2059	7.7105	0.034

Tl$_2$As$_2$SnS$_6$

Origin: synthetic
Method: differential pulse voltammetry, inverse mode
Electrolyte: 1 M KCl
Method parameters: E_{pulse} = 0.050 V t_{pulse} = 0.01 s
E_{step} = 0.004 V t_{int} = 0.50 s
E_{dep} = −1.700 V t_{dep} = 60 s

Results of peak search:

	E / V vs. Ag/AgCl	height / A E - 04	area /AV E - 06	W$_{1/2}$ / V
1	−0.730	1.419	6.483	0.033
2	−0.378	0.476	2.352	0.040

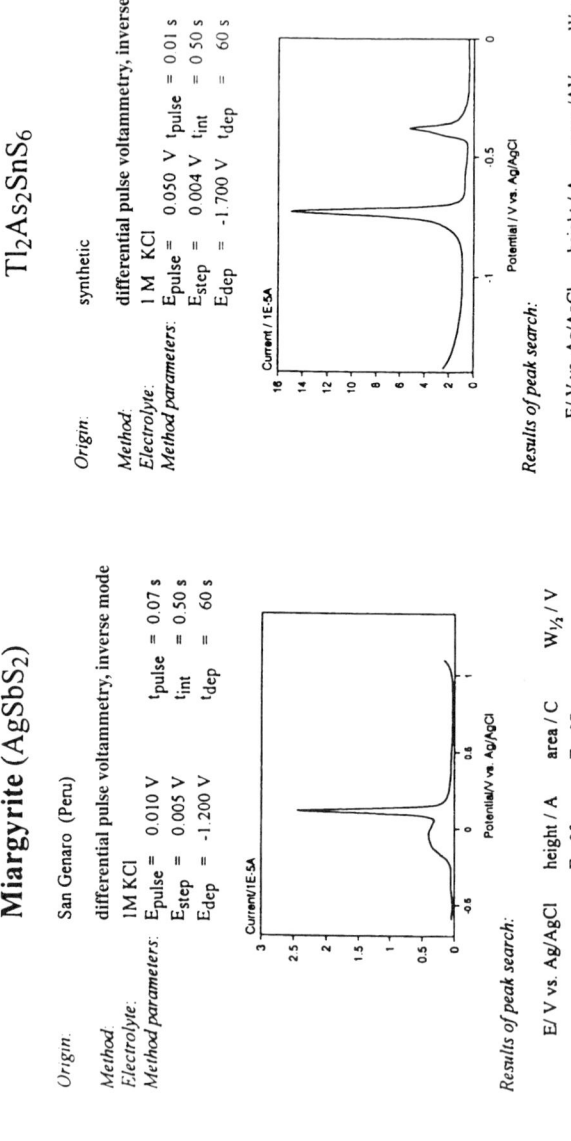

FIG. 24 Continued

Lorandite (TlAsS$_2$)

Origin: synthetic
Method: differential pulse voltammetry, inverse mode
Electrolyte: 1 M KCl
Method parameters: E_{pulse} = 0.050 V t_{pulse} = 0.01 s
E_{step} = 0.004 V t_{int} = 0.50 s
E_{dep} = -1.400 V t_{dep} = 60 s

α-Matildite (AgBiS$_2$)

Origin: synthetic
Method: differential pulse voltammetry, inverse mode
Electrolyte: 1 M KCl
Method parameters: E_{pulse} = 0.050 V t_{pulse} = 0.07 s
E_{step} = 0.005 V t_{int} = 0.50 s
E_{dep} = -1.500 V t_{dep} = 60 s

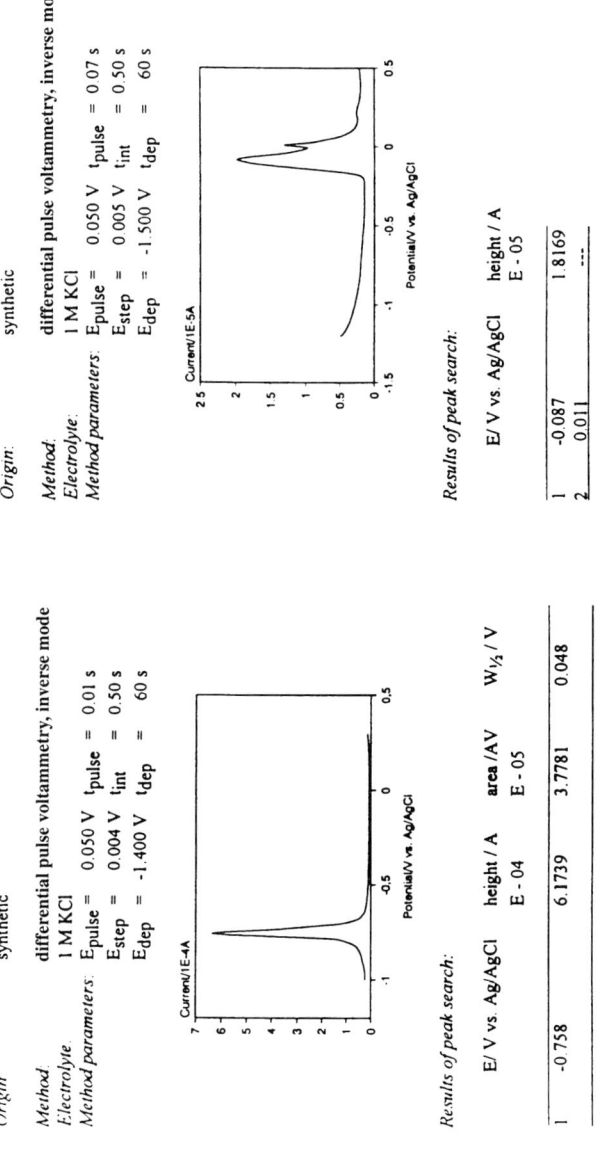

Results of peak search: (Lorandite)

	E / V vs. Ag/AgCl	height / A E - 04	area / AV E - 05	W½ / V
1	-0.758	6.1739	3.7781	0.048

Results of peak search: (α-Matildite)

	E / V vs. Ag/AgCl	height / A E - 05
1	-0.087	1.8169
2	0.011	---

FIG. 24 Continued

Cuprite (Cu$_2$O)

Origin: Gumerchewsk, Ural (Russia)

Method: differential pulse voltammetry, inverse mode
Electrolyte: 1 M KCl
Method parameters: E_{pulse} = 0.050 V t_{pulse} = 0.01 s
E_{step} = 0.004 V t_{int} = 0.50 s
E_{dep} = -1.400 V t_{dep} = 60 s

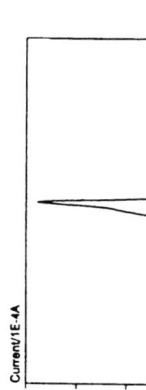

Results of peak search:

	E/V vs. Ag/AgCl	height / A E - 04	area /AV E - 05	W$_{½}$ / V
1	-0.068	7.1836	4.8064	0.055
2	0.255	2.3612	3.5910	0.132

Galena (PbS)

Origin: Neudorf, Harz

Method: differential pulse voltammetry, inverse mode
Electrolyte: 1 M KCl
Method parameters: E_{pulse} = 0.050 V t_{pulse} = 0.01 s
E_{step} = 0.004 V t_{int} = 0.50 s
E_{dep} = -1.200 V t_{dep} = 60 s

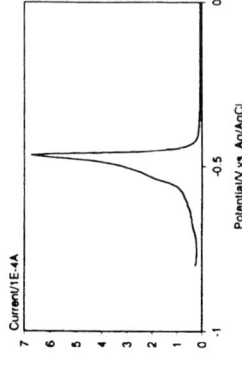

Results of peak search:

	E/V vs. Ag/AgCl	height / A E - 04	area /AV E - 05	W$_{½}$ / V
1	-0.467	6.6083	4.0254	0.037

FIG. 24 Continued

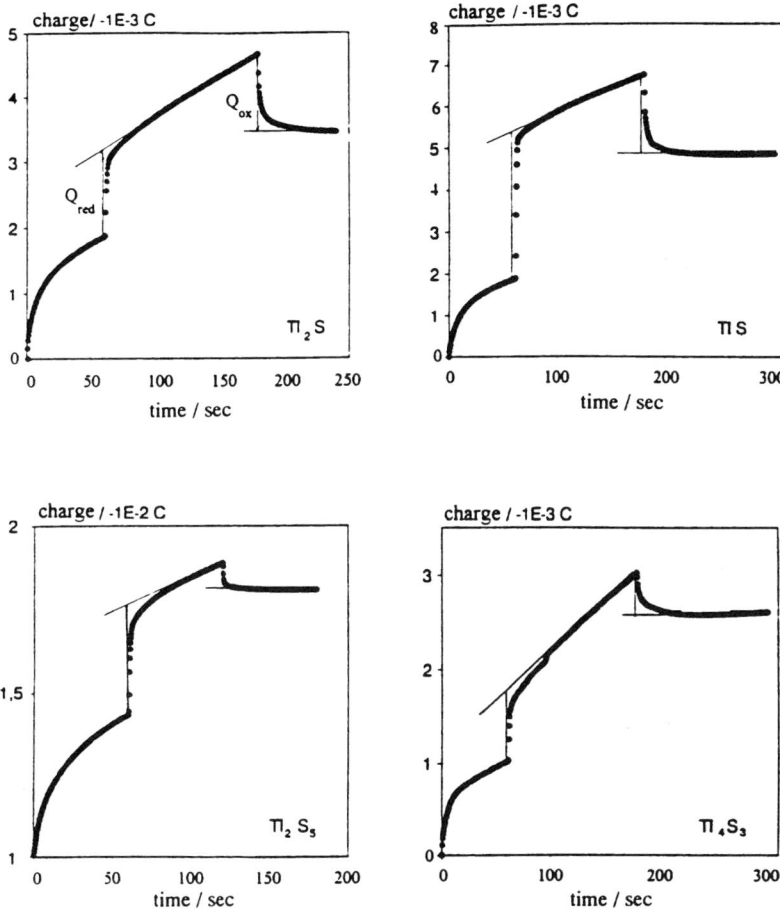

FIG. 25. Coulograms of microparticles of thallium sulfides in 2 M NaOH + 2 × 10^{-4} M $HgCl_2$. Potentials: Tl_4S_3 –1.0 V (60s), –1.32 V (120s), –0.5 V (120s); Tl_2S –1.0 V (60s), –1.4 V (60s), –0.5 V (60s); TlS –1.0 V (60s), –1.4 V (120s), –0.5 V (120s); Tl_2S –1.0 V (60s), –1.4 V (120s), –0.5 V (60s).

At the first potential (–1.0 V), mercury was plated from the stirred solution. After this period, the solution was quiescent. At the second potential, the thallium sulfide was reduced to thallium metal, and at the third potential, the thallium metal was oxidized to thallium(I) ions. (From Ref. 55.)

TABLE 3

Reactions and Ratios of Charges Necessary for the Reduction of a
Thallium Sulfide to the Charge for the Oxidation of Thallium

Reduction of sulfide	Oxidation of thallium	$q_{red}:q_{ox}$	
		Theory	Experiment (mean values)
$Tl_2S + 2e^- \rightarrow 2Tl + S^{2-}$	$2Tl \rightarrow 2Tl_+ + 2e^-$	2:2 = 1	1.18
$TlS + 2e^- \rightarrow Tl + S^{2-}$	$Tl \rightarrow Tl^+ + e^-$	2:1 = 2	2.0
$Tl_4S_3 + 6e^- \rightarrow 4Tl + 3S^{2-}$	$4Tl \rightarrow 4Tl^+ + 4e^-$	6:4 = 1.5	1.65
$Tl_2S_5 + 10e^- \rightarrow 2Tl + 5S^{2-}$	$2Tl \rightarrow 2Tl^+ + 2e^-$	10:2 = 5	4.68

Source: Ref. 55.

the same signal for the oxidation of the metallic thallium formed in a preliminary reduction. However, when the charge q_{red} for the reduction of the metal sulfides and afterwards the charge q_{ox} for the oxidation of the formed thallium are measured, one can distinguish the sulfides by their different ratios of q_{red}/q_{ox}. Figure 25 shows the chronocoulograms for their reduction and subsequent oxidation, and Table 3 gives the experimental values for q_{red}/q_{ox}. These experiments can only be performed with a mercury plating preceding the reduction of the thallium sulfides. Because the solution was not stirred, the mercury reduction current dropped to such small values that the reduction and oxidation steps could still be measured very well. When the experiment was performed without mercury, which obviously amalgamated with the thallium metal, it was observed that the thallium deposit, at a certain potential before its oxidation, was vigorously repelled from the electrode. This repulsion was observed as a black cloud rejected from the electrode surface.

In case of coprecipitates, VMP allows the detection of very small amounts of certain phases, provided they are electroactive. A good example is the detection of manganese(IV) oxides in calcium carbonate coprecipitations [89]. Finally, it should be stressed that the identification of mineral phases can be performed with field instrumentation, independent from any laboratory [90].

E. Structure Analysis

Electroanalysis is generally considered as a tool for the quantitative analysis of solutions. However, growing attention is paid to its potential for the

Voltammetry of Microparticles

structural characterization of dissolved species, especially in inorganic and organometallic chemistry. It is particularly cyclic voltammetry, even referred to as "electrochemical spectroscopy" [91], that is already routinely used for the study of new compounds. The simplicity of its use and low cost of instrumentation makes it very attractive. Therefore, it is interesting to see what the contribution of electrochemical methods is to the structural analysis of solid compounds. The papers already published on this aspect of solid state electroanalysis give a clear answer and allow one to expect even more from future applications.

Obtaining information on the structure of a solid compound by help of electrochemical measurements is still mainly based on empirical knowledge rather than on theoretical interpretation. As was done in the early days of spectroscopic techniques, the very first step is to search for an empirical correlation between electrochemical behavior and structure of redox centers. Besides these empirical correlations, some experiments can be interpreted on the basis of strict thermodynamic principles. This can be demonstrated for systems that form mixed crystal phases. In this way it is possible to unambiguously distinguish *phase mixtures* from *mixed phases*, as both differ in their thermodynamic properties. Without a shadow of doubt, this is important information regarding structure, which is even more attractive when it can be derived from microsamples (see above) and x-ray amorphous compounds. The entire discussion of the phase analysis of solids concerns in essence structure analysis, since the structure of a phase is its most important feature.

The electrochemistry of solid metal hexacyanometalates is well suited to demonstrate the potential of VMP for structure analysis. Figure 7 depicts a part of the lattice of Prussian blue type metal hexacyanometalates. Three different metal positions can be distinguished: (1) carbon coordinated metal ions; in the case of Prussian blue these are low-spin iron(II) ions of the hexacyanoferrate units; (2) nitrogen-coordinated metal ions; in the case of Prussian blue these are high-spin iron(III) ions; (3) metal ions in the holes of the cubes (so-called interstitial metal ions). Preferably, the latter are alkali metal ions, but there is evidence that the same metal ions which are nitrogen-coordinated can also be present in these positions. Figure 26 shows cyclic voltammograms of some metal hexacyanoferrates that were immobilized on the surface of a PIGE. They all show the hexacyanoferrate system, only manganese hexacyanoferrate having a second system due to the redox activity of the manganese ions coordinated to the nitrogen of the cyanide ions. Figure 27 illustrates very

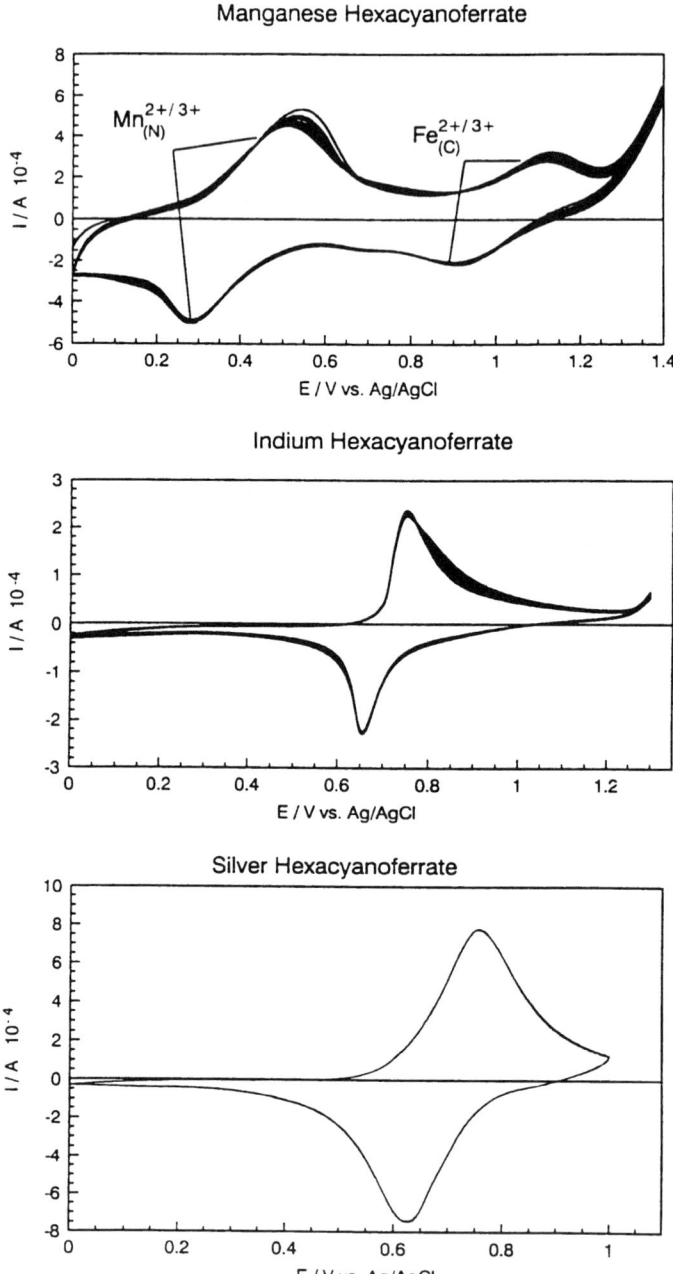

FIG. 26. Cyclic voltammograms of three metal hexacyanoferrates. Electrolyte: 0.1 M KNO_3; scan rate 0.1 V sec^{-1}.

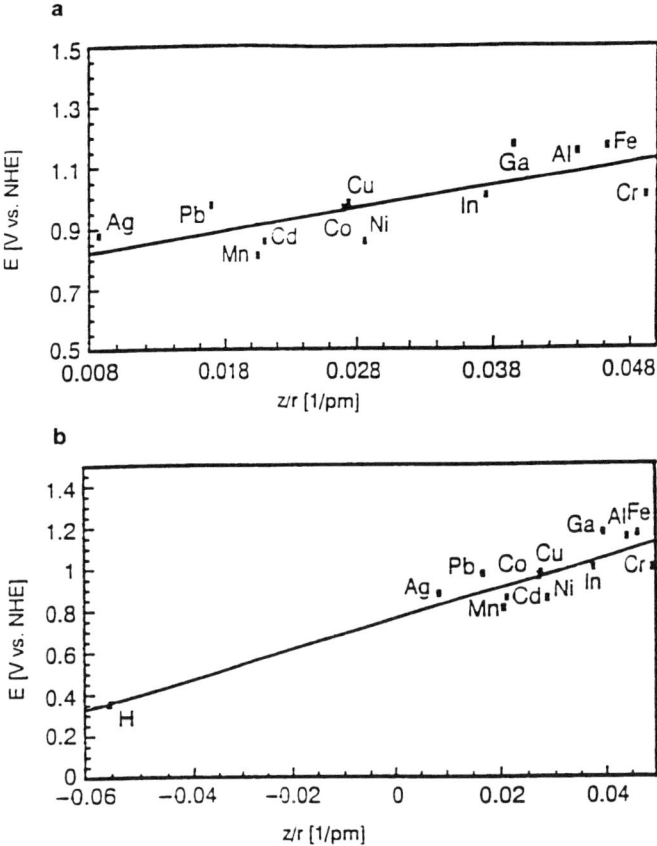

FIG. 27. (a) Plot $E_{s,f}^0$ of solid metal hexacyanoferrates M'[M"(CN)$_6$] versus the ratio z/r of the effective charge and radius of the metal ions M'. (b) The same plot as in (a) but with the standard potential E_{aq}^0 of the hexacyanoferrate ions in water as well as the ratio z/r for protons. (From Ref. 92.)

well how the peak potentials vary from one metal hexacyanoferrate to the other. It was shown that the formal potential of the hexacyanoferrate system depends on the ionic potential (formal charge divided by the effective radius) of the nitrogen-coordinated metal ions [92]. This study was performed to achieve a better understanding of the factors that control the thermodynamics of solid state electrochemical reactions, which proceed in the presence of an adjacent electrolyte solution. The authors suggest that the standard Gibbs energy of a redox process of a solid compound $\Delta G°_s$ can be related to the standard Gibbs energy $\Delta G°_{aq}$ of the same redox center in aqueous solution by the following equation:

$$\Delta G°_s = \Delta G°_{aq} + \Delta G°_{in} + \Delta G°_{it} \tag{5}$$

Here, $\Delta G°_{it}$ denotes the standard free energy of the transfer of the charge compensating counterion, and $\Delta G°_{in}$ is the intrinsic change of standard free energy caused by the different environment, i.e., an aqueous solution in the one case and a solid lattice environment in the other. Unfortunately, it is as impossible to fully independently determine $\Delta G°_{in}$ and $\Delta G°_{it}$ as it is impossible for the reaction to proceed without the compensating ion transfer in the case of the solid compound. The results [92] show that, in case of the hexacyanoferrates, $\Delta G°_{it}$ contributes less than $\Delta G°_{in}$. This tendency can be noted in Table 4. Changing the nature of the exchanged cation shifts the formal potentials of the hexacyanoferrate ions only by some 10 mV, whereas the chemical environment (z/r in Fig. 27) shifts it by several hundred millivolts.

VMP also helps to perform electrochemical measurements with

TABLE 4

Formal Potentials $E_f = (E_{p,a} + E_{p,c})/2$ of Metal Hexacyanometalates Immobilized on a Graphite Electrode and Immersed in 0.1 M Solutions of Alkali Metal Chlorides (scan rate 0.1 V sec^{-1})

Cu[Fe(CN)$_6$][a]	Ni[Fe(CN)$_6$] E_f[V vs. SHE]	Cd[Fe(CN)$_6$]	ΔG_h [kJ mol^{-1}][b]
Li$^+$ 0.848	0.693	0.664	−114.6
Na$^+$ 0.808	0.602	0.673	−89.7
K$^+$ 0.888	0.768	0.768	−73.5
Rb$^+$ 1.008	0.773	1.002	−67.5
Cs$^+$ 0.775[c]	0.533	0.923	−60.8

[a]The formulas M′[M″(CN)$_6$] indicate only the metal M′ coordinated to nitrogen and the metal M″ coordinated to carbon.
[b]Standard free energy of hydration of alkali metal ions.
[c]The voltammetric system was poorly developed.
Source: Ref. 92.

compounds which are thermodynamically unstable and which could not be synthesized on the surface of an electrode. Iron(III) hexacyanochromate(II) (Fe hcc) is such an example. This compound is known to isomerize at a temperature above 100°C to chromium(III) hexacyanoferrate(II) (Cr hcf), a compound that is thermodynamically stable at room temperature. Fe hcc could be synthesized by a bulk precipitation but it was not possible to achieve the electrochemical deposition of a Fe hcc film on an electrode. When Fe hcc was cyclically oxidized and reduced after mechanical immobilization on a PIGE, one observed an electrochemically initiated isomerization at room temperature while the kinetics of this process can be observed using cyclic voltammetry or chronocoulometry [93]. Figure 28 shows cyclic voltammograms recorded during isomerization. Figure 29 depicts the isomerization of iron hexacyanomanganate (Fe hcm) to manganese hexacyanoferrate (Mn hcf).

The voltammetry of microparticles is also a powerful tool for study-

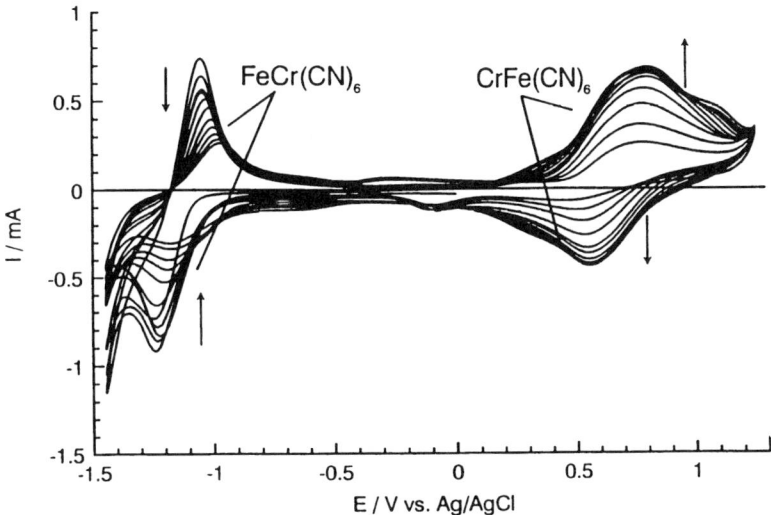

FIG. 28. Cyclic voltammograms of iron(II) hexacyanochromate(III) (Fe hcc) mechanically immobilized on a PIGE. The isomerization of Fe hcc to chromium(III) hexacyanoferrate(II) (Cr hcf) can be deduced from the simultaneous decrease of the hexacyanochromate system accompanied by the growing hexacyanoferrate system. Electrolyte 0.1 M KCl, scan rate 0.1 V sec^{-1}. (From Ref. 93.)

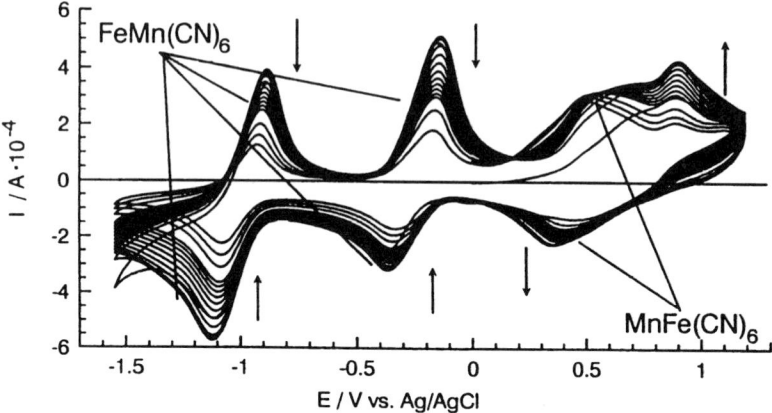

FIG. 29. Cyclic voltammograms of iron(II) hexacyanomanganate(III) (Fe hcm) mechanically immobilized on a PIGE. The isomerization of Fe hcm to manganese(II) hexacyanoferrate(II) (Mn hcf) can be deduced from the simultaneous decrease of the hexacyanomanganate signals accompanied by the growing of the hexacyanoferrate signal. Electrolyte 0.1 M KCl, scan rate 0.1 V sec^{-1}.

ing electrochemically initiated lattice substitution reactions of metal hexacyanometalates. The first example concerned the substitution of high-spin iron ions of Prussian blue by cadmium ions, a reaction that led to the formation of cadmium hexacyanoferrate [52]. Later it showed that these substitution reactions are based on an *insertion-substitution* mechanism [94]. At first, the foreign metal ion enters the lattice of the parent metal hexacyanoferrate during a reduction step. Then the inserted metal ion replaces the N-coordinated metal ion and leaves the replaced metal ion in the interstitial place. From this location, it is removed into the solution during the following oxidation. In this way it is possible to switch all equilibria onto the side of the new metal hexacyanoferrate which covers the parent compound as an epitaxial layer. Figure 30 gives a few examples of such reactions. It is clearly visible in the figure how the signal of the new hexacyanoferrate is growing while the signal of the parent hexacyanoferrate is decreasing simultaneously. The mechanism of the insertion-substitution reaction has been deduced from plots of peak current ratios of parent and daughter hexacyanoferrates versus the scan rate. If the daughter compound epitaxially covers the parent compound, then the diffusion layer

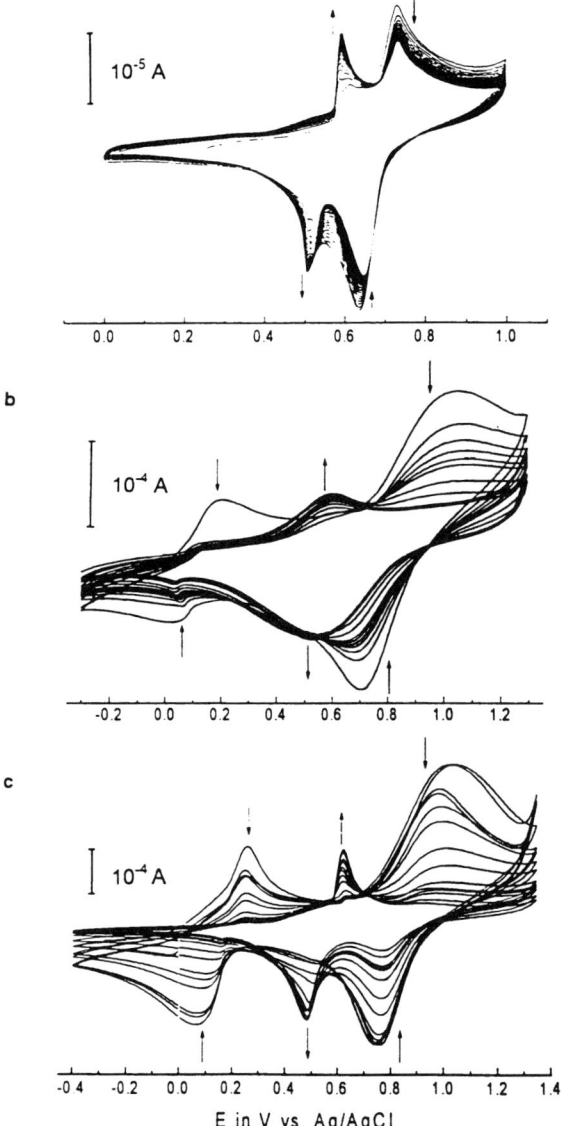

FIG. 30. Multiple cyclic voltammograms recorded during the formation of (a) cadmium hexacyanoferrate on silver hexacyanoferrate, (b) nickel hexacyanoferrate on Prussian blue, and cadmium hexacyanoferrate on Prussian blue. Scan rate: 0.1 V sec^{-1}. Electrolyte solutions for (a) and (c) 0.1 M cadmium nitrate; (b) 0.1 M nickel nitrate. (Adapted from Ref. 94.)

will advance first through the daughter compound layer and only subsequently through the parent compound core. Thus, the plot of the current ratio versus the scan rate will drop to zero. If the two compounds are equally accessible for the reaction, as in the case of a powder mixture, then this peak current ratio will be independent on the scan rate. Figure 31 illustrates this behavior for the nickel hexacyanoferrate layer on Prussian blue. Figure 32 provides a scheme depicting the electrochemically initiated substitution reactions which can be performed with hexacyanoferrates immobilized on an electrode surface. From a theoretical treatment of the progression of a solid state electrochemical reaction through the solid phase [95], it follows that the reaction layer starts to grow at the three-phase boundary electrode-solid-solution and expands from there with exponential borderlines (see Sec. III.H). This does not alter the principle of the reaction pathway shown in Fig. 32. The electrochemical conversion of silver(I) hexacyanoferrate(III) into Prussian blue was used as the working principle of an iron sensor [96].

Voltammetric studies of microparticles have been used to analyze

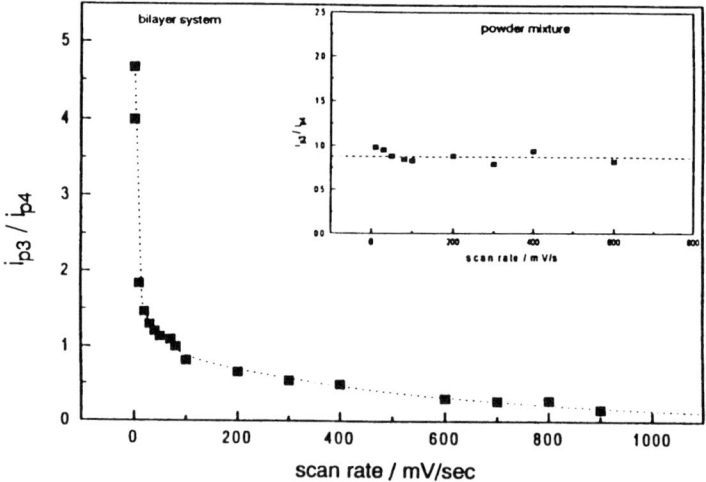

FIG. 31. Plots of the ratio of peak current (I_{p3}: peak current of Cd hcf, I_{p4}: peak current of Ag hcf) versus the scan rate for the bilayered system of cadmium hexacyanoferrate on silver hexacyanoferrate and for a powder mixture of these two compounds. (From Ref. 94.)

Voltammetry of Microparticles

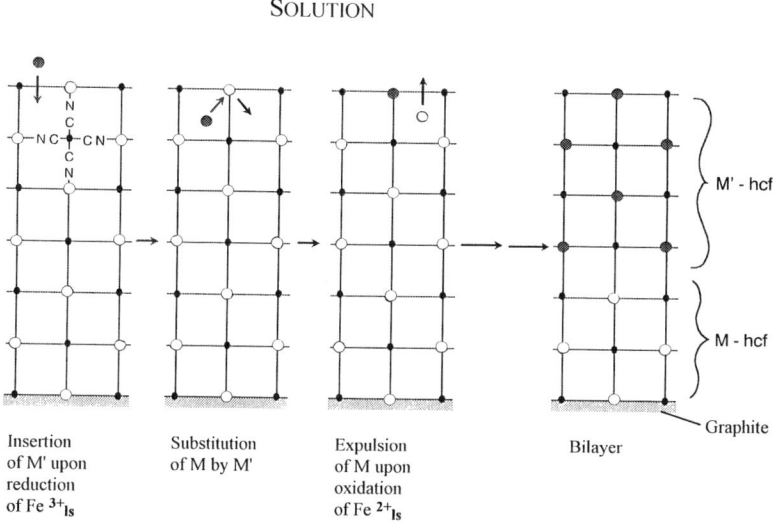

FIG. 32. Mechanism of the insertion-substitution reactions leading to the formation of a bilayered system of parent and daughter hexacyanoferrates, when the parent hexacyanoferrate is Prussian blue. (From Ref. 94.)

mixed hexacyanoferrates, i.e., solid solutions of one metal hexacyanoferrate in another. This is no simple analytical task. Other analytical methods do not give such unambiguous results as electrochemical measurements. X-ray analysis is difficult since hexacyanoferrates are often almost amorphous, and even if their crystallinity is sufficient, they will have very similar lattice constants. Then it becomes already difficult to decide whether a precipitate is a mixture of two phases or just one mixed phase, not to mention a quantitative analysis of the mixed phase. IR spectroscopy does not help, either. The solid state voltammetric analysis is superior to all these methods since it fully relies on the thermodynamics of the compounds. A good example is the formation of mixed nickel-iron hexacyanoferrates

[82]. For nickel and iron as N-coordinated metal ions, the position of the hcf signal differs remarkably (see Fig. 33). After the formation of a mixed phase, there is a random distribution of nickel and high-spin iron ions in these positions. The hcf ion "sees" this random distribution and attains a formal potential in between the values for the pure Fe hcf and the pure Ni hcf. In case of a phase mixture, both hexacyanoferrates behave indepen-

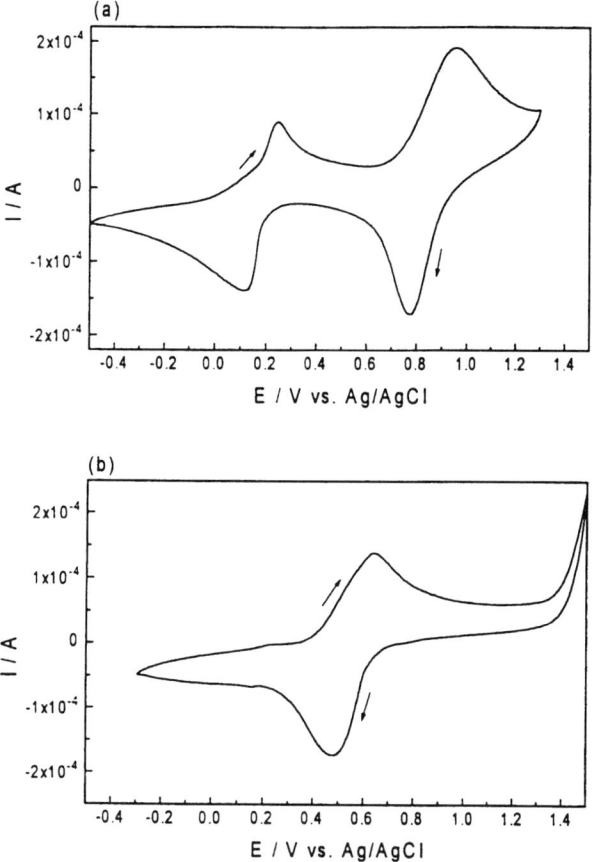

FIG. 33. (a) Multiple cyclic voltammograms of Prussian blue immobilized on a paraffin-impregnated graphite electrode (PIGE) immersed in 0.1 M KCl, scan rate 0.1 V sec^{-1}. (b) Cyclic voltammogram of nickel hexacyanoferrate immobilized on a PIGE immersed in 0.1 M KNO$_3$, scan number 7, scan rate 0.1 V sec^{-1}. (Adapted from Ref. 82.)

Voltammetry of Microparticles

dently and each of the phases contains hcf ions surrounded either by nickel or iron. Thus, one obtains both hcf systems. In the case of hexacyanoferrates, the excess free energy of mixing is obviously very small. The deviation from the linearity of a plot of formal potentials of the hcf system of the mixed compounds versus the mole ratio of nickel and iron is also very small (see Fig. 34) and presumably only due to the mixing entropy.

The voltammetry of microparticles has been proven to be very helpful in studying ion insertion into the host lattice of hexacyanoferrates and also in such cases where no substitution or isomerization occurs. Thus, it has been possible to explain the phenomenon that some metal ions will induce one of the voltammetric systems of Prussian blue to vanish. This was studied in detail for Tl^+, Rb^+, and NH_4^+ ions [97]. It was shown that during lattice contraction taking place when Everitts salt (the fully reduced form of Fe hcf) was oxidized to Prussian blue, the previously mentioned ions were immobilized in the Prussian blue and could not diffuse in this compound. This study also explains why Prussian blue is efficient in the detoxification after thallium poisoning [98] and also internal ^{137}Cs contamination [99]. Since the incorporation of thallium ions into Prussian blue is chemically irreversible when the potential range is confined to appropriate values, this effect can be used for a selective and sensitive preconcentration in stripping voltammetry [100] when graphite–paraffin–Prussian blue composite electrodes are used. The incorporated thallium ions can be reduced to metallic thallium at around –1.4 V vs. Ag/AgCl. During this reduction the

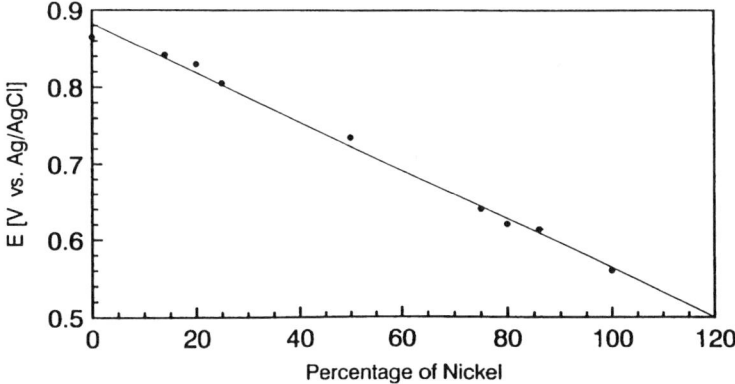

FIG. 34. Plot of formal potential vs. percentage of nickel in mixed hexacyanoferrates. (From Ref. 82.)

thallium ions are substituted by potassium ions from the solution. Reoxidation of the metallic thallium leads to a resubstitution of the potassium ion by the thallium ions.

Downard et al. [101] applied VMP in their studies of solid and water insoluble cobalt complexes with bis(2-(dimethylarsino)phenyl) methylarsine (mtas). These compounds gave electrochemically reversible responses according to the following reaction equations:

$$\{[Co(mtas)_2](ClO_4)_3\} + e^- \Leftrightarrow \{[Co(mtas)_2](ClO_4)_2\} + ClO_4^- \quad (XVI)$$

$$\{[Co(mtas)_2](ClO_4)_2\} + e^- \Leftrightarrow \{[Co(mtas)_2](ClO_4)\} + ClO_4^- \quad (XVII)$$

Figure 35 depicts typical multicyclic voltammograms of $\{[Co(mtas)_2](ClO_4)_2\}$. They show the expulsion of anions usually accompanying reduction processes to maintain the charge balance. When instead of measuring the perchlorates of the cobalt complexes, the attempt was made to measure the voltammograms of the tetraphenylborates, no response was obtained (unless the latter anions had been oxidatively destroyed at a very positive potential). This shows that tetraphenylborate ions are not released from the solid, presumably due to strong hydrophobic interaction between the organic parts of the anion and cation. Most remarkably, the authors observed that the formal potentials of the redox systems are almost identical for both acetonitrile solutions of the complexes and immobilized solids with an adjacent aqueous solution (see Table 5). The authors believed that this is probably not the result of compensating actions. They assumed that the ion transfer in the case of the reacting solid did not contribute to the overall thermodynamics of the reaction. They further assumed that the environment of the redox centers did not change considerably once the solid was dissolved. These results are very interesting in comparison to those obtained for the relatively small hexacyanoferrate ions (see discussion above).

F. Determination of Thermodynamic Data of Solid Compounds

The question whether the voltammetry of microparticles lends itself to determining thermodynamic constants of solid compounds deserves special attention, as it is of great importance to solid state chemistry. Thermodynamic data can be derived from measurements of characteristic potentials,

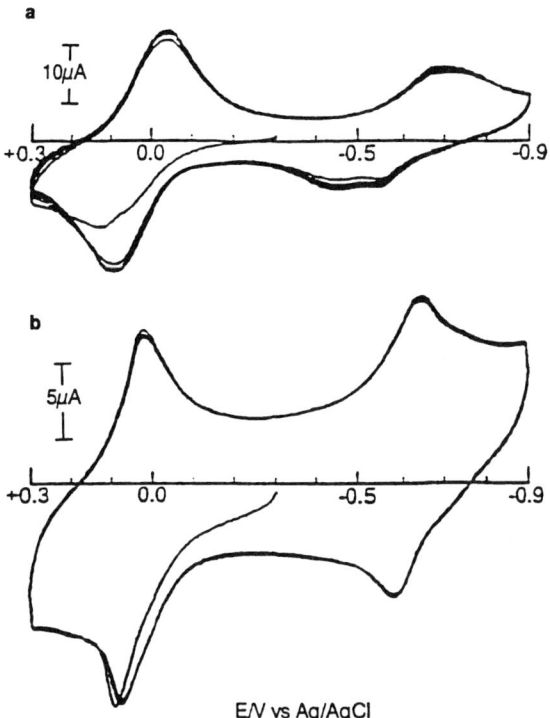

FIG. 35. Cyclic voltammograms obtained in aqueous (0.1 M LiClO$_4$) electrolyte for the solid [Co(*mtas*)$_2$](ClO$_4$)$_2$ which was mechanically attached to a basal plane pyrolyte graphite electrode (scan rate = 100 mV sec^{-1}): (a) first six cycles: (b) cycle after 40 repeat cycles. (From Ref. 101.)

e.g., peak potentials in cyclic or linear sweep voltammetry, especially when the rate of the electrochemical process is not kinetically controlled. Thus, electrochemical reversibility is the most reliable prerequisite for a thermodynamic interpretation of potentials.

As discussed in Sec. III.B, electrochemical reactions coupled with an ion insertion exhibit in many cases a reversibility that allows a thermodynamic interpretation of the data. For correctly understanding the formal potentials of these systems, serious problems still remain. First, the formal potentials determined as $E_f = 1/2\ (E_{pa} + E_{pc})$ of the peak system in cyclic voltammetry characterize both an electron and ion transfer equilibrium.

TABLE 5
Voltammetric Data (scan rate = 100 mV sec^{-1}) for Cobalt Complexes[a]

[Co(mtas)$_2$]X$_n$					$E_{1/2}$/V[c]		
X	n	Phase	Electrolyte (0.1 M)	Electrode[b]	Co(III)/Co(II)	Co(II)/Co(I)	Co(I)/Co(0)
BF$_4^-$	3	Solution	CH$_3$CN/Bu$_4$NClO$_4$	gc	0.04 (60)	−0.61 (60)	−1.85 (66)
BPh$_4^-$	2	Solution	CH$_3$CN/Bu$_4$NPF$_6$	gc	0.04 (62)	−0.61 (62)	−1.86 (74)
BPh$_4^-$	2	Solution	CH$_2$Cl$_2$/Bu$_4$NPF$_6$	gc	0.14 (86)	−0.57 (78)	−2.05[d]
ClO$_4^-$	3	Solid	H$_2$O/LiClO$_4$	bpg	0.05 (54)	−0.61 (90)	
ClO$_4^-$	2	Solid	H$_2$O/LiClO$_4$	bpg	0.04 (50)	−0.61 (60)	
BPh$_4^-$	2	Solid	H$_2$O/LiClO$_4$	bpg	0.03 (36)	−0.63 (60)	
BF$_4^-$	3	Solution	H$_2$O/LiCl	bpg	−0.13 (60)		
					−0.06 (40)[c]		
BF$_4^-$	3	Solution	H$_2$O/LiCl	gc	−0.13 (72)	−0.60 (120)[c]	

[a] 1 mM complex for solution studies.
[b] gc = Glassy carbon; bpg = basal plane graphite.
[c] $E_{1/2} = (E_p^{ox} + E_p^{red})/2$; Corrected to V vs. SCE (nonaqueous solvents), V vs. Ag/AgCl (3 M NaCl) (aqueous electrolytes); values in parentheses indicate ΔE_p^{red} for irreversible reduction.
[d] Adsorption controlled response.

Source: Ref. 101.

Voltammetry of Microparticles

The two equilibrium constants cannot be separately determined. If ox denotes the oxidized form and red the reduced form of a redox center in the solid compound, and if Me⁺ is the cation inserted into the solid compound during reduction, the electrode reaction, with its standard potential $E°_{s,f}$, can be formulated according to the following:

$$\{ox\} + e^- + Me^+ \Leftrightarrow \{redMe\} \quad \text{(XVIII)}$$

This electrode reaction can be separated into two independent equilibria:

$$\{ox\} + e^- \Leftrightarrow \{red^-\} \quad \text{(XIX)}$$

with the standard potential $E°_s$, and

$$\{redMe\} \Leftrightarrow \{red^-\} + Me^+ \quad \text{(XX)}$$

Equilibrium (XX) is in itself an electrochemical reaction because it describes the distribution of a charged particle between two phases. For the equilibrium (XIX), we can formulate the Nernst equation:

$$E = E°_s + \frac{RT}{nF} \ln \frac{f_{\{ox\}}}{f_{\{red^-\}}} + \frac{RT}{nF} \ln \frac{c_{\{ox\}}}{c_{\{red^-\}}} \quad (6)$$

where $E°_s$ is the standard potential of the redox system in the solid matrix. Equilibrium (XX) can be described by the constant defined in the following:

$$K = \frac{a_{\{red^-\}} a_{Me^+}}{a_{\{redMe\}}} = \frac{f_{\{red^-\}} c_{\{red^-\}} a_{Me^+}}{f_{\{redMe\}} c_{\{redMe\}}} \quad (7)$$

The combination of Eqs. (6) and (7) yields:

$$E = E°_s + \frac{RT}{nF} \ln \frac{f_{\{ox\}}}{f_{\{red^-\}}} + \frac{RT}{nF} \ln \frac{K}{f_{\{red^-\}}} + \frac{RT}{nF} \ln \frac{c_{\{ox\}}}{c_{\{redMe\}}} + \frac{RT}{nF} \ln a_{Me^+} \quad (8)$$

where the first three terms can be regarded as the formal potential $E°_{s,f}$ of reaction (XVIII). Unfortunately, a separate experimental determination of $E°_s$ and K is not possible.

From the preceding equations, it follows that a formal potential, which is per se defined for equal amounts of the reduced and oxidized forms, can be determined by voltammetric measurements. However, the standard potential is not yet defined as there is no agreement upon a standard state for redox centers in a solid. Thus, it has not yet been possible to define and determine the activity coefficients of the subspecies ox, red, and redMe of the solid compound. Nevertheless, the formal potentials of redox species in solid matrices can be determined and compared with each other to study the influence of the solid environment on the redox properties of the redox centers [92] as long as the structures of these compounds are similar.

In some of the published examples, thermodynamic data were determined with a solid compound immobilized on an electrode surface, but the data obtained referred to the *dissolved* compound. The solid sample only served to provide a constant source of dissolved species at the electrode surface. This was an interesting approach for compounds that have a very low solubility. Bond and Scholz [81,102] studied the electrochemical behavior of lead and mercury dithiocarbamates. When these compounds were immobilized on the surface of a graphite electrode, they gave rise to a slowly decreasing response of the electrochemical system:

$$Me(dtc)_2 + 2e^- \Leftrightarrow \{Me\} + 2dtc^- \tag{XXI}$$

as can be seen in Fig. 36. The formal potential $E_f = 1/2\ (E_{pa} + E_{pc})$ of the peak system in cyclic voltammetry was determined by the conditional overall stability constant β_2 of the metal dithiocarbamates in the electrolyte solution that was used. This can be easily seen from a comparison between previously reported values of some lead and mercury compounds with the values determined by Bond and Scholz [81,102], based on this assumption. Provided that the studied pair of peaks in cyclic voltammetry is due to the equilibrium (XXI), the conditional stability constant β_2 of the metal dithiocarbamate can be calculated according to the following equations:

$$E = E^\circ_{Me^{2+}/Me} + \frac{RT}{2F} \ln a_{Me^{2+}} \tag{9}$$

$$\beta_2 = \frac{a_{Me(dtc)_2}}{a_{Me^{2+}} a^2_{dtc^-}} \tag{10}$$

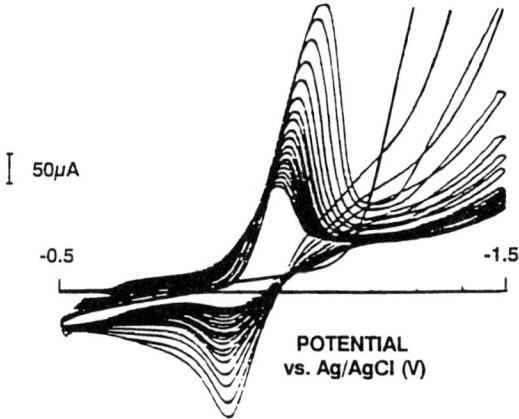

FIG. 36. First 20 cycles of cyclic voltammograms for reduction of Pb(i-Pr$_2$dtc)$_2$ mechanically transferred to a PIGE with aqueous 0.1 M KNO$_3$ as electrolyte: initial potential = –0.50 V vs. Ag/AgCl; scan rate = 100 mV sec^{-1}. (From Ref. 102.)

$$E = E^{\circ}_{\text{Me}^{2+}/\text{Me}} + \frac{RT}{2F}\ln\frac{1}{\beta_2} + \frac{RT}{2F}\ln\frac{a_{\text{Me(dtc)}_2}}{a^2_{\text{dtc}^-}} \tag{11}$$

$$E^{\circ}_f = \frac{E_{pa} + E_{pc}}{2} = E^{\circ}_{\text{Me}^{2+}/\text{Me}} + \frac{RT}{2F}\ln\frac{1}{\beta_2} + \frac{RT}{2F}\ln\frac{f_{\text{Me(dtc)}_2}}{f^2_{\text{dtc}^-}} \tag{12}$$

Table 6 gives a comparison of conditional stability constants of some dithiocarbamates and dithiophosphates as determined by the method outlined above with values reported in literature. Since the values are in total agreement with each other, the authors [102] determined the conditional overall stability constants of 28 different dithiocarbamates of lead and mercury. This study demonstrated the potential of VMP bearing in mind that such measurements could be performed in a very simple and quick manner, which is just the opposite of what can be said about the usual potentiometric procedures applied to determine stability constants. However, it must be remarked that the method is limited to systems exhibiting a reversible or at least quasi-reversible electrochemistry. In two cases of the studied dithiocarbamates, peak separation in cv approached or exceeded 300 mV and the calculated stability constants were definitely incorrect, at

TABLE 6

Comparison of Stability Constants Determined by Voltammetric Measurements on Solids Attached to a Graphite Electrode with Literature Values

Complex	log β_2	
	Present paper values	Literature values
Hg(Et$_2$dtp)$_2$ in 0.3 M KNO$_3$, 40% EtOH	28.3 ± 0.4	27.7
Hg (Et$_2$dtc)$_2$ in aqueous 0.1 M KNO$_3$	38.2 ± 0.2	38.07 ± 0.3
Pb (Et$_2$dtc)$_2$ in aqueous 0.1 M KNO$_3$	17.7 ± 0.2	17.7 (polarography) 13.3 (extraction)
Pb(pydtc)$_2$ in aqueous 0.1 M KNO$_3$	17.2 ± 0.7	17.1 (polarography) 16.8 (extraction)

Source: Ref. 81.

least for the latter case, because a deviating value is reported in literature. But in all other cases, with peak separations around 100–150 mV, reliable determinations of the stability constants were possible. Finally, it should be remarked that the authors were enabled by this comprehensive study to discover an unknown linear dependence of the logarithms of stability constants on the molecular weight of the complexes [102]. This new correlation could only be found, for there were many data of one series of complexes determined where the chelating group was practically unchanged, whereas the organic tail of the ligands varied considerably. Figure 37 shows this dependence for the studied complexes.

Since the observed signals of lead and mercury dithiocarbamates originated from the dissolved species, the formal potentials of the system (XXI) were strongly correlated with the values obtained by conventional solution cyclic voltammetry in dichloromethane [102].

Another type of electrochemical reaction which was studied with respect to thermodynamics was the reduction of some sulfide minerals [103,104]. These sulfide minerals exist in two different modifications in the solid state. Examples are xanthoconite and proustite, both of the composition Ag_3AsS_3, and pyrostilpnite and pyrargyrite, both of the formula

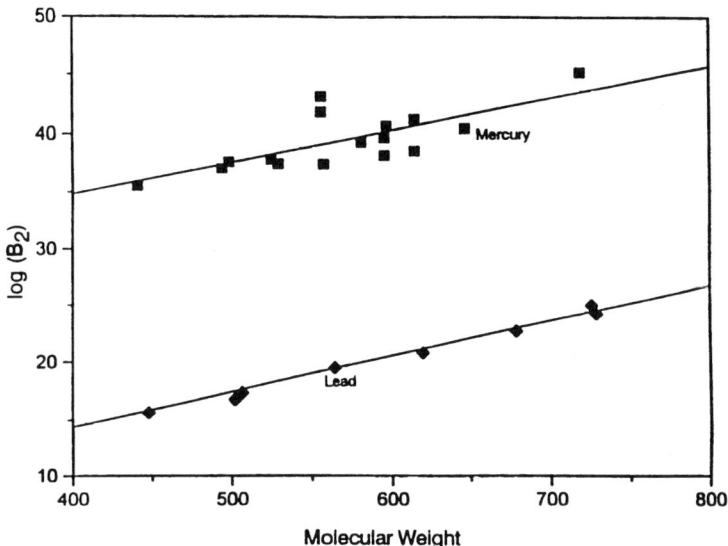

FIG. 37. Plot of log β_2 in aqueous 0.1 M KNO_3 for mercury and lead dithiocarbamate complexes vs. molecular weight. (From Ref. 102.)

Ag_3SbS_3. The electrochemical behavior under AbrSV conditions showed that the following reactions proceeded:

$$\{Ag_3AsS_3\} + 3e^- \Leftrightarrow 3\{Ag\} + AsS_3^{3-} \qquad (XXII)$$

$$\{Ag_3SbS_3\} + 3e^- \Leftrightarrow 3\{Ag\} + SbS_3^{3-} \qquad (XXIII)$$

Interestingly, these reactions are electrochemically reversible, provided that the time scale of the measurement is sufficiently short. This is necessary because the soluble thioarsenate and thioantimonate anions will undergo decomposition reactions if the time is sufficient. Since each pair of modifications is reduced to one and the same dissolved anion and solid silver, the differences in their reduction peak potentials reflect the free energy difference of the two modifications. Thus, it became possible to determine the Gibbs free energy of transformation of these two pairs of mineral modifications.

G. In Situ and Ex Situ Combinations of the Voltammetry of Solid Microparticles with Other Analytical Techniques

The technique of mechanically immobilizing solid particles on electrode surfaces allows the application of all electrochemical techniques. But this will certainly never suffice for a complete study of the processes. Therefore, it is very important to combine the electrochemical measurements with other independent analytical techniques. In situ combinations are preferable but, unfortunately, not always possible. Table 7 gives a survey of the described combinations.

Scanning electron microscopy equipped with an energy dispersive x ray detector (SEM/EDX) is of great value for the study of the composition of immobilized particles before and after electrochemical reactions. To perform an SEM/EDX measurement, it is necessary to excise the top 2 mm of the graphite electrode with a suitable saw in such a way as to obtain a disc. This disc can be easily mounted on the sample table of the microscope. Although SEM/EDX can only be used ex situ, it allows one to determine exactly the elemental composition of immobilized microparticles, thus helping considerably in the elucidation of reaction pathways.

A quartz microbalance can be applied for in situ studies. Until now, only gold electrodes have been used. The immobilization of microparticles was achieved with the help of a cotton swab.

X-ray diffraction is of special interest, but an in situ application is accompanied by several problems. Conventional x-ray tubes have radia-

TABLE 7

In Situ and Ex Situ Combinations of Voltammetry of Microparticles with Other Analytical Techniques

Technique	In or ex situ	Ref.
X-ray diffraction	In situ	68
X-ray diffraction	Ex situ	49
Electron microscopy/EDX	Ex situ	59, 52, 73, 74 97, 101
Quartz microbalance	In situ	48, 49, 52, 53
Light microscopy	In situ	71
Diffuse reflectance spectroscopy	In situ	105
Diffuse reflectance spectroscopy	Ex situ	101
Infrared spectroscopy	Ex situ	47, 116
UV-VIS spectroscopy	In situ	115

tion intensities that do not allow the electrolyte film between the cell window and the electrode to be thicker than a few tenths of a mm. Further, the sample should give a strong diffraction pattern, which sometimes needs larger amounts of sample. All these restrictions limit the application of this technique to suitable examples.

In situ incident light microscopy is very valuable for electrochromic systems. With the help of a conventional diode array spectrometer coupled to the microscope using a light guide, an in situ recording of diffuse reflection spectra is possible [105]. Figure 38 shows the two spectra of silver oc-

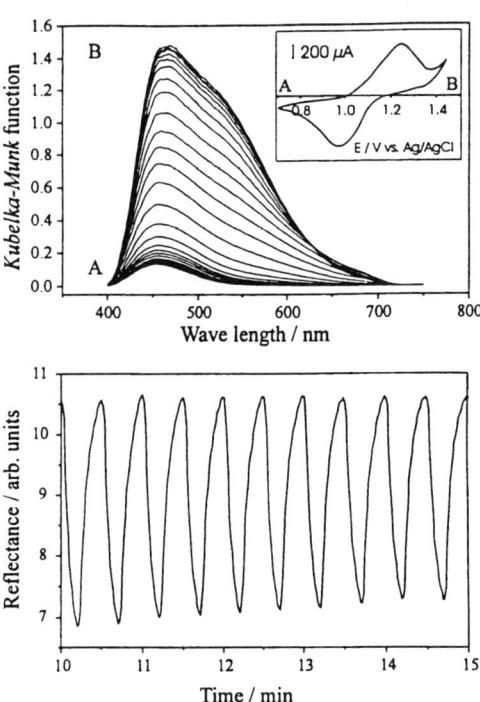

FIG. 38. (Top) *Kubelka-Munk* function of silver octacyanomolybdate(V) (B) and silver octacyanomolybdate(IV) (A) as measured under a light reflection microscope equipped with a diode array spectrometer. The insert shows the cyclic voltammogram of solid silver octacyanomolybdate(IV/V); electrolyte 0.1 M $AgNO_3$, scan rate 50 mV sec^{-1}. (Bottom) Absolute light reflection intensity of solid silver octacyanomolybdate(IV/V) immobilized on the surface of a graphite electrode during continuous cyclic voltammetry. Wavelength 560 nm, electrolyte 0.1 M $AgNO_3$, scan rate: 50 mV sec^{-1}.

tacyanomolybdate(V) and -(IV) together with a cyclic voltammogram of the system:

$$\{Ag_4[Mo^{(IV)}(CN)_8]\} \Leftrightarrow \{Ag_3[Mo^{(V)}(CN)_8]\} + Ag^+ + e^- \quad (XXIV)$$

Figure 38 also depicts the reflection trace at 560 nm during the continuous cyclic voltammetry of the solid compound between 0.7 and 1.45 V vs. Ag/AgCl in a 0.1 M $AgNO_3$ solution.

H. Problems of a Theoretical Description of the Voltammetry of Microparticles

The vast majority of experimental setups in classical electrochemistry exhibits electrodes with a well-defined interface of two phases. In VMP the electrode is always a multiphase system. In the simplest case, the electrode will consist of at least three phases, i.e., one liquid and two solid phases. One phase is the solid support, serving both as the electrical conductor and also to keep the solid sample particles immobilized. The second solid phase is the sample. The third phase is the electrolyte solution, which is necessary for the ion exchange accompanying every electrochemical reaction to ensure charge neutrality in the phases. The electronic conductivity of the sample phase will determine the potential difference between the sample phase and the electrolyte solution. If this conductivity is very high, the potential difference will be equal to that existing between the support (e.g., graphite) and the solution. In this case the electrode reaction can, if not otherwise inhibited, proceed on the entire surface of the sample particles. Metal and alloy particles will react in this way.

However, when the conductivity of the sample is very low, i.e., when it is an electronic insulator, one may wonder how an electrochemical reaction can proceed at all. Experimental evidence for such electrochemical reactions (e.g., 47, 48, 49, 53, 59, 77) would indicate that the only possible reaction place is the three-phase boundary between the electronically conducting support, the sample, and the electrolyte solution (Fig. 39). At this three-phase boundary, electrons can be exchanged between the support and the sample and ions between the sample and the solution. In a recent study [95] the propagation of an electrochemical reaction through a solid was considered for the case of a mixed electronic and ionic conductivity of the solid. A coupled diffusion of electrons and ions within the crystal lattice is separately treated with two differential equations. The redox reaction is initiated

Voltammetry of Microparticles

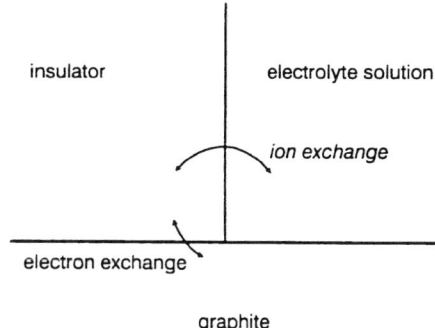

FIG. 39. Three-phase electrode supporting both an ion transfer between the solid compound and the electrolyte solution and also the electron transfer between the solid compound and the electron conductor.

by the polarization of the three phase boundary, where the crystal is in contact with both the electrode and the solution. From this contact line the redox reaction advances on the surface and into the crystal body by the diffusion of ions and the conductance of electrons. The theoretical treatment of the very simple model of the three-phase boundary, which exists when a crystal is attached to a solid electrode surface, allows the following conclusions:

1. The three-phase boundary is always the starting line for the reaction front, independent of the geometry of the particle and its conductivity.
2. The reaction will be surface-confined if the diffusion of ions into the bulk of the crystal is impossible or very slow.
3. Generally, the net current is the sum of surface current and bulk current.
4. In case that both the surface and the bulk reaction proceed at comparable rates, the reaction front expands from the three-phase boundary as shown in Fig. 40.
5. The surface current will be negligible in cases where the bulk reaction is dominating.

The voltammetric behavior of azobenzene microcrystals is an example where the electrode reaction is confined to the surface of the solid materials [106].

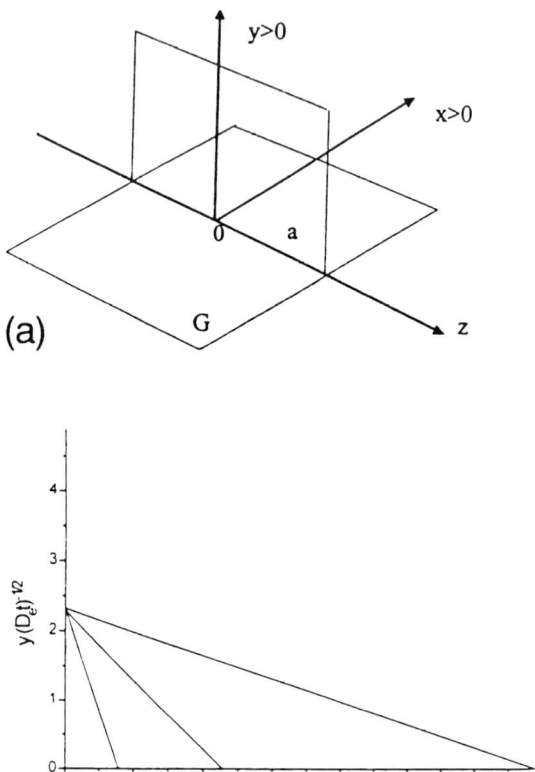

FIG. 40. (a) Coordinate system with a part of the crystal/solution interface, a, and a part of the graphite surface, G. (b) Isoconcentration lines $c_B = c_{B,x=0,y=0}/10$ in the dimensionless volume of the solid particle for $D_e = 9D_c(1)$, $D_e = D_c(2)$ and $D_e = D_c/9(3)$. D_e is a diffusion coefficient of electrons and D_c is a diffusion coefficient of ions. x and y are the coordinates as in (a). (From Ref. 95.)

Voltammetry of Microparticles

Electrochemical reactions at a three-phase boundary—solid electrode (e⁻ conductor)/electrolyte solution (ion conductor)/organic liquid—were described for the electrolysis of emulsions [107] (although in most cases, the electrolysis of microemulsions proceeds as a result of solution equilibria [108]). It is well known that such three-phase boundaries as electrode/solution/gas play an important role in electrochemical reactors [109]. Electrochemical studies of insulators are still seminal, but from what has been published, it seems rather unlikely that direct electron or hole injection can play any significant role in voltammetry of microparticles because the voltages necessary for such reactions are not accessible in these experiments [110].

Another problem for a theoretical treatment is the morphology of the sample distributed on the electrode surface. Microcrystals of the sample are scattered on a usually solid support, e.g., the graphite electrode. These microcrystals have a certain size distribution and may have a preferential orientation due to the immobilization procedures. The microcrystals will expose their different faces to the solution and also to the solid support. Thus, their polycrystallinity can decisively influence their electrochemical properties.

Further difficulties arise from the great variety of different mechanisms of the electrode reactions studied in VMP. This can lead to the involvement of additional solid phases, including the direct conversion of one solid phase into another [49,68]. Thus, the number of phases can vary during the electrochemical reaction, and the interfacial areas are no longer independent of time. Instead of well-defined interfaces, it is also possible that a continuous transformation from one phase to the other occurs through mixed-phase formation. Thus, in case of silver halides being reduced to silver metal, the voltammograms can be simulated with a model assuming the formation of mixed crystals between the silver halide and silver metal [111], an assumption supported by independent data.

The transformation of one solid phase into another solid phase cannot always proceed along a continuous series of mixed crystal phases. This happens when the system of the starting phase, e.g., {ox}, and the product phase, e.g., {$C_n$red}, exhibits a miscibility gap. This means that mixed crystals $\{(ox)_x(C_n red)_{1-x}\}$ for all possible values of $0 \leq x \leq 1$ do not exist. Such an immiscibility gap of the solid phases can lead to a considerable splitting of the anodic and cathodic peaks in cyclic voltammetry when there is also no "solution pathway" for the reaction to proceed upon [112]. Thus, the miscibility gap between the oxidized and reduced phases leads to an inert zone as described, for example, for TCNQ [49]. Figure 41 depicts

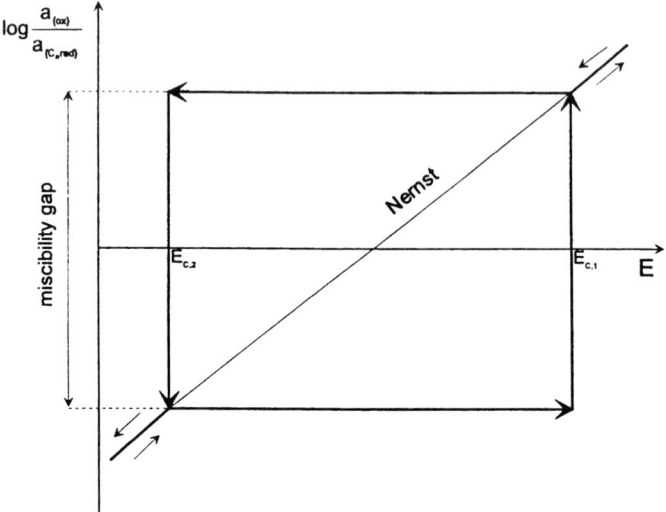

FIG. 41. Schematic plot of the logarithm of the acitvity ratios of the oxidized to the reduced form in the solid state as a function of electrode potential. (From Ref. 112.)

the change of the composition of the solid phases in the course of an electrochemical transformation with a miscibility gap. Figure 42 shows a simulated cyclic voltammogram for a certain case of immiscibility. Whenever a miscibility gap exists the formation of the new phase will be associated with nucleation-growth kinetics. Then, the latter can be deduced from the voltammetric signal as discussed in Sec. III.B.5 for the case of TCNQ. The occurrence of an inert zone (splitting of the voltammetric signals with a potential zone without faradaic current in between) as a result of immiscibility is not a purely thermodynamic phenomenon. It will occur only when there is no formation of an oversaturated mixed phase which could continuously disproportionate into the border phases. However, especially when the structures of the two stable border phases $\{(ox)_x(C_n red)_{1-x}\}$ and $\{(ox)_y(C_n red)_{1-y}\}$—with x being near 1 and y being near 0—are very different, it may not be possible to enter the range between x and y, preventing a disproportionation into the stable border phases, although thermodynamics would demand it.

As far as the inherent heterogeneity of the electrode is concerned, the

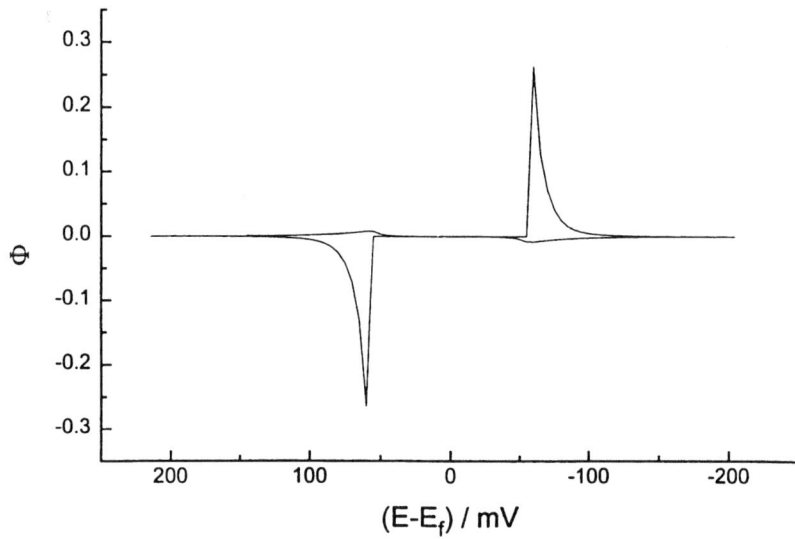

FIG. 42. Staircase cyclic voltammetry of a small hemispherical solid electroactive microparticle with a dimensionless radius $5r_0 v^{1/2} (D\Delta E)^{-1/2} = 20$. Reversible redox reaction influenced by the limited miscibility of components: $Z_{red/ox} = Z_{ox/red} = 0.1$. (From Ref. 112.)

question is raised as to what the rate of the electron transfer between different solid particles is. This question was studied in the case of the electrochemical reduction of Cu_2Se particles immobilized on a rough glassy carbon electrode [50]. In the presence of thiocyanate ions, a remarkable shift in the reduction signal toward more positive potentials was observed accompanied by an increase in peak currents and a decrease in the half-width of peaks. The shift in peak potentials followed an adsorption isotherm of thiocyanate ions. These results led to the assumption that adsorbed thiocanate ions increase the rate of electron transfer due to a bridging of Cu_2Se particles and the glassy carbon surface (Fig. 43). In this case the action of the thiocyanate ions resembled its role in the well-known electrocatalytic reduction of nickel and other ions [113,114]. The electron transfer between different solid particles has to be definitely studied in much more detail in the future.

We are still far away from having a sufficient theoretical understanding of the processes associated with the electrochemistry of immobilized

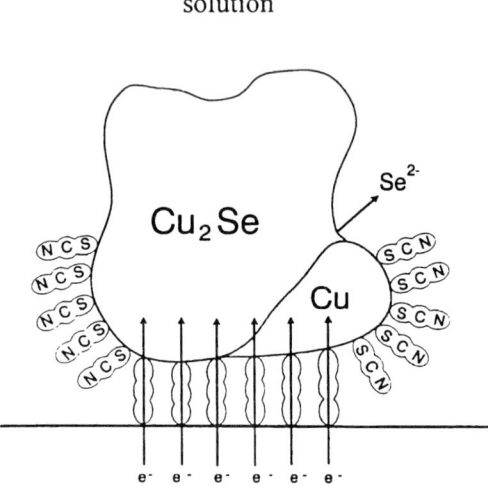

FIG. 43. Schematic situation at the interface between the glassy carbon electrode and mechanically immobilized copper(I) selenide particles in the presence of thiocyanate ions. (From Ref. 50.)

solid particles. The growing amount of experimental results will certainly help in developing models for them.

IV. CONCLUSIONS

The voltammetry of microparticles has been developed to win access to the electrochemistry of solid materials, regardless of their individual physical and chemical properties. In the short time that has elapsed since its introduction, this technique has proved successful in studying a great variety of different solid materials. These studies have provided new insight into the mechanisms of electrochemical reactions of solid compounds. Electrochemical measurements using immobilized solid microparticles have the potential to become a standard technique for the investigation of solid materials. They can assist in the development of new electrode materials, sensors, electrochromic devices, batteries and accumulators, new plating systems, and many other fields of application of solid materials in electronics. In addition, the voltammetry of microparticles will expand our insight into the structure-reactivity relations of solid compounds, making electrochemistry a versatile tool for the characterization of solids. It will deepen our understanding of electrochemical reaction mechanisms at multiphase electrodes and will contribute to the theory of electrochemistry.

ACKNOWLEDGMENTS

It is a great pleasure to acknowledge co-workers and Ph.D. and diploma students who contributed to the development of the voltammetry of microparticles. One of them, Dr. Birgit Meyer, co-authors this review. All others shall be thanked here as well: Dipl.-Chem. Aleš Dostal, Dipl.-Chem. Michael Hermes, Dipl.-Chem. Heike Kahlert, Dr. Günter Kauschka, Dr. Georg Kubsch, Dipl.-Chem. Britta Lange, Dr. Lutz Nitschke, Dipl.-Chem. Fazil Rabi, Dipl.-Chem. Uwe Schröder, and Dr. Song Zhang.

Further thanks must go to: Prof. Dr. Hans-Joachim Bautsch, Dr. Ferdinand Damaschun, and Dr. Gerd Wappler (all of them from the Museum für Naturkunde, Humboldt-Universität zu Berlin), Dr. Wolf-Dieter Müller (Charité Berlin), and Prof. Dr. Günter H. Moh (Mineralogisch-Petrographisches Institut, Universität Heidelberg, deceased).

The following guest scientists are particularly thanked for their share in the development of the voltammetry of microparticles: Dr. Šebojka Komorsky-Lovrić (Zagreb, Croatia), Dr. Milivoj Lovrić (Zagreb, Croatia), Dr. Nina Fedorovna Zakharchuk (Novosibirsk, Russia), Prof. Dr. Anna Brainina (Ekaterinburg, Russia), Prof. Dr. S. Jayarama Reddy (Tirupati, India), Dr. Alexander Jaworski (Warsaw, Poland), and his Ph.D. supervisor, Prof. Dr. Zbigniew Stojek (Warsaw, Poland).

It is my most sincere wish to finally thank Prof. Dr. Alan M. Bond (Melbourne, Australia) for his long-lasting friendship, enthusiastic support and for all his contributions to the development of the voltammetry of immobilized particles. Deutsche Forschungsgemeinschaft and Fonds der Chemischen Industrie are acknowledged for their generous funding of the project, as well as Humboldt-Universität for providing the basic facilities for research.

REFERENCES

1. D. O. Raleigh, in *Electroanalytical Chemistry*, Vol. 6 (A. J. Bard, ed.), Marcel Dekker, New York, 1973, pp. 87–186.
2. H. Rickert, *Electrochemistry of Solids, An Introduction*, Springer-Verlag, Berlin, 1982.
3. P. G. Bruce, ed., *Solid State Electrochemistry* Cambridge University Press, Cambridge, 1995.
4. J. Lipkowski and P. N. Ross, eds., *Electrochemistry of Novel Materials, Frontiers of Electrochemistry*, VCH, New York, 1994.
5. M. D. Ward, *Electrochemical Aspects of Low-Dimensional Molecular Solids*, in *Electroanalytical Chemistry*, Vol. 16 (A. J. Bard, ed.), Marcel Dekker, New York 1966, p. 181.
6. F. Scholz and B. Lange, Trends Anal. Chem. *11*:359 (1992).
7. F. Scholz and B. Meyer, Chem. Soc. Rev. *23*:341 (1994).
8. K. Z. Brainina, E. Y. Neyman, and V. V. Slepushkin, *Inversionnye elektroanaliticheskie metody*, Khimiya, Moskva, 1988.
9. H. Kaesche, *Die Korrosion der Metalle*, Springer-Verlag, Berlin, 1990.
10. F. Scholz, L. Nitschke, and G. Henrion, Naturwissenschaften *76*:71 (1989).
11. F. Scholz, L. Nitschke, G. Henrion, and F. Damaschun, Naturwissenschaften *76*:167 (1989).
12. F. Scholz, GIT Fachz. Lab. *36*:917 (1992).
13. K. Brainina and E. Neyman, *Electroanalytical Stripping Methods*, John Wiley & Sons, New York, 1993.
14. K. Z. Brainina and M. B. Vidrevich, J. Electroanal. Chem. *121*:1 (1981).
15. V. A. Chanturiya and V. E. Vigdergauz, *Elektrokhimiya sulfidov, teoriya i praktika flotatsii*, Nauka, Moskva, 1993.
16. V. V. Slepushkin, Zh. analit. khimii *42*:606 (1987).
17. V. V. Slepushkin, B. M. Stifatov, and E. J. Neyman, Zh. analit. khimii *49*:911 (1994).
18. T. Kuwana and W. G. French, Anal. Chem. *36*:241 (1964).
19. V. I. Belyi, T. P. Smirnova, and N. F. Zakharchuk, Appl. Surf. Sc. *39*:161 (1989).
20. W. Gruner, J. Kunath, L. N. Kalnishevskaja, J. V. Posokin, and K. Z. Brainina, Electroanalysis *5*:243 (1993).
21. N. F. Zakharchuk, B. Meyer, H. Hennig, F. Scholz, A. Jaworski, and Z. Stojek, J. Electroanal. Chem. *398*:23 (1995).
22. K. Kalcher, J.-M. Kauffmann, J. Wang, K. Vytras, C. Neuhold, and Z. Yang, Electroanalysis *7*:5 (1995).
23. D. Bauer and M. P. Gaillochet, Electrochim. Acta *19*:597 (1974).
24. M. Lamache and D. Bauer, Anal. Chem. *51*:1320 (1979).
25. F. Chouaib, O. Cauquil, and M. Lamache, Electrochim. Acta *26*:325 (1981).
26. M. Eguren, M. L. Tascón, M. D. Vázques, and P. Sánchez Batanero, Electrochim. Acta *33*:1009 (1988).

27. P. Sánchez Batanero, M. L. Tascón Garcia, and M. D. Vázques Barbado, Quimi. Anal. *8*:393 (1989).
28. I. M. Kolthoff and J. T. Stock, Analyst *80*:860 (1955).
29. K. Micka, in *Adv. in Polarography, Proceedings of the 2nd Int. Congress on Polarography, Cambridge 1959* (I. S. Longmuir, ed.), Pergamon Press, New York, 1959, pp. 1182–1190.
30. R. Kalvoda, in *Adv. in Polarography, Proceedings of the 2nd Int. Congress on Polarography, Cambridge 1959* (I. S. Longmuir, ed.), Pergamon Press, New York, 1959, pp. 1172–1181.
31. M. R. Dausheva and O. A. Songina, Uspekhi Khimii *42*:323 (1973).
32. M. Heyrovský and J. Jirkovský, Langmuir *11*:4288 (1995).
33. M. Heyrovský, J. Jirkovský, and B. R. Müller, Langmuir *11*:4293 (1995).
34. M. Heyrovský, J. Jirkovský, and M. Štruplová Bartácková, Langmuir *11*:4300 (1995).
35. M. Heyrovský, J. Jirkovský, and M. Štruplová Bartácková, Langmuir *11*:4309 (1995).
36. R. A. Mackay and J. Texter, eds., *Electrochemistry in Colloids and Dispersions*, VCH Publishers, Inc., New York, 1992.
37. G. Inzelt, in *Electroanalytical Chemistry*, Vol. 18 (A. J. Bard, ed.), Marcel Dekker, Inc., New York, 1994, pp. 90–241.
38. K. Doblhofer, in *Electrochemistry of Novel Materials, Frontiers of Electrochemistry* (J. Lipkowski and P. N. Ross, eds.), VCH, New York, 1994, pp. 141–205.
39. T. T. Wooster, M. L. Longmire, H. Zhang, M. Watanabe, and R. W. Murray, Anal. Chem. *64*:1132 (1992).
40. T. C. Franklin, J. Darlington, R. Nnodimele, and R. C. Duty, in *Electrochemistry in Colloids and Dispersions* (R. A. Mackay and J. Texter, eds.), VCH Publishers, Inc., New York, 1992, pp. 319–329.
41. P. J. Kulesza and Z. Galus, J. Electroanal. Chem. *269*:455 (1989).
42. P. J. Kulesza, G. Roslonek, and L. R. Faulkner, J. Electroanal. Chem. *280*:233 (1990).
43. P. J. Kulesza, L. R. Faulkner, J. Chen, and W. G. Klemperer, J. Am. Chem. Soc. *113*:379 (1991).
44. P. J. Kulesza, M. A. Malik, S. Zamponi, M. Berrettoni, and R. Marassi, J. Electroanal. Chem. *397*:287 (1995).
45. A. M. Bond, A. Bobrowski, and F. Scholz, J. Chem. Soc., Dalton Trans. *411*:(1991).
46. F. Scholz, B. Lange, A. Jaworski, and J. Pelzer, Fresenius' J. Anal. Chem. *340*:140 (1991).
47. A. M. Bond, R. Colton, F. Marken, and J. N. Walter, Organometallics *13*:5122 (1994).
48. S. J. Shaw, F. Marken, and A. M. Bond, J. Electroanal. Chem. *404*:227 (1996).
49. A. M. Bond, S. Fletcher, F. Marken, S. J. Shaw, and P. G. Symons, J. Chem. Soc. Faraday Trans. *92*:3925 (1996).

50. B. Lange, M. Lovrić, and F. Scholz, J. Electroanal. Chem. *418*:21 (1996).
51. D. Blum, W. Leyffer, and R. Holze, Electroanalysis *8*:296 (1996).
52. A. Dostal, B. Meyer, F. Scholz, U. Schröder, A. M. Bond, F. Marken, and Sh. J. Shaw, J. Phys. Chem. *99*:2096 (1995).
53. S. J. Shaw, F. Marken, A. M. Bond, Electroanalysis *8*:732 (1996).
54. F. Scholz, L. Nitschke, G. Henrion, and F. Damaschun, Fresenius' Z. Anal. Chem. *335*:189 (1989).
55. S. Zhang, B. Meyer, G. Moh, and F. Scholz, Electroanalysis *7*:319 (1995).
56. T. Grygar, Coll. Czech. Chem. Commun. *61*:93 (1996).
57. T. Grygar, J. Šubrt, J. Boháček, Collect. Czech. Chem. Commun. *60*:950 (1995).
58. T. Grygar, J. Electroanal. Chem. *405*:117 (1996).
59. A. M. Bond, R. Colton, F. Daniels, D. R. Fernando, F. Marken, Y. Nagaosa, R. F. M. Van Steveninck, and J. N. Walter, J. Am. Chem. Soc. *115*:9556 (1993).
60. F. Scholz, L. Nitschke, G. Henrion, Electroanalysis *2*:85 (1990).
61. F. Scholz, F. Rabi, and W.-D. Müller, Electroanalysis *4*:339 (1992).
62. N. A. Madigan, T. J. Murphy, J. M. Fortune, C. R. S. Hagan, and L. A. Coury, Jr., Anal. Chem. *67*:2781 (1995).
63. Š. Komorsky-Lovrić, M. Lovrić, and A. M. Bond, Anal. Chim. Acta *258*:299 (1992).
64. T. Grygar, Coll. Czech. Chem. Commun. *60*:1261 (1995).
65. T. Grygar, J. Solid State Electrochem., *1*: 77 (1997).
66. F. Scholz, L. Nitschke, E. Kemnitz, T. Olesch, G. Henrion, D. Hass, R. N. Bagchi, R. Herrmann, N. Pruss, and W. Wilde, Fresenius Z. Anal. Chem. *335*:571 (1989).
67. S. Scheurell, F. Scholz, T. Olesch, and E. Kemnitz, Supercond. Sc. Technol. *5*:303 (1992).
68. B. Meyer, B. Ziemer, and F. Scholz, J. Electroanal. Chem. *392*:79 (1995).
69. R. Schöllhorn, Angew. Chem. *92*:1015 (1980); Angew. Chem. Int. Ed. Engl. *19*:983 (1980).
70. P. M. S. Monk, R. J. Mortimer, and D. R. Rosseinsky, *Electrochromism, Fundamentals and Application*, VCH, Weinheim, 1995.
71. U. Schröder, B. Meyer, and F. Scholz, Fresenius' J. Anal. Chem. *356*:295 (1996).
72. R. E. Dueber, A. M. Bond, and P. G. Dickens, J. Electrochem. Soc. *139*:2363 (1992).
73. A. M. Bond, J. B. Cooper, F. Marken, and D. M. Way, J. Electroanal. Chem. *396*:407 (1995).
74. A. M. Bond, R. Colton, P. J. Mahon, and W. T. Tan, J. Solid State Electrochem. *1*: 53 (1997).
75. M. D. Ward, in *Electroanalytical Chemistry*, Vol. 16 (A. J. Bard, ed.), Marcel Dekker, Inc., New York, 1989, pp. 181–312.

76. R. L. Deutscher and S. Fletcher, J. Electroanal. Chem. *277*:1 (1990).
77. A. M. Bond and F. Marken, J. Electroanal. Chem. *372*:125 (1994).
78. Š. Komorsky-Lovrić, J. Electroanal. Chem. *397*:211 (1995).
79. F. Scholz, W.-D. Müller, L. Nitschke, F. Rabi, L. Livanova, C. Fleischfresser, and C. Thierfelder, Fresenius' J. Anal. Chem. *338*:37 (1990).
80. F. Scholz and B. Lange, Fresenius' J. Anal. Chem. *338*:293 (1990).
81. A. M. Bond and F. Scholz, Langmuir *7*:3197 (1991).
82. S. J. Reddy, A. Dostal, and F. Scholz, J. Electroanal. Chem. *403*:209 (1996).
83. B. Meyer, S. Zhang, and F. Scholz, Fresenius' J. Anal. Chem. *356*:267 (1996).
84. S. J. Jayarama Reddy, M. Hermes, and F. Scholz, Electroanalysis *8*:955 (1996).
85. G. Gandolfi, Miner. Petrogr. Acta *13*:67 (1967).
86. B. Meyer, Ph.D. thesis, Berlin, 1995.
87. B. Lange, F. Scholz, A. Weiß, G. Schwedt, J. Behnert, and K.-P. Raezke, Int. Lab. *23*:23 (1993).
88. F. Scholz, L. Nitschke, and G. Henrion, Fresenius Z. Anal. Chem. *334*:56 (1989).
89. Š. Komorsky-Lovrić, J. Bartoll. R. Stößer, and F. Scholz, Croat. Chim. Acta, in press.
90. A. M. Bond and F. Scholz, J. Geochem. Explor. *42*:227 (1992).
91. J. Heinze, Angew. Chem. *96*:823 (1984).
92. F. Scholz and A. Dostal, Angew. Chem. *107*:2876 (1995); Angew. Chem. Int. Ed. Engl. *34*:2685 (1995).
93. A. Dostal, U. Schröder, and F. Scholz, Inorg. Chem. *34*:1711 (1995).
94. A. Dostal, M. Hermes, and F. Scholz, J. Electroanal. Chem. *415*:133 (1996).
95. M. Lovrić and F. Scholz, J. Solid State Electrochem. *1*: 108 (1997).
96. H. Kahlert and F. Scholz, Electroanalysis, *9*: 922 (1997).
97. A. Dostal, G. Kauschka, S. J. Reddy, and F. Scholz, J. Electroanal. Chem. *406*:155 (1996).
98. W. J. Meggs, R. S. Hoffman, R. D. Shih, R. S. Weisman, and L. R. Goldfrank, Clin. Toxicol. *32*:723 (1994).
99. D. R. Melo, J. L. Lipsztein, C. A. N. de Oliveira, and L. Bertelli, Health Physics *66*:245 (1994).
100. H. Kahlert, Š. Komorsky-Lovrić, M. Hermes, and F. Scholz, Fresenius' J. Anal. Chem. *356*:204 (1996).
101. A. J. Downard, A. M. Bond, L. R. Hanton, and G. A. Heath, Inorg. Chem. *34*:6387 (1995).
102. A. M. Bond and F. Scholz, J. Phys. Chem. *95*:7460 (1991).
103. B. Lange, F. Scholz, H.-J. Bautsch, F. Damaschun, and G. Wappler, Phys. Chem. Minerals *19*:486 (1993).
104. B. Meyer and F. Scholz, Phys. Chem. Minerals *24*:50 (1997).
105. U. Schröder and F. Scholz, J. Solid State Electrochem. *1*: 62 (1997).

106. Š. Komorsky-Lovrić, J. Solid State Electrochem. *1*: 94 (1997).
107. H. Feess and H. Wendt, Ber. Bunsenges. Phys. Chem. *85*:914 (1981).
108. J. F. Rusling, in *Electroanalytical Chemistry*, Vol. 18 (A. J. Bard, ed.), Marcel Dekker, Inc., New York, 1994, pp. 1–88.
109. J. O'M. Bockris and A. K. N. Reddy, *Modern Electrochemistry*, Vol. 2, Plenum Press, New York, 1970, pp. 1382–1385.
110. F. Willig, Adv. Electrochem. Electrochem. Eng. *12*:1 (1981).
111. A. Jaworski, Z. Stojek, and F. Scholz, J. Electroanal. Chem. *354*:1 (1993).
112. F. Scholz, M. Lovrić, and Z. Stojek, J. Solid State Electrochem. *1*: 134 (1997).
113. Ya. I. Turyan, J. Electroanal. Chem. *338*:1 (1992).
114. L. Pospíšil and R. de Levie, J. Electroanal. Chem. *25*:245 (1970).
115. A. M. Bond, F. Marken, E. Hill, R. G. Compton, and H. Hügel, J. Chem. Soc. Perkin Trans. 2: 1735 (1977).
116. A. M. Bond, R. Colton, F. Marken, and J. N. Walter, Organometallics *16*: 5006 (1997).

ANALYSIS IN HIGHLY CONCENTRATED SOLUTIONS: POTENTIOMETRIC, CONDUCTANCE, EVANESCENT, DENSOMETRIC, AND SPECTROSCOPIC METHODOLOGIES

Stuart Licht

Technion Israel Institute of Technology, Haifa, Israel

I. Introduction 88
II. Potentiometric Analysis 89
   A. Conventional pH measurement 89
   B. Colorimetric estimates of pH in concentrated acids and bases 91
   C. Calculation of pH in concentrated acids and bases 91
   D. Potentiometric measurement of pH in concentrated alkaline solutions 93
   E. Theoretical and experimental tools for concentrated solution potentiometric analysis 101
III. Evanescent (Solvent) Activity Analysis 103
   A. Solvent activity 103
   B. Determination of mean solution activity from solvent activity 104
   C. Traditional analysis of activity 105
   D. Evanescent methodology for analysis of solvent activity 106
IV. Conductometric Analysis of Concentrated Solutions 113
   A. Conventional conductometric titration 113
   B. Conventional conductometric titration of concentrated solution 115
   C. Differential conductometric methodology 118
   D. An example of differential conductometric methodology 119
V. Differential Densometric Analysis 121
   A. Differential densometric principles 121
   B. Differential densometric methodology 123

C. Differential densometric analysis example   123
   D. Differential densometric benefits and disadvantages   127
VI. Submicrometer Path UV/VIS/IR Spectroscopy   127
   A. Conventional absorption spectroscopy limits   127
   B. Submicrometer path length cells   129
   C. Submicrometer path length absorption spectroscopy   129
   D. Submicrometer path length examples   132
VII. Final Comments   137
   References   137

I. INTRODUCTION

In industrial, environmental, and many synthetic and research processes, highly concentrated solutions are prevalent. Yet, conventional spectroscopic, electrochemical, and chromatographic methodologies are often not suitable to probe chemistry in the concentrated solution phase. In order to analyze concentrated solutions, most conventional methodologies require an initial solution dilution step. However, this dilution step precludes measurement of interesting complexation, association, or solvent-driven equilibria which can occur in the concentrated phase. There are few direct or in situ analytical methodologies for analysis of these highly concentrated electrolytes. Instead, analysis of these solutions is often accomplished with extraction and dilution steps. From a practical perspective, dilution steps can lengthen analysis time, add contaminants, and provide only incomplete information regarding chemical interactions in a concentrated solution.

Solution phase methodologies have focused on analysis of dilute solutions for both pragmatic and fundamental reasons. The magnitude of conventional spectroscopic extinction coefficients dictates that concentrated solutions tend to be opaque, precluding conventional absorption spectroscopy of solutions containing molar concentrations. Electrochemical methods are limited in concentrated solutions due to uncertainties including ill-defined liquid junction potentials, electrode instability, and solution activity. From a fundamental perspective, Debye Hückel theory breaks down, and activity coefficients can deviate substantially from unity in concentrated solutions.

A principal subject of this chapter is three methodologies, each electrochemical in nature, which are suitable for analysis of concentrated solutions. These methodologies, variations of electroanalytical or ionic methods, are concentrated solution potentiometric analysis, evanescent (solvent) activity analysis, and differential conductometric analysis. Section II probes fundamental and practical issues related to potentiometric analysis (ion-selective electrode methodologies) including solutions over 10 molal (m) in concentration, and compares theoretical derivation and experimental determinations of aqueous pH values of over 17. Section III presents a dynamic technique to determine activities of concentrated electrolytes. This evanescent analysis correlates induced evaporative mass loss of solution with solvent and solution activity. Ionic conductivity, which is enhanced, rather than obscured, with increasing electrolyte, provides a precise probe of concentrated electrolytes. This is the framework for differential conductometric analysis in Sec. IV, in which speciation is determined by comparing the conductivity of isolated reactants with that of an equilibrated homogeneous mixture.

In addition to the three electrochemical methodologies, two chemical methodologies are investigated techniques which can be appropriate for the analysis of concentrated media (solutions of 0.1 to over 10 molar (m) in analyte). The variation of several physical chemical phenomena are accentuated in increasingly concentrated solutions and, in principal, may be the basis for analytical probes and methodologies in these solutions. Differential densometric analysis (Sec. V) takes advantage of such a physical chemical property (density) [1,2]. In concentrated solutions, the solute comprises a significant mole fraction of the electrolyte, and density variations can be used to gauge equilibria. A final methodology, submicrometer path UV/Vis/IR spectroscopy, discussed in Sec. VI, uses a specialized spectroscopic cell in which the cell path length can be shorter than the wavelength of incident light through the cell [2,3].

II. POTENTIOMETRIC ANALYSIS

A. Conventional pH Measurement

The conventional domain of aqueous pH analysis generates and evaluates pH values in the range of 0 to 14. pH is traditionally defined from the hydrogen ion activity, a_{H+}:

$$pH = -\log(a_{H+}) \qquad (1)$$

In solutions dominated by hydroxide ions, it is convenient to rewrite Eq. (1) with inclusion of the equation describing the water dissociation constant, K_w:

$$K_w = \frac{a_{H^+} \gamma_{OH^-} m_{OH^-}}{a_w}; \quad K_w(25°C) = 14.00 \quad (2)$$

where at the hydroxide molal concentration, m_{OH^-}, the hydroxide activity coefficient is γ_{OH^-} and the water activity coefficient is a_w, (a_w and $\gamma_{OH^-} = 1$ in 55.508m (pure) water). Upon substitution of Eq. (2), Eq. (1) yields:

$$pH(25°C) = 14.00 + \log\left(\frac{\gamma_{OH^-} m_{OH^-}}{a_w}\right) \quad (3)$$

Individual redox half cells cannot be fully isolated, and therefore individual ion activity coefficients cannot be precisely measured. Hence, measurement of pH in accord with Eqs. (1) or (3) is problematic. However, the potential of a full cell may be measured, for example, using a potentiometric cell consisting of a sensing electrode and a reference electrode immersed in electrolyte:

$$\text{pH electrode} \mid \text{sample solution} \mid E_{reference} \text{electrode} \quad (4)$$

In addition to the desired response to pH, the potential of the cell expressed by Eq. (4) can have uncertainty from the possible response to interfering ions, E_c, at the pH electrode and to liquid junction potentials, E_j, at the reference electrode interface:

$$E_{H^+} = E°_{H^+} + S\log(a_{H^+}) = E_j - E_c \quad (5)$$

leading to the further necessity of an operational definition of pH according to the convention of Bates [4]:

$$pH_{measured} = pH_{standard} + \left(\frac{E_{H^+} - E_{ref}}{S}\right) \quad (6)$$

where $(E_{H^+} - E_{ref})$ is the differential voltage measured between the pH and the reference electrode. The slope, S, is the measured differential voltage response with pH of the electrode (typically, $S(\text{volts}) = 17.64/T(K)$ per pH unit or approximately 0.059 V at 25°C). $pH_{standard}$ is the defined pH of a standardizing solution used to calibrate the potential response of the cell; several pH standardizing solutions are available [4].

pH measurement is one of the most widely used analytical techniques, and the conventional cell configuration of Eq. (4) is found in chemistry, biology, and environmental laboratories throughout the world. This generally consists of a single probe containing both a glass pH electrode and a reference electrode in close proximity, with their potential difference measured by a high-input impedance voltammeter, necessary due to the high resistivity of the glass ion selective electrode.

B. Colorimetric Estimates of pH in Concentrated Acids and Bases

In highly concentrated (over 1 N) acidic or alkaline solutions, pH values are expected to range beyond the conventional bounds of $0 < \text{pH} < 14$. Colorimetric techniques have been employed to estimate pH in concentrated acid and concentrated alkaline solutions [5,6], and generally employ absorption spectroscopy to evaluate an equilibrium, which is influenced by hydrogen ion activity. By studying a relatively dilute indicator species, these techniques bypass the problems of high absorbance expected if the concentrated analyte (in this case H^+ or OH^-) was directly observed. However, these techniques are limited by problems arising from solution oxidation/reduction of the dilute indicator, overlapping absorbance by other species or colloidal suspensions [7]. In addition, these techniques are indirect. Rather than measure pH they measure chromophores whose concentration are affected by pH. For identical solutions, the different colorimetric techniques can been shown to estimate hydrogen ion activities, which vary by several orders of magnitude in both acidic and basic media [8] depending on the indicator utilized, and generally are related to the correct pH only by an indeterminate constant [9].

C. Calculation of pH in Concentrated Acids and Bases

The theoretical foundation for understanding of pH becomes increasingly problematic in concentrated solutions. The fundamental definition of pH [Eq. (1)] is based on the hydrogen ion activity. However, single ion activities are experimentally inaccessible and are expected to deviate substantially from unity in concentrated electrolytes.

Whereas single ion activities are experimentally inaccessible, this is not the case for mean solution activities, which have been well characterized for acidic and alkaline solutions. These experimentally reproducible, fundamental values for solution activity provide a basis for a modified definition of pH, in which Eqs. (1) and (3) are modified to incorporate these measurable parameters. For example, with alkali hydroxide salt solutions

[Eq. (3)] may be modified to calculate pH from known mean solution activities, $a_{\pm} = a_{KOH} = \gamma_{\pm}MOH^m$ MOH; $\gamma_{\pm}MOH \equiv$ activity coefficient in solution of concentration m_{MOH}, and at 25°C [8]:

$$pH_{calculated} = 14.00 + \log\left(\frac{\gamma_{\pm MOH} m_{MOH}}{a_w}\right) \qquad (7)$$

pH values substantially beyond 14 can exist in aqueous solutions. Debye-Hückel theory indicates that activity coefficients approach unity at low concentrations. However, very concentrated aqueous alkali hydroxide solutions are known to have mean solution activity coefficients which can reach more than an order of magnitude higher; indeed, nearly saturated, highly concentrated sodium or potassium hydroxide solutions have mean solution activities greater than 30 [8]. In these solutions the majority of the water is

TABLE 1

The Calculated pH[a] of Aqueous KOH Solutions at 25°C

m_{KOH}	$\gamma_{\pm KOH}$	a_w	$\log(a_{KOH}^2/a_w)$	$pH_{calculated}$
0.000100	0.99	1.00	−8.01	10.00
0.00100	0.97	1.00	−6.03	10.99
0.0100	0.91	1.00	−4.04	11.96
0.100	0.99	1.00	−2.20	12.90
0.300	0.74	0.99	−1.30	13.35
1.00	0.75	0.96	−0.23	13.89
3.00	1.06	0.872	1.06	14.66
4.50	1.51	0.790	1.77	14.93
6.00	2.22	0.700	2.40	15.28
7.50	3.29	0.604	3.00	15.61
9.00	4.86	0.508	3.58	15.94
10.5	7.10	0.419	4.12	16.25
12.0	10.2	0.342	4.64	16.55
13.5	14.5	0.278	5.14	16.85
15.0	20.2	0.221	5.62	17.14
16.0	25.0	0.187	5.93	17.33
17.0	30.9	0.157	6.24	17.52
17.5	34.4	0.143	6.40	17.62

[a]Determined from Eq. (7) using data from Ref. 8 and 10. The mean solution activities, $a_{\pm} = a_{KOH}$, are determined as the product of $\gamma_{\pm MOH}$ and $\mathbf{m}_{MOH}$.

bound within the sphere of ionic hydration and the water activity drops to less than 0.2. Thus the value of $\gamma_{\pm MOH}/a_w$ can be larger than 100, and from Eq. (7) for solutions over 10 m, pH values larger than 17 are expected.

Table 1 presents calculated pH values for a range of potassium hydroxide electrolytes. For example, at 25°C in respective 3.0 **m** and 17.5 **m** KOH solutions, a_{KOH} increases from 3.18 to 602 and a_w decreases from 0.872 to 0.143 [10]. In accordance with Eq. (7), this leads to predicted pH values of 14.66 and 17.62, respectively, in 3.0 **m** and 17.5 **m** KOH solutions. Equation (7) provides a consistent fundamental basis to predict pH of highly concentrated alkaline solutions. Unusually high pH values are expected, and, as demonstrated in the next section, these can be experimentally measured and verified.

D. Potentiometric Measurement of pH in Concentrated Alkaline Solutions

Several obstacles must be considered in potentiometric measurement of concentrated electrolytes. These are (1) competitive ion interference, (2) liquid junction potential errors, and (3) chemical attack of the sensor electrode. Effects of competitive ion interference will occur as cation errors in alkali pH measurement. High cation concentrations, for example, K$^+$ in KOH or Ba^{2+} in BaOH$_2$, are necessary to introduce high hydroxide concentration. In increasingly concentrated alkali solutions, as the solvated proton concentration diminishes these other cation concentrations increase and can increasingly influence the potential of the pH electrode.

The sensing element of commercially available pH electrodes is comprised of glasses of various compositions. The magnitude of the cation error for these pH electrodes varies with glass composition and cation type in solution. However, for example, in a sodium hydroxide solution, it typically generates on the order of a 1–2 pH unit underestimate of the pH in a 1 **m** NaOH solution. Various commercial ion-selective electrode manufacturers have specific glass sensor compositions comprised to minimize the magnitude of cation error in alkaline solutions [4,7]. Cation errors can be substantially less in alkaline potassium solutions compared to sodium electrolytes, providing an effective medium to determine and monitor concentrated pH [8]. Other commercial glass compositions have been chosen to minimize the response to hydrogen ion and maximize the response to alkali or alkali hydroxide cations to create cation-sensing ion-selective electrodes. In concentrated alkaline solutions, these electrodes effectively respond exclusively to the cation salt activity (competitive ion interference due to the extremely low hydrogen ion activity is then not significant).

When a cation-selective potentiometric electrode is available, the interference on a pH electrode due to this cation can be quantified. For example, the potassium ion error at a pH electrode may be determined by potentiometric measurement of the cell:

$$E_{potassium} \text{ electrode I sample solution I pH electrode} \tag{8}$$

Whereas the potentials of either pH or potassium electrode cells contain uncertainties in individual ion acitivities and liquid junction potentials, [Eqs. (5) and (9)], this is not the case for the potential of the combined cell [Eq. (8)], which is denoted $E_{K^+} - E_{H^+}$.

$$E_{K^+} = E°_{K^+} + S \log(a_{K^+}) - E_j \tag{9}$$

As derived from Eqs. (2), (5), and (9), the potential of cell (7) is expected to vary with log (a_{KOH}^2/a_w) [8]. The nonlinear variation of $E_{K^+} - E_{H^+}$ with log (a_{KOH}^2/a_w) provides a measure of the cation error on the pH electrode, E_c, for this cell:

$$E_{K^+} - E_{H^+} = a \text{ constant} + S \log\left(\frac{a_{KOH}^2}{a_w}\right) + E_c \tag{10}$$

Figure 1 presents the potential difference measured between pH and K$^+$ selective glass electrodes in both dilute and highly concentrated potassium hydroxide solutions. As seen in the figure, using this commercial low cation error pH electrode, voltage deviations from linearity are small for concentrations up to 6 m KOH (log (a_{KOH}^2/a_w) = 2.4), but then rises to 0.05 V, equivalent to an 0.8 pH unit cation error in a 17.5 m KOH (log(a_{KOH}^2/a_w) = 6.4) electrolyte.

The pH-measuring cell represented by Eq. (4) contains a reference electrode, which can present a second uncertainty to pH measurement of concentrated electrolytes. A conventional reference electrode, such as either the saturated calomel or the silver chloride reference electrodes commonly employed, contains an internal solution separated from the analyte solution by either a ceramic frit, gel, membrane, small orifice, or other flow impeding mechanism. The ion activity differential between the sample and internal reference solutions gives rise to a liquid junction potential, E_j, which is expected to vary as different sample solutions are probed by the pH cell of Eq. (4). The calculation of E_j is highly problematic, as evi-

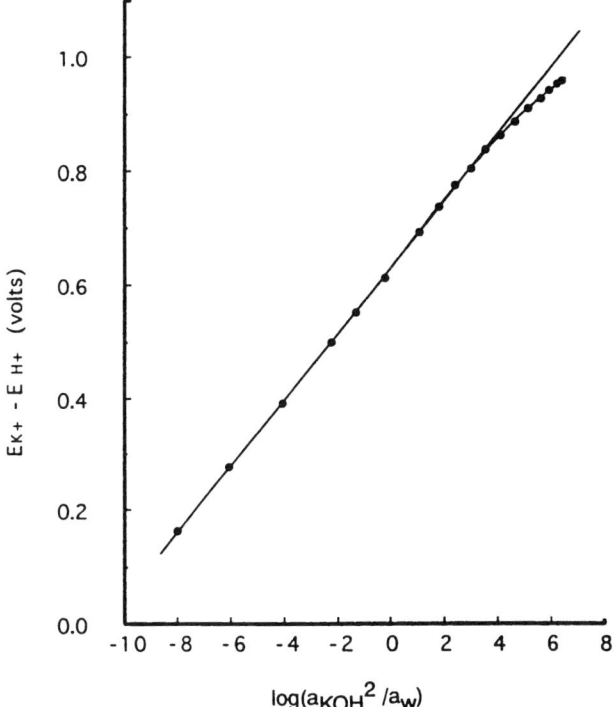

FIG. 1. The voltage of a liquid junction potential free potassium hydroxide cell as a function of log (a_{KOH}^2/a_w) at 25°C. The cell is comprised of a Radiometer G202BH pH electrode measured versus a Beckman 39317 cation selective electrode immersed in solutions varying from 10^{-4} to 17.5 m KOH. Cell potential is measured with a conventional high input impedance voltammeter. (From Ref. 8.)

denced by an order of magnitude difference in predicted liquid junction potentials even between simple monovalent salt solutions [4,11–13]. In one estimate, Picknett observes a correspondence of E_j to solution specific conductivity [12]. His model does not require the additional assumption of the Henderson model [4], which had assumed a specific ionic distribution in the liquid junction region. Picknett estimates E_j variation between KOH and KCl solutions to vary linearly with concentration and to generate a small potential of approximately 1 mV per decade differential in concentration [12].

Figure 2 presents the potential response, E_{H^+}, of a low alkali error pH cell as a function of concentration in KOH solutions at 25°C. In the figure, the upper dashed line represents measured E_{H^+} in a saturated calcium hydroxide solution with known pH of 12.454, further calibrated with a 0.0100 **m** borax solution (with a defined primary pH standard of 9.180). At high pH the measured cation interference can be used to correct the measured pH, and the operational definition of pH may be modified to incorporate this cation error potential:

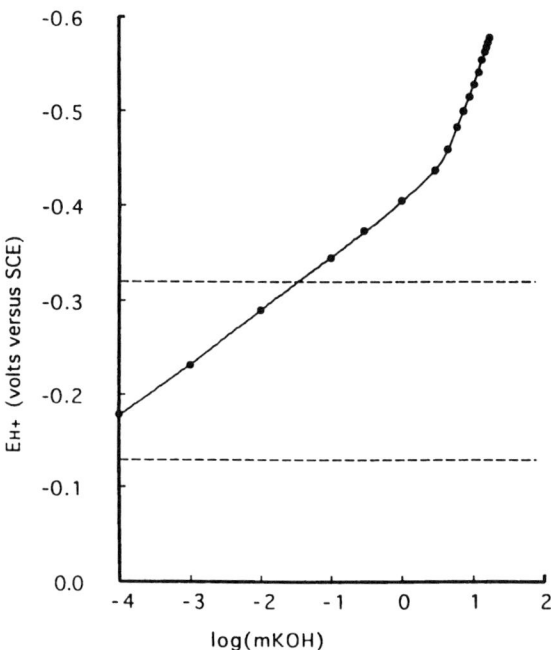

FIG. 2. The voltage response of a low cation error pH cell as a function of potassium hydroxide concentration at 25°C. The cell is comprised of a Radiometer G202BH pH electrode measured versus a Radiometer K401 saturated calomel electrode immersed in solutions varying from 10^{-4} to 17.5 **m** KOH. Cell potential is measured with a conventional high input impedance voltammeter. The upper and lower dashed lines are the potentials measured, respectively, in saturated calcium hydroxide and 0.1 **m** borax solutions. (From Ref. 8.)

$$pH_{measured} = pH_{standard} + \left(\frac{E_{H^+} - E_{ref}}{S} \right) + \left(\frac{E_c}{S} \right) \quad (11)$$

Using the measured values of E_{H^+} presented in Figure 2, values of pH$_{measured}$ have been determined without cation correction from Eq. (6) and also determined with the cation correction values of E_c from Figure 1 and Eq. (10). Figure 3 presents the resultant comparison of pH$_{measured}$ to theoretical pH values calculated and summarized in Table 1 as pH$_{calculated}$. Closed circle points are for pH$_{measured}$ prior to cation correction; open circle points include the cation correction of E_c/S from Eq. (11). In this figure, it is seen that pH$_{measured}$ provides an highly linear estimate of the solution pH. With appropriate cation correction of E_c/S, the maximum deviation for pH$_{measured}$ is ~0.1 pH units. This small deviation is consistent with the low values of liquid junction potential error, which would be predicted by the Picknett model [12].

Due to the relatively small ion interferences, potassium electrolytes

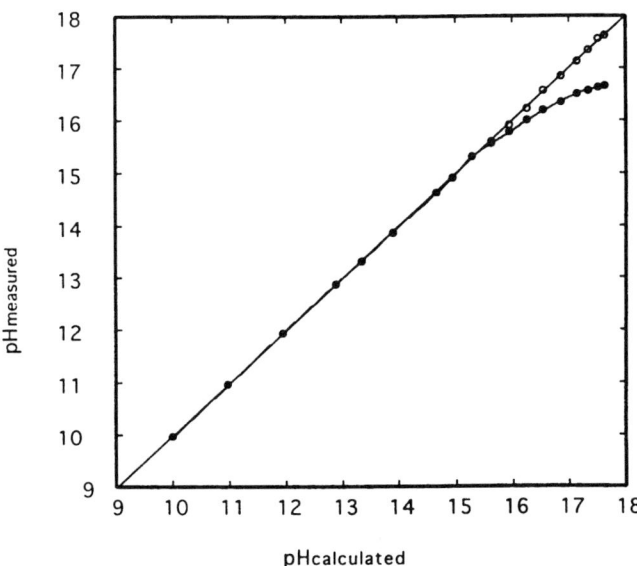

FIG. 3. Potentiometric pH measurement in concentrated aqueous KOH solutions at 25°C. Predicted pH, pH$_{calculated}$, is compared to measured pH before (solid data points) and after (open data points) correction for cation error. (From Ref. 8.)

provide an attractive media for the measurement of pH in highly alkaline solutions. However, it is also of interest to measure the magnitude of other cation interferences. Figure 4 presents measured sodium ion errors at 25°C for high pH values. Various concentrations of sodium chloride were added to 1.0, 4.5, 9.0, and 17.5 **m** potassium hydroxide solutions and the pH measured with a low cation error commercial pH electrode. In the figure, the dashed lines represent the pH of the sodium salt–free solutions. Analogous experiments were done using lithium chloride to determine the magnitude of the pH depression, which results from the presence of lithium ions. These results are shown in Fig. 5. Lithium ion interference errors are larger than sodium ion er-

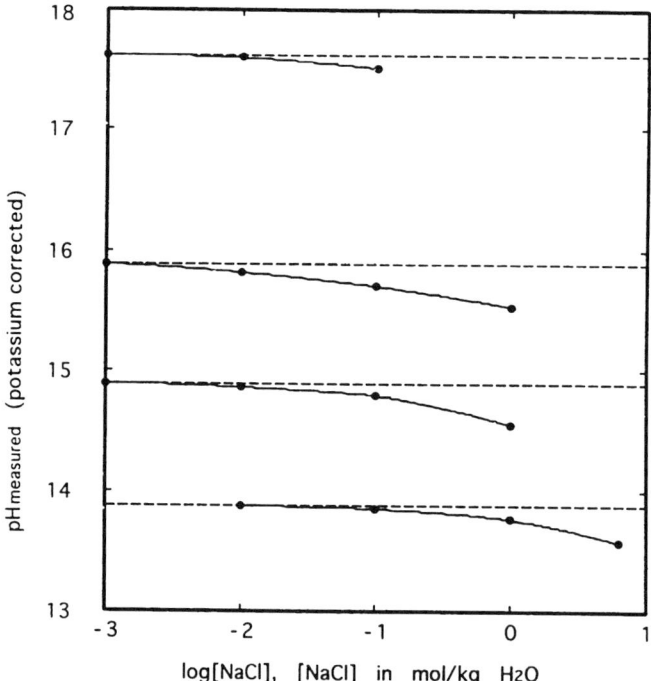

FIG. 4. Measured sodium errors for a Radiometer G202BH pH electrode in 1.0 (lowest curve), 4.5, 9.0 (lowest curve), 17.5 **m** KOH solutions as a function of added NaCl at 25°C. The pH cell is as described for Fig. 2. The dashed lines represents the sodium free solutions. (Modified from Ref. 8.)

rors, which in turn are substantially larger than potassium ion errors. It is evident from Figs. 4 and 5 that these pH errors, due to the presence of relatively high known concentrations of lithium (up to 0.1 m) and sodium (up to 1 m), may be compensated for even in highly alkaline solutions if the primary cation is potassium and a low cation error pH electrode is used.

Despite the known alkali attack on glass, pH measurement can be shown to be stable on the order of hours in highly concentrated alkaline solutions. In glass, alkalis break the silicon oxygen bonds in the bridge network resulting in the dissolution of water-soluble silicates [14,15]. At higher temperature (80°C), significant dissolution of pH sensitive glass has been observed to occur in 1 N sodium, lithium, potassium, barium, and

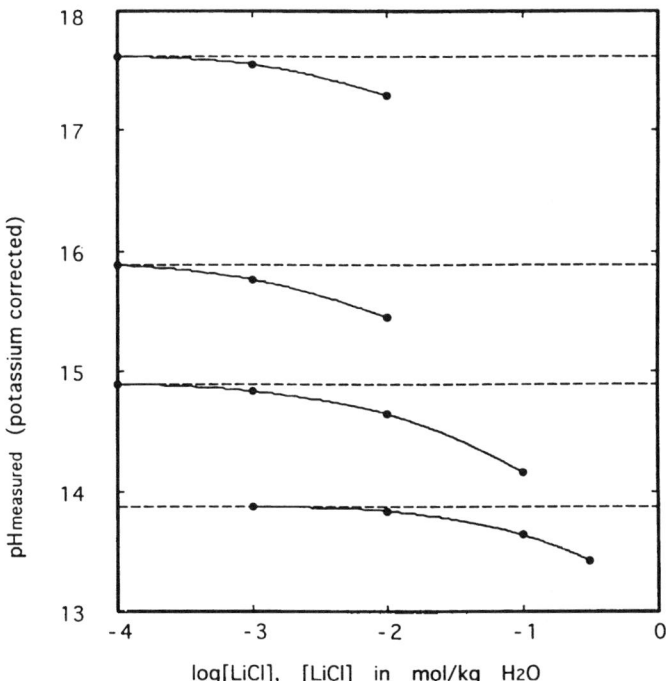

FIG. 5. Measured lithium errors for a Radiometer G202BH pH electrode in 1.0 (lowest curve), 4.5, 9.0 (lowest curve), 17.5 m KOH solutions as a function of added LiCl at 25°C. The pH cell is as described for Fig. 2. The dashed lines represents the lithium free solutions. (Modified from Ref. 8.)

ammonium hydroxide solutions. The greatest rate of dissolution was for the sodium solution and followed the order Na > Li > K > Ba > NH_4 [16]. No studies have been previously reported regarding the attack of pH-sensitive glass in solutions of higher alkalinity or lower temperature. Figure 6 shows the measured pH in 17.5 m KOH at 25°C with a commercial pH electrode as a function of time. As seen in the figure, the pH for this electrode (a Radiometer G202BH electrode) remained relatively constant for the first 6 hours of immersion in concentrated KOH, followed by a decrease of 2.1 pH units by the hundredth hour. However, the pH returned to within ~0.1 pH

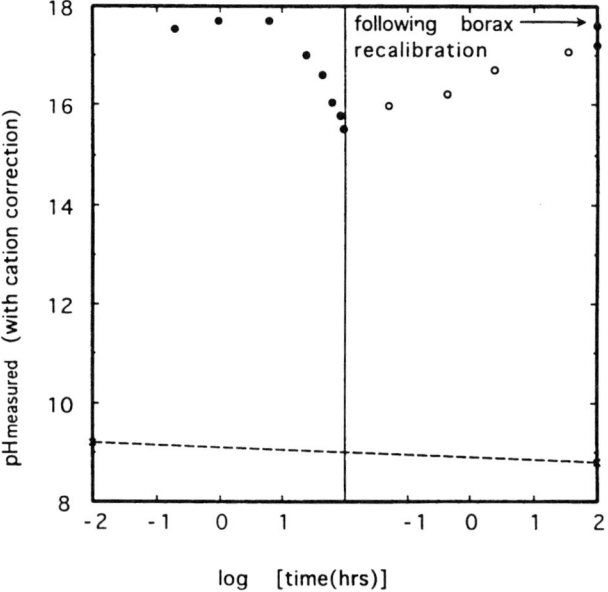

FIG. 6. pH measurement in concentrated 17.5 m KOH solutions as a function of time at 25°C. The pH cell is as described for Fig. 2. The open circles represent individual measurements made during continual immersion of the pH electrode in the concentrated KOH solution. The reference electrode was only placed in the solution during pH measurement, and otherwise was in a saturated KCl solution. Closed circles represent individual subsequent measurements performed after soaking the pH electrode in distilled water for the indicated time. The dashed line represents the pH of an 0.01 m borax solution measured at the beginning and then at the conclusion of the experiment. (Modified from Ref. 8.)

units of the original value, as measured versus a 0.01 **m** borax reference after the electrode was rinsed and then soaked for 100 hours in distilled water.

E. Theoretical and Experimental Tools for Concentrated Solution Potentiometric Analysis

The systematic use of mean solution activities, rather than colorimetric indicators or single ion activities, provides a consistent formalism for the prediction of solution interactions in concentrated electrolytes. Applying this formalism to pH, as shown for concentrated alkali solutions using Eq. (7), aqueous pH values substantially outside the conventional pH range of 0–14 are expected. For potassium hydroxide solutions, these pH values range to a pH of >17, as summarized in Table 1. The pH of these concentrated solu-

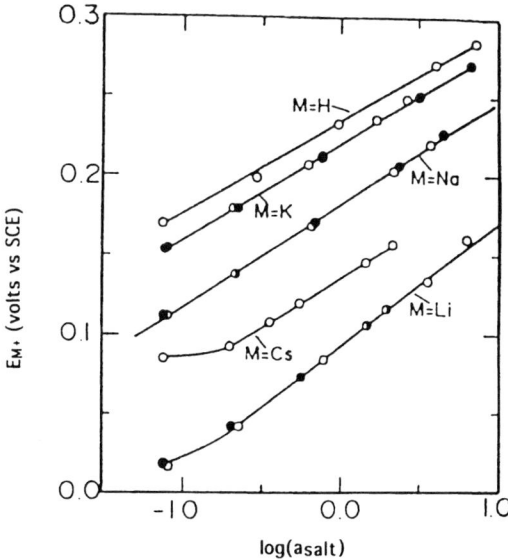

FIG. 7. The voltage response of a Beckman 39137 cationic selective electrode, measured versus a Radiometer K401 saturated calomel electrode, in aqueous alkali chloride and hydroxide solutions at 25°C. Data for 0.1, 0.3, 1.0, and 4.5 **m** halide (open circle) and hydroxide (solid circle) solutions are generally presented. Due to solubility limitations, data for 4.0 **m** LiOH is presented, and an additional data point for 0.6 **m** CsCl is included. (From Ref. 8.)

tions can be experimentally probed using glass ion-selective electrode methodologies if proper consideration is given to expected errors arising from competitive ion interference, liquid junction potential errors, and chemical attack of the sensor electrode. The feasibility of pH measurement of extremely concentrated alkaline solutions has been exemplified with potassium hydroxide electrolytes, and reproducible pH values can be measured with a room temperature stability of several hours. As shown in Fig. 7, ion-selective glass electrodes are also available, which generate logarithmic response to activity of a variety of cations. Under controlled potentiometric conditions, using an approach analogous to the high pH measurements discussed in this chapter, this can permit cation determination in highly concentrated salt solutions and has been applied to the potentiometric determination of potassium activities in highly concentrated potassium sulfide solutions as shown in Fig. 8. These measurements indicated that the second acid dissociation constant of hydrogen sulfide was considerably smaller (by five orders of magnitude) than generally had been thought and could only be measured in extremely concentrated electrolytes. This would substantially alter the sol-

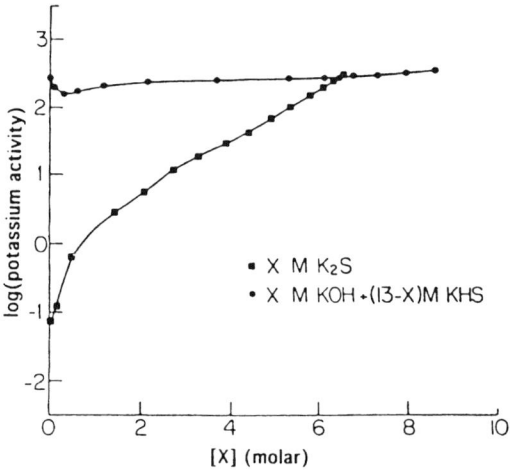

FIG. 8. The measured potassium activity at 25°C of aqueous solutions × molarity in either K_2S (lower curve) or in a solution containing a constant 13 M concentration of K^+ in which 13 M KHS is titrated by 13 M KOH (upper curve). (From Ref. 17.)

ubility products of a wide variety of metal sulfides [18]. In order to verify this, other methods of concentrated electrolyte analysis were sought.

III. EVANESCENT (SOLVENT) ACTIVITY ANALYSIS

A. Solvent Activity

Another approach to exploring the chemistry of concentrated solutions is to infer information regarding the chemical interactions of the solute species by measuring properties of the solvent. "Pure" solvent (without added salt) has an activity of unity. In dilute solutions many bulk solvent properties are substantially unaffected by the dissolved salt. However, in increasingly concentrated solutions a growing fraction of the solvent will be directly influenced by solute/solvent interactions. For example, as discussed in Sec. II.C, in concentrated aqueous solutions most of the water can be strongly bound within the solvation sphere of ionic hydration and the water activity can drop by an order of magnitude. The number of water molecules associated with the hydration sphere of a dissolved ion varies severalfold depending on the experimental technique employed to make the measurement. However, several techniques indicate that the primary hydration spheres surrounding Li^+, Na^+, K^+, and Cs^+ cations contain, respectively, approximately 9, 5, 3, and 0 water molecules. A concentrated solution can contain several moles of salt per 55.51 moles (1 kg) of water. Even these approximate values for alkali cation hydration sphere indicate that a substantial fraction of water in concentrated lithium, sodium, or potassium salt solutions will be bound in hydration spheres. This is readily evident in the vapor pressure above these solutions, which falls proportionally to the drop in water activity.

The water activity, a_w, of aqueous solutions has been characterized and parameterized with a single coefficient, of α_o, for each specified salt expressing the variation of water activity with molal solution concentration, C. At a constant temperature, T, this results in the general expression [19]:

$$\alpha_0 = -127.8 \left(\frac{\partial \log a_w}{\partial C} \right) T \tag{12}$$

Table 2 lists the reported water activity values of a variety of saturated solutions at 25°C and, where available, the values for α_o [19,20].

The theory of the interrelationship of the solvent activity and the mean

TABLE 2
Variation of Water Activity, a_w, and a_o [Eq. (12)] for Saturated Solutions in Equilibrium at 25°C of the Indicated Salt

Salt	Solubility (m)	α_o	a_w
None			1.000
K_2CrO_7	0.6		0.980
KNO_3	3.8	0.74	0.925
KBr	5.8	2.25	0.807
NaCl	6.2	3.45	0.753
$NaNO_2$	13.		0.65
NaBr	9.2	25.6	0.577
$Mg(NO_3)_2$	4.9	12.1	0.529
K_2CO_3	8.3		0.428
$MgCl_2$	5.9	20.9	0.330
$CaCl_2$	7.5	11.1	0.284
KOH	20.8		0.080

Source: Refs. 19 and 20.

solution activity is well established and permits a mathematical vehicle to determine solute activity from measured solvent activity. This theory is reviewed in the next section, and in the following section an experimental methodology for the rapid determination of solvent activity will be presented.

B. Determination of Mean Solution Activity from Solvent Activity

Compared to the thermodynamic standard state of a pure solvent s, $a_s^\circ \equiv 1$, the activity of the solvent component in an electrolytic solution, a_s, is given by:

$$a_s = \frac{P_s}{P_s^o} \qquad (13)$$

where P_s and P_s° are the respective partial pressures (or more precisely the fugacity) for the solution or pure solvent. In a solution containing one or more electrolytes, k_1, k_2, k_3 ..., quantitative variations of solvent activity from unity are emphasized by a parameter defined as the osmotic coefficient, ϕ:

$$j = \frac{-1000 \ln a_s}{M_s (\Sigma_k n_k m_k)} \qquad (14)$$

Analysis in Highly Concentrated Solutions

where M_s is the molar mass of the solvent and 1 mole of electrolyte gives v_k moles of ions in solution.

A precise knowledge of solvent activity, such as a_w for aqueous solutions, reveals direct information of a variety of physical chemical properties including osmotic pressure, freezing point depression, and boiling point elevation. Additionally Gibbs, Duhem, and others have shown that each of the activity coefficients of a multicomponent system may be calculated given the activity coefficient variation of any one component as a function of system composition [21a]. Hence, mean solution and solute activity may be determined from solvent activity when this solvent activity has been determined over a wide electrolyte concentration range. In concentrated electrolytes, the mass fraction or molal fraction of solute becomes significant compared to that of the solvent. In this case, a method to monitor solvent activity may reveal significant variations in electrolyte speciation.

In aqueous solution a_w, solvent activity, is often summarized by osmotic coefficient from Eq. (14):

$$\varphi = \frac{-55.51 \ln a_w}{\Sigma_k v_k m_k} \quad (15)$$

In accordance with Gibbs and Duhem, the mean solution activity, $\gamma_\pm$, is intimately related to φ (and therefore solvent activity), and at a given concentration, C [21a]:

$$\ln \gamma_\pm(m) = \varphi(m) - 1 + \int_{c=0}^{m} [\varphi(C) - 1] C^{-1} dC \quad (16)$$

Solute activity may be determined from Eqs. (15) and (16) when solvent activity have been determined over a wide concentration range of electrolyte. Inversely, the osmotic coefficient and therefore the water activity may be determined when the mean solution activity is known:

$$\varphi(m) = 1 + m^{-1} \int_{c=0}^{m} C d \ln \gamma(C) \quad (17)$$

C. Traditional Analysis of Activity

In 1917 Bousfield introduced the concept of "isopiestic solutions" to determine the concentrations of solutions having the same vapor pressure [22]. In isopiestic methods two open solutions, one of known solvent activity and the other of unknown solvent activity, are placed in an enclosed container, which permits free solvent exchange between the solutions. The solutions are allowed to come to thermal equilibrium resulting in a fixed vapor pressure in the container. During this equilibration process, there is a mass gain or loss from the known solution during the condensation or distillation from the unknown solution. This mass change is then measured and related to solvent activity of the known solution. This process may be iterated until the solvent activity in the unknown solution is determined by choice of a known solution which undergoes no mass change and therefore no net solvent exchange. Isopiestic methods were modified several times in the early 1900s and used extensively in the first half of this century. However, these methods require days for a single measurement [23–25].

Electrolyte activities may be determined by any method that either directly or indirectly probes the chemical potential of solvent or solute in solution. These methods include isopiestic [21b], potentiometric [21c], freezing point, boiling point, osmotic pressure, relative speciation, solubility, and transport properties including conductivity and diffusion [21d]. The solvent and mean solution activities of several hundred single salt solutions have been determined. Of these, the majority have been studied by the isopiestic method [21d,23–27], which is also more applicable for highly speciated equilibria, in which potentiometric (Nernstian) analysis would be complex or not suitable [21c]. We therefore sought a contemporary methodology to speed up the lengthy measurement time required for traditional isopiestic methods. This method has been termed "evanescent" (evanescent $\int$ transitory, dissipating like a vapor), and is discussed in the following section.

D. Evanescent Methodology for Analysis of Solvent Activity

Evanescence methodology digitally monitors the real-time mass variation of a solution to determine the rate of driven evaporative loss as a function of time in a controlled environment. In this method, the dynamic mass loss via solvent evaporation, dm_{solnh}/dt, from the liquid phase can be used as a probe to determine solvent activity to a high degree of precision. At constant temperature, the evaporation rate of solvent from solution will be related to the differential of the solvent partial pressure in the ambient gas phase and the thermodynamic partial pressure limited by the solvent activity in solution. In this method of analysis, the partial pressure of solvent in the ambient gas is effectively kept fixed and the resultant dynamic mass loss is monitored and related to solvent activity. The effect leads to accelerated solvent activity measurements compared to previous conventional isopiestic measurements. This is accomplished by introducing a reproducible gas boundary layer consisting of a relatively nonturbulent regulated flow of gas across a fixed geometry of the solution under analysis. The gas is introduced with a fixed partial pressure of solvent, and the time variation of the re-

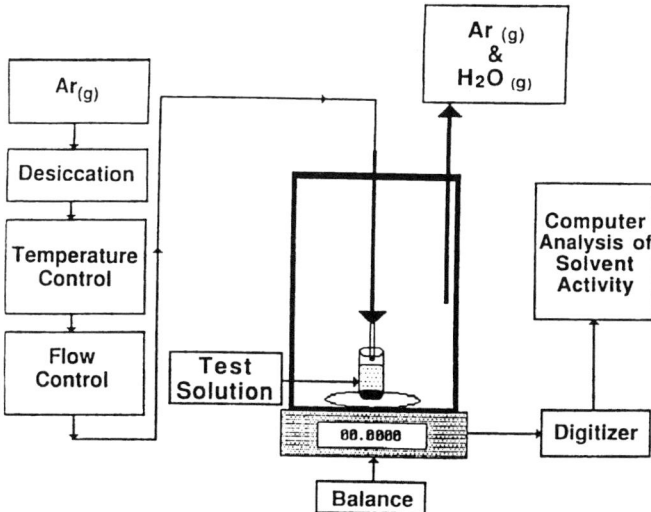

FIG. 9. Schematic representation of the evanescent determination of solvent activity for a "test solution." An on-line computer compares the rate of solvent loss to calibrated evanescent loss of solutions of known solvent activity, and converts the digitized mass variation to solvent activity of the test solution. In the evanescent solvent activity apparatus used, all control gases were initially dehydrated and temperature controlled to 25 ± 0.3°C by passing through an extended stainless steel tubular coil immersed in a 25°C water bath. Output gas flow rate, typically maintained at 25 cm^3/sec, was controlled by valve and monitored using a conventional Fisher Scientific Precision bore flowrator 01 150/S-51801 flow meter. Mass measurements were determined with a Fischer XA-200 electroanalytical balance with 0.1 mg resolution. The balance, interfaced through an RS-232C serial port, internally accumulates and digitizes measured mass every 2.5 seconds. TRUE BASIC software was developed to create algorithms for on-line mass collection, the calculation of differential mass, and analysis of solvent activity.

sultant solution mass is monitored. Differential solution mass, dm_{soln}/dt, is calculated in real time and related to calibrated $dm_{standard}/dt$ for solutions of known activity, resulting in determination of sample solvent activity.

Evanescent solvent activity applied to the determination of water activity is summarized in Fig. 9. Highly dried gas, such as argon passed over a dessicator, is introduced at fixed temperature and flow rate into a confined electroanalytical balance chamber. In the chamber, the gas flows over a "test solution," the analyte solution, at a flow rate optimized to maximize solution evaporation without inducing excessive solution turbulence. The rate of evaporation of solvent from the test solution is monitored as mass loss by the electroanalytical balance, digitized, and interfaced to a computer, which continuously determines differential mass loss as a function of time. The differential mass is compared to the mass loss induced from standard solutions of known solvent activity and converted to solvent activity of the test solution. As shown in Fig. 10, in which the inlet gas humidity is varied, the time response for evanescent analysis is on the order

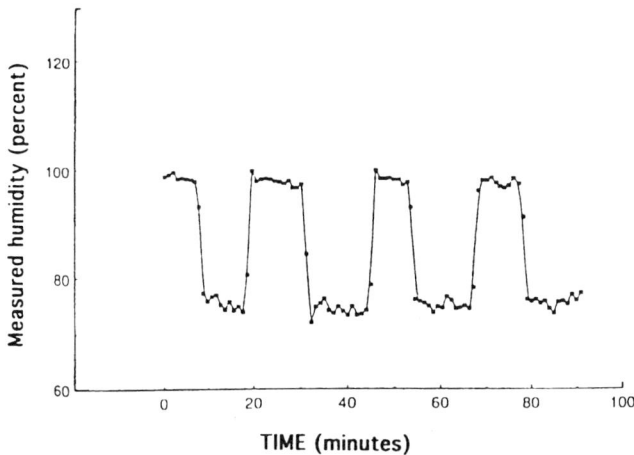

FIG. 10. The time response of evanescent analysis. Evanescent measurement conducted using a fixed sample solution as described in Fig. 9, in which the inlet gas is not dry argon, but rather cycled between argon fixed with either 100% or 75% relatively humidity at 25.0°C. (Further experimental details are presented in Ref. 28.)

Analysis in Highly Concentrated Solutions 109

of minutes, rather than the time scale of days associated with conventional isopiestic measurement.

Evanescent activity measurements were examined by bringing argon in contact with a variety of standard solutions of known solvent activity. Evanescent calibration with KOH standard solutions is described herein. Further experimental results summarizing evanescent measurements in a variety of different salt solutions are found in Refs. 3 and 28. Figure 11 presents the evanescent mass loss of a variety of concentrated aqueous KOH test solutions at 25°C. From the data in this figure, the inset of Fig. 11 summarizes the measured value of $dmass_{soln}/dt$ as a function of KOH solution concentration. Under a fixed gas flow and fixed cell geometry conditions, the mass transport rate of solvent across the gas/liquid inter-

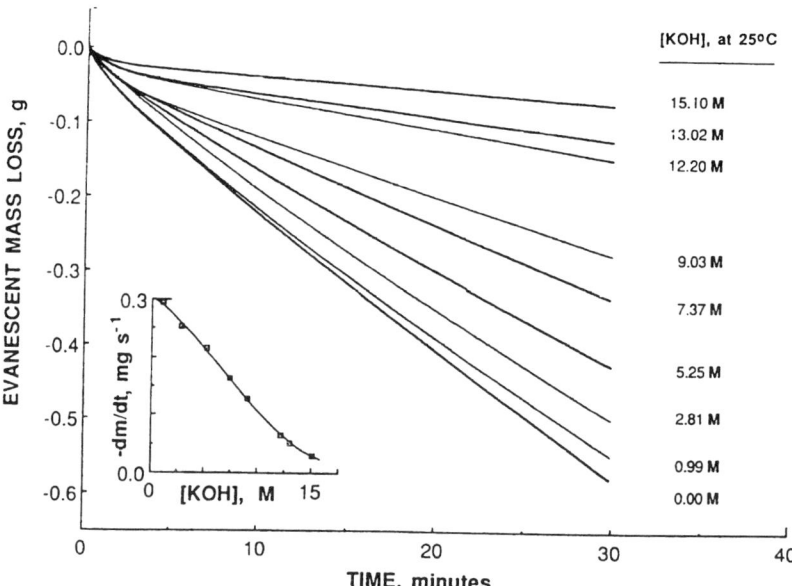

FIG. 11. The evanescent mass loss of solvent from a variety of aqueous KOH solutions. Evaporation of solvent from each KOH solution to desiccated Ar_{gas} is monitored by the mass loss of water at 25.0°C, using the experimental configuration a described for Fig. 9 and as further described in Refs. 3 and 28. Figure inset: Evanescent time evolved rate of mass loss, $-dm/dt$, as a function of KOH molar solution concentration.

face, $dmass_{soln}/dt$, is related to the differential between the solution solvent activity and the solvent partial pressure in the flow gas, p_w:

$$\frac{dmass_{soln}}{dt} \infty\, a_w = p_w \qquad (18)$$

There are limits to which the rate, $dmass_{soln}/dt$, will remain proportional to the solvent activity. Limiting the choice of analytes to nonvolatile solutes, including most conventional salts in aqueous solution, constrains the solution's mass variation to solvent, and simplifies interpretation of $dmass_{soln}/dt$. From a thermodynamic perspective, the ideal case occurs at $dmass_{soln}/dt = 0$. In this case, when the flow gas contains a solvent content, p_w, equivalent to the solution's solvent activity, no net exchange of solvent occurs across the solution/gas interface, and equilibrium activities are probed. Equilibrium is also approached as the gas flow rates approaches zero. The system then has infinite time to equilibrate the gas partial pressure to the solvent activity. At higher flow rates, a kinetic process (evaporative losses) is being used to probe thermodynamic information (solvent activity), and flow conditions will perturb the system from equilibria. In particular, the enthalpy of vaporization of most solutions is significant, and cooling due to evaporation will perturb the solution activity. At room temperature, each °C increases the vapor pressure of water by 8%. From a practical perspective, during measurement the gas flow rate is kept low to ensure that the solution temperature is affected by less than 0.1°C.

The linear response of the evanescent technique to probe a range of solvent activity is demonstrated with well-characterized KOH solutions (Tables 1 and 2). The variations of these known (standard) water activities as a function of measured evanescent solvent loss, $dmass_{soln}/dt$, are summarized in the inset of Figure 12. In accordance with Eq. (18), the variation of $dmass_{soln}/dt$ with a_w is highly linear to a first-order approximation, but small deviations to linearity are observed at both the low- and high-solvent activity extremes. Deviations may occur with imperfect temperature control, initial residual solvent in the carrier gas, or turbulence at the gas/liquid interface. Compensations for small deviations from linearity are incorporated by fitting a third order polynomial for a_w. The resultant polynomial acts as a calibrating standard, an algorithm for on-line conversion of measured $dmass/dt$ to solvent activity. This method provides a measurement of water activities in Fig. 12 to an average reproducibility of 1.0% over the activity range varying from 0.86 to 0.20, and 1.6% over the full 0–1.0 a_w range.

Analysis in Highly Concentrated Solutions

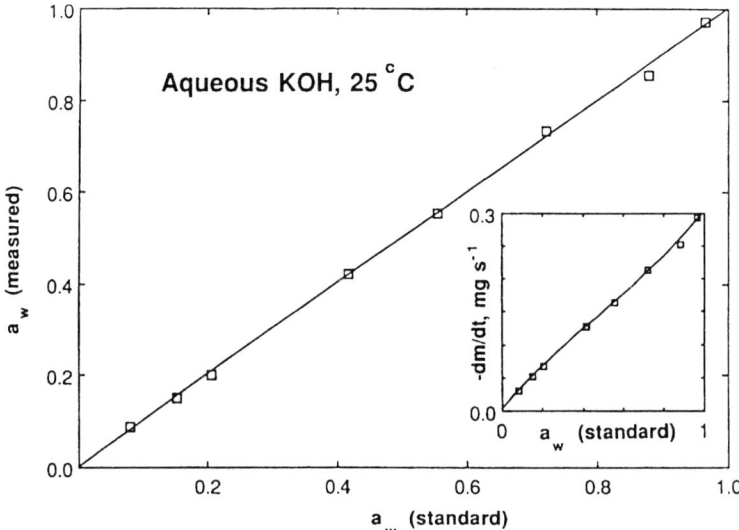

FIG. 12. Comparison of known (standard) solvent activities in aqueous KOH solutions with those regenerated from evanescent measurements at 25°C. Values for measured activity are determined from measured evanescent values of dm/dt using a third-order polynomial fit of dm/dt versus a_w data summarized in the figure inset. Inset: The variation of solvent mass loss, $-dm/dt$, with known water activity in aqueous KOH solutions. The curve represents a best third-order polynomial fit of the data.

Figure 13 summarizes solutions for mass losses, $dmass_{soln}/dt$, from aqueous K_2S solutions at 25°C over a wide concentration range. Potassium sulfide solutions may be nominally described as either K_2S solutions or as solutions equimolar in KOH and KHS. Molal and molar concentrations of equivalent solutions are presented in the first two columns of Table 3 and are consistent with recent density determinations [1]. The measured value of solvent mass loss, $dmass_{soln}/dt$, for each of the K_2S solutions as a function of solution molarity is summarized in the inset of Fig. 13. The direct conversion of dmass/dt to a_w for K_2S is quantified in the calibration presented in the inset of Fig. 12. The resulting analyzed values of a_w for aqueous K_2S solutions are presented in column 3 of Table 3.

Solution osmotic coefficients for sulfide solutions, ϕ, are presented

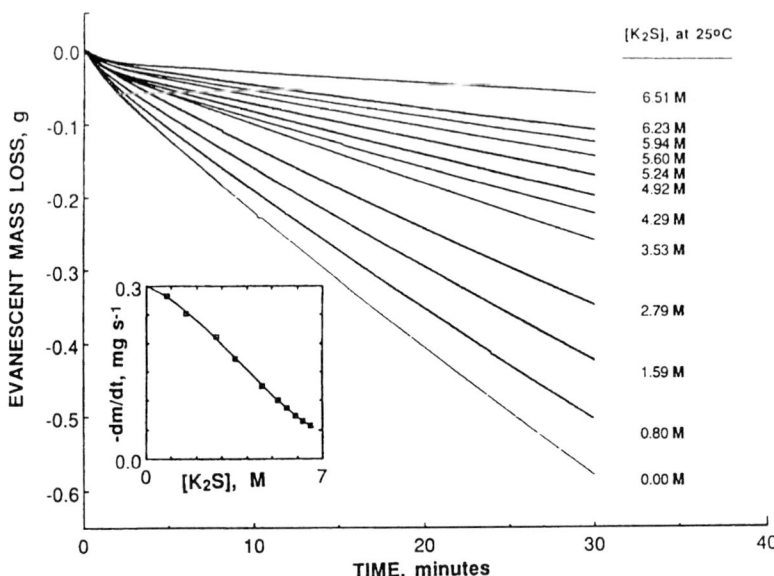

FIG. 13. The evanescent mass loss of solvent from a variety of aqueous K_2S solutions. Evaporation of solvent from each K_2S solution is monitored by the mass loss of water at 25.0°C, using the experimental configuration as described for Fig. 9 and as further described in Ref. 28. Figure inset: Evanescent time evolved rate of mass loss, dm/dt, as a function of K_2S molar solution concentration.

in column 5 of Table 3 as determined from Eq. (15) using the evanescent water activities summarized in column 3 of this table. Substitution of these osmotic coefficients into Eq. (15), and a conventional trapezoidal integration of the concentration variation of Eq. (16), results in the mean molal activity coefficient summarized in column 6 of Table 3. From these values in the high alkaline sulfide concentration domain, the evanescent technique provides in column 4 the first measurements for mean solution activity in sulfide solution, $a_\pm$, pertinent to an understanding of the sulfide second acid dissociation constant, which are summarized. As seen in this column, and analogous to concentrated potassium hydroxide solutions, very high solution activities dominate in this high concentration domain [3].

The evanescent (driven evaporation) technique provides a contem-

TABLE 3
Solvent and Mean Solution Activities of Aqueous Potassium Sulfide Solutions at 25°C

[K⁺], molar	[K⁺], molal	a_w	$a_\pm$	ϕ	$\gamma_\pm$
0.00	0.00	1.00	0.00	—	1.00
0.99	1.01	0.96	0.76	0.99	0.76
1.60	1.65	0.95	1.17	0.95	0.71
3.18	3.44	0.86	3.20	1.20	0.93
5.58	6.50	0.73	8.51	1.37	1.31
7.06	8.65	0.59	18.2	1.69	2.10
9.21	12.3	0.41	47.8	2.01	3.89
10.5	14.8	0.32	79.1	2.13	5.34
11.2	16.4	0.28	103.	2.18	6.29
11.9	17.9	0.23	135	2.24	7.50
12.5	19.4	0.20	170.	2.30	8.75
13.0	20.9	0.17	211.	2.34	10.1

a_w is measured using the evanescent technique, as described in the text. ϕ and $\gamma_\pm$ are calculated in accordance with Eqs. (15) and (16).

porary method to probe and broaden the understanding of solvent activity in concentrated electrolyte. The measured values of solvent activity provide analyte information by determination of solution activity in accordance with Eqs. (15) and (16). The methodology determines the rate of driven evaporative loss as a function of time in a controlled environment and evaporation rate correlates with solvent activity for a wide variety of known solutions. As opposed to conventional isopiestic methods, which can take days to determine solvent activity and are therefore often not useful in highly unstable, volatile, caustic, or reactive media evanescent methodology is demonstrated to determine solvent activity within minutes to a precision of better than 2%.

IV. CONDUCTOMETRIC ANALYSIS OF CONCENTRATED SOLUTIONS

A. Conventional Conductometric Titration

Two general instrumental techniques have been utilized for conductometric analysis of solutions. The first, common to academic and

TABLE 4

The Equivalent Conductivity of Hydrogen and Hydroxide Ions in Water

Temp (°C)	pK_w	λ, cm^2 ω^{-1} equiv^{-1}		
		λ^{oH+}	λ^{oOH-}	$\lambda^{oH+} + \lambda^{oOH-}$
0	14.938	224.1	117.8	341.9
5	14.727	250.0	133.6	383.6
10	14.528	275.6	149.6	425.2
15	14.340	300.9	165.9	466.8
18	14.233	315.8	175.8	491.6
20	14.163	325.7	182.5	508.2
25	13.995	350.1	199.2	549.3
30	13.836	374.0	216.1	590.1
35	13.685	397.4	233.0	630.4
40	13.542	420.0	250.1	670.1
45	13.405	442.0	267.2	709.2
50	13.275	463.3	284.3	747.6
55	13.152	483.8	301.4	785.2
60	13.034	503.4	318.5	821.9
65	12.921	522.0	335.4	857.4
70	12.814	539.7	352.2	891.9
75	12.711	556.4	368.8	925.2
80	12.613	572.0	385.2	957.2
85	12.520	586.4	401.4	987.8
90	12.431	599.6	417.3	1017
95	12.345	611.6	432.8	1044
100	12.264	622.2	448.1	1070
125	11.911			1196
150	11.637			1323
175	11.431			1425
200	11.288	824	701	1525
225	11.207			1581
250	11.192			1636
275	11.251			1676
300	11.406	894	821	1715
325	11.705			1737
350	12.295			1759
374	15.641			1780

Source: Ref. 31.

Analysis in Highly Concentrated Solutions 115

environmental laboratory, immerses two electrodes of fixed geometry in solution. An alternating potential is applied at the electrodes and the solution conductance component portion of the impedance signal is electronically determined. The second technique, often found in industrial applications, is electrodeless conductivity in which torroidal electrodes are situated on each side, but external, to the solution being investigated. An oscillating potential is applied to the first and induced in the second torroid, and the solution conductance is determined [29]. A variety of calibrating solutions have been determined for analysis of solution conductance [30].

Conductometric analytical methodologies make use of the different specific conductance of ions to determine which species are present in solution. These methodologies are particularly suitable for distinguishing between covalent, single, and multiple charge species, between soluble and insoluble species (as in solubility product determination), and acid dissociation constants. In aqueous media the hydrogen ion and hydroxide ion mobilities are severalfold higher than all other ions [31] and reflect the unique hopping/tunneling conductance mechanisms available to only H^+ or OH^- ion/solvent interactions. Therefore, conductometric measurement of H^+ or OH^- during acid/base titration presents an easily discerned and widely used indicator to probe the degree of acid protonation in muliprotonic acid, as in our study of ferrocyanic and ferricyanic acid [32]. The equivalent conductivity of hydrogen ion and hydroxide ion is presented in Table 4 for water from the melting through critical points.

B. Conventional Conductometric Titration of Concentrated Solution

Conductometric titrations conventionally utilize a titrant substantially more concentrated than the analyte, with the equivalence point and equilibrium constant related to the derivative of the conductivity variation. When this is not feasible, as in the study of highly concentrated analytes, an equally valid approach is to use equimolar (equally concentrated) analyte and titrant solutions and to model the results to expected conductance variation for the equilibrium under consideration. An example of this approach is presented here in which the high mobility of hydroxide is utilized to probe the extent of free hydroxide in KHS solution to probe the second acid dissociation constant, K_2, of H_2S. If during the course of the titration added hydroxide does not react, then solution conductance is expected to rise. However, this will not be the case if added OH⁻ reacts with free HS⁻ according to the hydrolysis reaction associated with K_2:

$$HS^- + OH^- \rightleftharpoons H_2O + S^{2-} \; ; \; K_{2b'} = \frac{K_2}{K_w} = \frac{a_w a_{S^{2-}}}{a_{HS^-} a_{OH^-}} \tag{19}$$

Figure 14 summarizes the measured conductometric titration of 6.5, 9.5, or 12.83 molar KHS solutions during titration by equimolar (respectively 6.5, 9.5, or 12.83 M) KOH solutions at 25°C. The figure inset presents two idealized variations of titrated KHS with added KOH, as the solution conductivity varies from that of the pure KHS electrolyte. The dotted curve indicates the expected variation in conductivity if there is no interaction between hydrosulfide and hydroxide. Hence, as seen at 50% added KOH, the conductivity acts as the simple average of the KHS and the higher KOH solution conductivities. In the second situation, the dashed curve indicates the expected variation in conductivity if the interaction of each hydrosulfide and hydroxide reacts fully to produce a free sulfide. In this situation, the point of 50% added KOH reflects a solution containing only K_2S and cannot contain the highly mobile hydroxide species. Each free sulfide is assumed to have equivalent conductivity substantially less than that of hydroxide and approximately that of free hydrosulfide.

In Fig. 14, the measured conductivity variation in the titration of KHS by KOH can be compared to the expected variation, as shown in the inset. In the 6.5 and 9.5 **M** titrations, the linear increase in conductivity with increasing KOH provides substantial evidence that added OH⁻ remains unassociated, and there is little free sulfide in solution. However, the observed similarity of the 12.83 **M** titration curve to the dashed curve in Fig. 14 indicates diminished free hydroxide in this solution and the formation of substantial quantities of free sulfide. The extent of these variations can be quantified by calculation of the distribution of species in solution as a function of the extent of titration, and reflect that little free S^{2-} exists in all but the most concentrated alkaline environments, expressed by the second acid dissociation constant of approximately $K_2 = 10^{-17}$. However, more precise measurements of K_2 can be determined using the differential conductometric methodology presented in the next section.

Analysis in Highly Concentrated Solutions

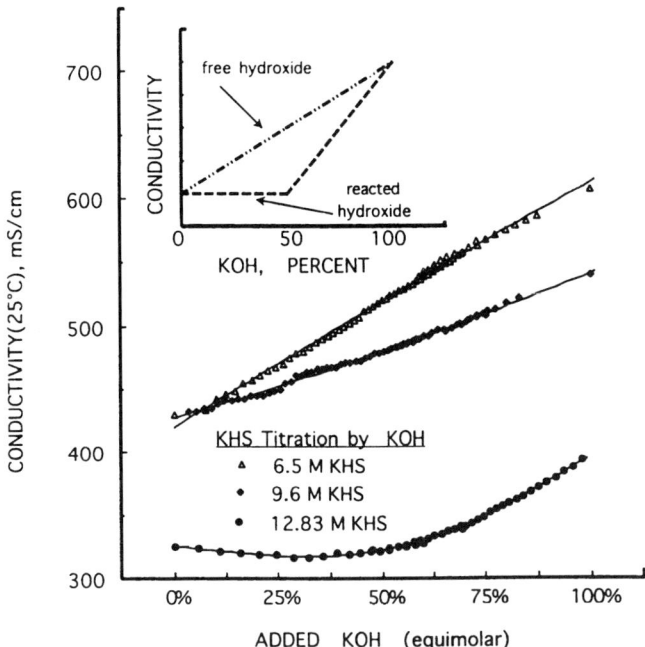

FIG. 14. The measured conductivity of 6.5, 9.5, or 12.83 M KHS solutions during titration by equimolar (respectively, 6.5, 9.5, or 12.83 M) KOH solutions at 25°C. Inset: The dotted curve indicates the expected variation in conductivity if the interaction of hydrosulfide and hydroxide produces no free sulfide; the dashed curve indicates the expected variation in conductivity if the interaction of each hydrosulfide and hydroxide reacts fully to produce a free sulfide, and assumes sulfide equivalent conductivity approximates hydrosulfide equivalent conductivity. Electrolytic conductivity was measured with a YSI 35 conductance meter and a Balsbaugh conductance probe with a cell constant of 10 cm^{-1}. The probe has a Noryl body and graphite electrodes minimizing both alkali attack and sulfide poisoning of the electrodes. The probe was calibrated with 1.00 M KCl solutions. Further experimental details are presented in Ref. 32.

C. Differential Conductometric Methodology

Differential conductometric analysis compares the specific conductances of isolated and equilibrated phases in liquid phase systems [33]. In concentrated solutions containing reagents A and B, the specific conductance change should act as a probe to gauge the equilibrium distribution of species in solution. The differential conductance, $\Delta\sigma_p$, measures the difference in the specific conductance of solutions containing the isolated reactants, σ_A and σ_B, compared to the specific conductance in the equilibrated system, σ_p, described in the equilibrium:

$$A + B \rightleftarrows P \qquad (20)$$

The differential conductance varies as a function of the given concentration, C, of isolated or combined equilibrated reactants:

$$\Delta\sigma_P(C) = 0.5(\sigma_A(C) + \sigma_B(C)) - \sigma_P(C) \qquad (21)$$

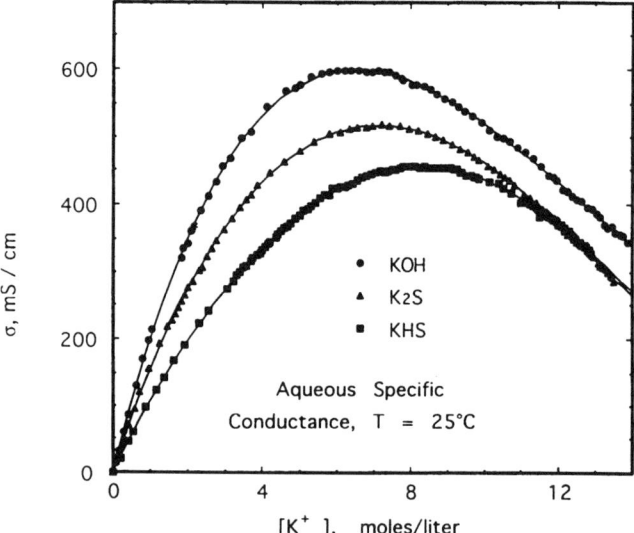

FIG. 15. The conductivity of concentrated KOH, K_2S, or KHS aqueous electrolytes at 25°C. [K$^+$] refers to the molar concentration of K$^+$ in an aqueous solution containing only the potassium salt indicated. Further details are presented in Ref. 32.

D. An Example of Differential Conductometric Methodology

The variation of aqueous KOH, KHS, and K_2S specific conductance at 25°C, and over a wide range of concentrations, is presented in Fig. 15. K^+ and HS^- have comparable limiting equivalent conductances ($\lambda°_{K^+}$ = 74 S cm^2 equiv^{-1} and $\lambda°_{HS^-}$ = 67 S cm^2 equiv^{-1}, respectively, at 25°C). The KOH specific conductance is dominated by the exceptionally high equivalent conductance of hydroxide, $\lambda°_{OH^-}$ = 199 S cm^2 equiv^{-1} at 25°C.

In Figure 16, the magnitude of the $\sigma_K 2S$ deviation is reported as a differential conductance, $\Delta\sigma_K 2S$, determined as:

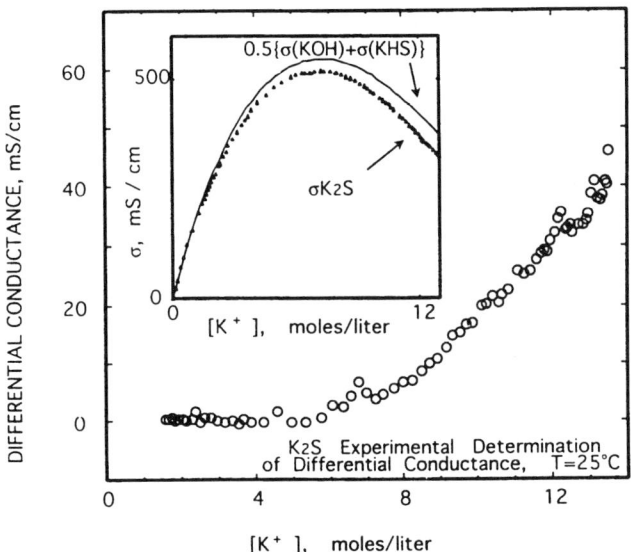

FIG. 16. Differential conductivity, $0.5(S_{KHS} + S_{KOH}) - S_K 2S$, in aqueous concentrated alkaline potassium sulfide solutions at 25°C. Differential conductivity is determined as the magnitude of the K_2S conductivity deviation from the average of the KHS and KOH conductivities. The nominal molar concentration of each solution may be described as $[K_2S] = 0.5[K^+]$ or alternatively as a solution comprised of $[KHS] = [K^+]$ and $[KOH] = [K^+]$. Inset: Aqueous K_2S conductivity compared to the average of the KHS and KOH conductivity as a function of concentration. $0.5(S_{KHS} + S_{KOH})$ is determined by a fifth order polynomial fit of experimental conductivity, as summarized in Table 2.

$$\Delta\sigma_{K_2S} = 0.5(\sigma_{KHS} + \sigma_{KOH}) - \sigma_{K_2S} \tag{22}$$

The variation of the aqueous potassium sulfide specific conductance, σ_K2S, can be separated into two distinct regions when compared to KHS and KOH electrolytes of similar concentration. As shown by the differential conductance measurements reported in Fig. 16, aqueous K_2S electrolytes containing less than 6 **M** K^+ act in a manner similar to an average of equimolar separate solutions of KOH and KHS. This variation in conductance reflects the distribution of species within these solutions. At sufficiently low concentrations of K_2S, there is virtually complete hydrolysis. At sufficiently low concentrations of K_2S, there is virtually complete hydrolysis, no free sulfide exists, and these electrolytes act as simple solutions containing equimolar free OH^- and HS^-, in accordance with the left side of Eq. (19). These conditions can only be met for a second acid disso-

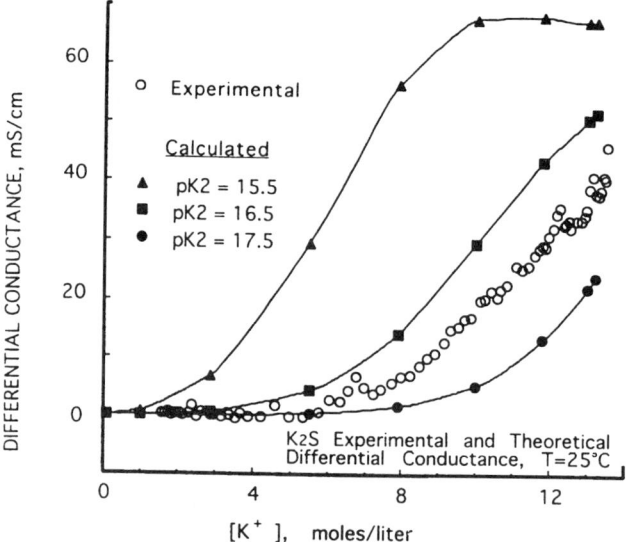

FIG. 17. Predicted and experimental differential conductivity of aqueous K_2S solutions at 25°C. Predicted differential conductivity for K_2S is calculated from Eq. (22), using the relative distribution of ions in solution, Eq. (19), and utilizes the experimental conductivities of equivalent KOH and KHS solutions.

Analysis in Highly Concentrated Solutions

ciation constant of H_2S, which is several orders of magnitude smaller than commonly reported.

In aqueous K_2S electrolytes containing greater than 6 M K^+, the increase in differential conductance parallels the depletion of free OH^- and the formation of free S^{2-}. Figure 17 compares predicted and experimental differential conductivity of aqueous K_2S solutions at 25°C. K_2 on the order of 10^{-12}, as had been reported, would lead to a much sharper rise in differential conductance than observed. In Fig. 17 the predicted differential conductance for K_2S is calculated based on K_2 values varying from $10^{-15.5}$ to $10^{-17.5}$. A best fit of $pK_2 = 16.9 \pm 0.2$ models the measured differential conductance for the complete range of K_2S concentrations studied [33].

V. DIFFERENTIAL DENSOMETRIC ANALYSIS

A. Differential Densometric Principles

In dilute solutions, the molecular volume change associated with chemical reactions has little effect on the density of the bulk solution. However, in concentrated solutions the solute constitutes a significant fraction of the solution, and changes in molecular volume change associated with chemical reactions will be reflected by significant changes in the density of the bulk solution.

The combined volume of the solute in concentrated electrolytes can be significant compared to the total solvent volume. Indeed, the volume encompassed by dissolved hydrated species in approximately 10 M aqueous salt solutions can be larger than that encompassed by the volume of free water. The combined volume of individual solute species may be substantially affected by equilibrium displacement, including that occurring by simple addition, as exemplified by Eq. (20) or by charge transfer from reactants A, B to form products C, D:

$$A^- + B^- \rightleftharpoons C^{2-} + D \tag{22}$$

Differential densometric analysis compares the densities of isolated and equilibrated phases in a system. Figure 18 illustrates the variation in volume that can occur between such isolated and equilibrated phases in a system. In concentrated solutions of reactants A and B, the cumulative volume change should act as a probe to gauge the equilibrium distribution of species in solution. The differential density, $\Delta d_{A,B}$, measures the difference in the density of solutions containing the isolated reactants, d_A and d_B, compared to the density

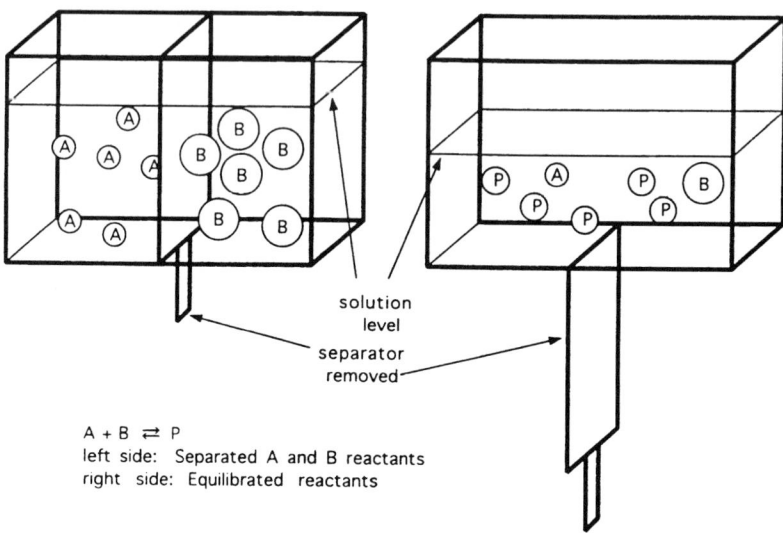

FIG. 18. Schematic of decrease in bulk solution volume occurring when the molecular volume of product species, P, is smaller than the combined molecular volumes of reactants A and B.

in the equilibrated system, $D_{A,B}$. The differential density varies as a function of the given concentration, C, of isolated or combined equilibrated reactants:

$$\Delta d_{A,B}(C) = d_{A,B}(C) - 0.5(d_A(C) + d_B(C)) \qquad (23)$$

Δd_{AB} varies with the relative solvated volume of product and reactant species and is proportional to the distribution of ions in solution. The concentration variation of $\Delta d_{A,B}(C)$ compared to that at a fixed concentration, $Dd_{A,B}(C_{fixed})$, acts as a measure of the relative concentration of products in these electrolytes. Hence, in accordance with Eqs. (20) and (23), for a solution that contains a total concentration, C = [A] + [B] + [P]:

$$\frac{\frac{[P]}{C}}{\frac{[P]_{fixed}}{C_{fixed}}} \cong \frac{\Delta d_{A,B}(C)}{d_{A,B}(C_{fixed}) - 0.5(d_A(C_{fixed}) + d_B(C_{fixed}))} \qquad (24)$$

Equation (24) becomes exact in the limit in which $[P]_{fixed}$ approaches C_{fixed}, i.e., complete reaction of A and B.

B. Differential Densometric Methodology

Differential densometric analysis requires precise measurement of differences in solution density. Specific gravity is readily measured by a Westphal balance in which the solution to be analyzed is situated on a balance; a plummet is suspended into the solution, displacing a fixed volume. The resultant measured mass change, relative to the displaced fixed volume, is the specific gravity of the solution. However, commercially available hydrometers and Westphal balances did not exhibit a desired relative density resolution of better than 0.0005 g/cm^3.

Specific gravity was measured with a contemporary in house–built Westphal balance, utilizing a ~15 cm^3 glass plummet externally suspended into a cell situated on a Fisher XA 200 0.0001 g resolution electroanalytical balance RS-232C interfaced to a 8086 microcomputer. Specific gravity was measured by the loss in weight of the glass plummet brought about by its buoyant suspension in approximately 45 cc of the liquid to be measured. The cell was maintained at a constant temperature of 25.0 ± 0.3°C, and the plummet was repeatedly calibrated upon immersion in deionized distilled water. In water, the ratio of the ±0.0003 g accuracy of the balance compared to the plummet mass yields a theoretical specific gravity reproducibility of 1.00000 ± 0.00002. Temperature uncertainty in the specific gravity is comparable to a theoretical specific gravity resolution for water of 1.00000 ± 0.00008. Actual reproducibility of the measured specific gravity was better than 1.0000 ± 0.0002.

In this study, differential densometric analysis is defined in terms of molar concentration units. The variation of density and molarity are both related by the spatial constraints of the solution, whereas molality is constrained by solvent mass. Similarly, the hydrolysis reaction [Eq. (19)] is a molar consistent, molal inconsistent, equation (as it involves solvent reaction). However, the required precision of the density measurements in this study dictated that molal concentrations be employed. Molar concentrations were subsequently determined from the solution density, d (for a solution comprised of a pure salt of molecular weight W, the molarity = d/(molality^{-1} + 0.001·W)).

C. Differential Densometric Analysis Example

As discussed in Sec. III, a K_2S solution can be considered a combined equimolar solution of KOH and KHS. The concentration of HS$^-$, OH$^-$, and S^{2-} in such solutions is constrained as in equilibrium Eq. (19). The variation of aqueous KOH, KHS, or K_2S specific gravity at 25°C, and over a wide range of molal concentrations, is presented in Fig. 19. The specific gravity was determined relative to unity specific gravity at infinite dilution, i.e., the specific gravity$_{25°C}(H_2O) \int 1$. The density is compared to the known density of water, i.e., $d_{25°C}(H_2O) = 0.99702$ g/cm^3, and the resultant molar variation of density is presented in the inset of Fig. 19. The functional variation of density is substantially linear with changes in concentration. A closer inspection of the (significant) contributions from higher order terms in the concentration variation of the density is achieved by first removing this large linear concentration dependence. This is derived by defining an equivalent density, scaled by concentration and similar to the concept of an equivalent conductance:

$$\text{Equivalent density} \equiv C^{-1}\{d(C) - d_{25°C}(H_2O)\} \tag{25}$$

where C is the concentration of K$^+$ in gram equivalents per liter. Figure 19 inset presents the variation of KOH, KHS, or K$_2$S equivalent density with concentration.

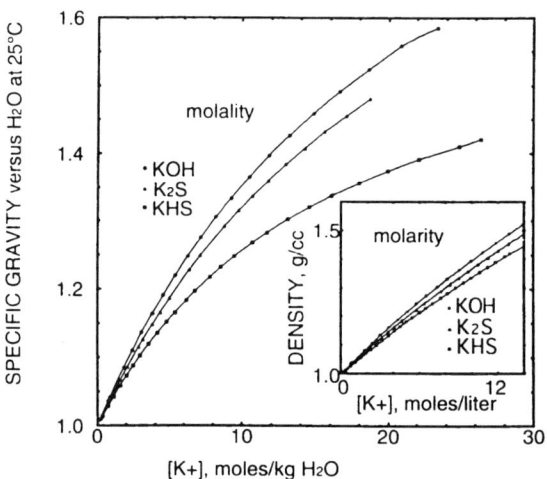

FIG. 19. The measured specific gravity of concentrated KOH, K$_2$S, or KHS aqueous electrolytes at 25°C. m_K^+ refers to the **molal** concentration of K$^+$ in an aqueous solution containing only the potassium salt indicated. Inset: The density of concentrated KOH, K$_2$S, or KHS aqueous electrolytes at 25°C. [K$^+$] refers to the **molar** concentration of K$^+$ in an aqueous solution containing only the potassium salt indicated.

Analysis in Highly Concentrated Solutions

In accordance with Eqs. (19) and (23), the differential density of a K_2S solution is determined upon comparison to the densities of separate KOH and separate KHS solutions:

$$\Delta d_{K_2S}([K^+]) = d_{K_2S}([K^+]) - 0.5(d_{KHS}([K^+]) + d_{KOH}([K^+])) \qquad (26)$$

where $[K^+]$ (moles/liter solution), is the molar concentration of K^+ in an aqueous solution nominally containing only the salt indicated.

The concentration variation of $\Delta d_K 2S$, calculated according to equation (26), is presented in Fig. 20. For aqueous electrolytes, nominally comprised of up to 3 M K_2S, it is evident that the solution density is comparable to solutions containing 50% isolated KOH and 50% isolated

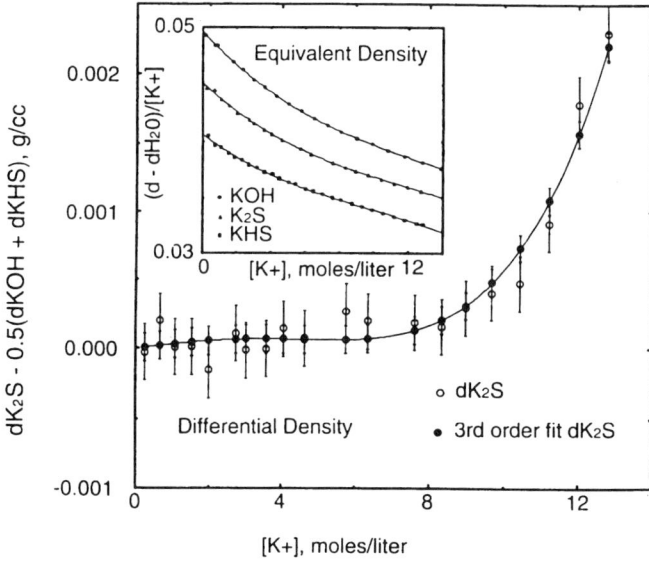

FIG. 20. Differential density in aqueous concentrated alkaline potassium sulfide solutions at 25°C. The nominal molar concentration of each K_2S solution may be described as $C_K2S = 0.5\,[K^+]$ or alternatively as a solution containing both $C_{KHS} = [K^+]$ and $C_{KOH} = [K^+]$. Δd_K2S open circle points are calculated using direct experimental values of $d_K2S([K^+])$, presented in Fig. 1 inset. The Δd_K2S solid circle points are calculated from a polynomial fit of experimental $d_K2S([K^+])$ values. Inset: The equivalent density of concentrated KOH, K_2S, or KHS aqueous electrolytes at 25°C, calculated from the measured density in accordance with Eq. (24).

KHS, with no evidence of volume changes due to the reaction of HS⁻ and OH⁻ to form free sulfide. At concentrations increasing above 3 **M** K_2S, a significant increasing deviation of solution density, in excess of the solution density of the isolated reactants, is evident. The observed increase in density is consistent with the formation of a more highly charged, and therefore more densely packed ion, such as S^{2-}.

The relative free sulfide in solution may be estimated by densometric comparison of K_2S solutions to pure equimolar solutions of KHS and KOH. Figure 21 presents the concentration variation of these predicted relative differential densities, assuming a range of hydrogen sulfide second acid dissociation constants varying from 12 to 18. These calculated values

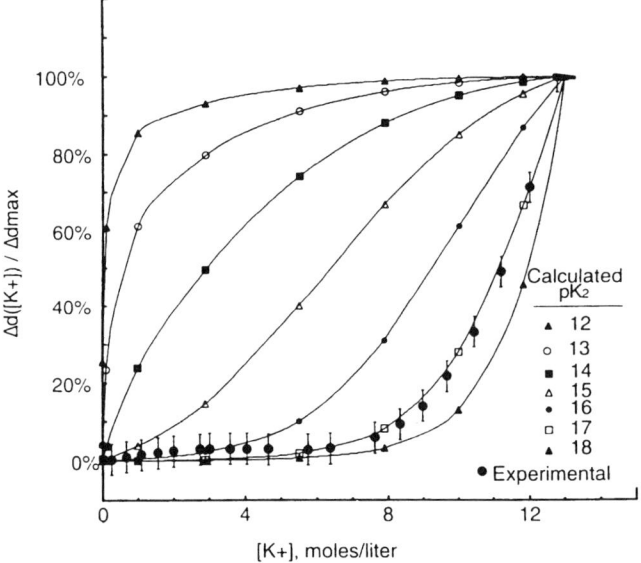

FIG. 21. The calculated and experimental concentration dependence of the relative differential density of aqueous K_2S solutions at 25°C. Calculated and experimental values are compared to a maximum value of Dd_K2S determined for 6.4 **M** K_2S. [K⁺] refers to the molar concentration of K⁺ in an aqueous solution of K_2S. Calculated relative differential density is determined using $K_w = 1.00 \times 10^{-14}$, values of a_w and activity from Table 1, and assuming a hydrogen sulfide second acid dissociation constant varying from $pK_2 = 12$ to 18.

Analysis in Highly Concentrated Solutions 127

for $(\Delta d_{K_2S})_{relative}$ can be compared to experimental values for $(\Delta d_{K_2S})_{relative}$, determined using the measured values of $\Delta d_{K_2S}([K^+])$ and $\Delta d_{K_2S}([K^+]$ = 12.8 M) as summarized in Fig. 20. In Fig. 21, upon comparison of idealized and measured values of $(\Delta d_{K_2S})_{relative}$, it can be seen that a best fit of $pK_2 = 17.1 \pm 0.3$ provides an excellent simulation of the measured relative differential density for the complete range of K_2S concentrations studied.

D. Differential Densometric Benefits and Disadvantages

Densometric analysis, unlike potentiometric and spectroscopic analytic techniques, cannot provide an absolute identification of the species being monitored. However, as opposed to other analytic techniques, which can exhibit increasing uncertainty with concentration, the physical analysis of solution density is particularly suited to concentrated solutions. The accuracy of density measurements is essentially concentration independent. Therefore, in Fig. 21, the effective signal-to-noise ratio of densometric analysis increases with increasing concentration. Combined with other techniques that qualitatively identify the relevant species in a concentrated liquid phase equilibrium, differential densometric analysis can provide a precise analysis of the equilibrated distribution of species. The example used illustrates application to unusually concentrated media and demonstrates the advantage of this technique to probe liquid phase equilibria in this regime.

VI. SUBMICROMETER PATH UV/VIS/IR SPECTROSCOPY

A. Conventional Absorption Spectroscopy Limits

The molar absorptivity of chromophores generally limits analysis by absorption spectroscopy to concentrations of less than 1 M. Higher concentrations result in absorption, A, greater than 2 and beyond the limit of conventional spectrophotometers. Direct path absorption studies are contingent on the simple Lambert-Bouguer linear relationship between light absorption and the path traversed, l. This combines in Beer's 1852 relationship between absorption and concentration, C, to yield for a species of molar absorptivity, ε:

$$A = \log\left(\frac{I_0}{I}\right) \propto \varepsilon l \; ; \quad A = \varepsilon l C \qquad (27)$$

Spectroscopic absorption phenomena have been investigated over a tremendous range of spatial dimensions varying from astronomic distances down to molecular dimensions. In the smallest spatial domains the spectroscopy of adsorbed monolayers and Langmuir-Blodgett films of thickness on the order of angstroms has been widely probed. There has been a growing interest in the spectroscopy of thin films by techniques using physical properties arising at surface interfaces, including diffraction [34] and reflectance and polarization enhancement [35,36].

In purely solution phase, there has been no analogous spectroscopic studies to the small dimensions used in thin film studies, although it is noted that liquid jets can have micrometer path lengths and that total internal reflectance techniques (using evanescent wave) can be used to probe the solution/prism interface at the submicrometer level. Direct path ultraviolet (UV), visible (VIS), or infrared (IR) absorption spectroscopy has been conducted with cells of path length down to several micrometers in the study of rapid kinetic measurements [37] and concentrated analytes [38], but due to material constraints there are few reports of liquid phase cells below one micrometer. Until recently, no systematic investigation of liquid phase absorption cells with dimensions smaller than the wavelength of incident radiation had been reported [3], although a submicrometer thickness thin film UV absorption spectrum having a quite diffuse spectral structure (to a resolution of 900 cm^{-1}) has been reported [39]. This cell path length domain is intriguing for near-ultraviolet, visible, and infrared absorption spectroscopy. It is in this domain that the cell path length, and light pathway, approach and actually become smaller than the dimensions of the wavelength of incident radiation.

Although literature references to submicrometer path length liquid phase absorption spectroscopy are scarce, a variety of schemes have been investigated for producing subwavelength illumination to deliver visible light to confined locations. These include channeling excitons to the point of illumination [40] or by propagation of photons through submicrometer apertures [41,42]. The significance of the micrometer domain in spectroscopy is evident in a Raman study in which spectroscopic cells of micrometer spatial resolution are presented to address molecular effects in single living cells [43]. Alternative techniques to investigate in situ liquid phase effects with a high degree of spatial resolution have been developed. These include x-ray diffraction to explore the structure of the electrochemical double layer [34] and microelectrode probes to map the movement of as few as 40,000 discrete ions [44].

B. Submicrometer Path Length Cells

Cells of micrometer or submicrometer dimensions were fabricated by sandwiching appropriate plates (as described below) in a modified 0186 0072 Perkin Elmer cell mount. Cells were assembled in a laminar flow chamber filled with prefiltered Ar. Solutions were prefiltered by injection through a syringe filter.

The 0.08 μm path length cell contained two EDMUND B43,400 precision single surface flats, fused quartz windows (1/20 wave (31 nm) flatness, 12.7 mm diameter, 3.18 mm thickness). A 0.015 μm pore, 25 mm diameter membrane unit was used to filter both Ar and liquids. Two ESCO E210450 precision flat, fused quartz windows (1/10 wave (53 nm) flatness, 25.4 mm diameter, 6.35 mm thickness) were used for the 0.28 μm path length cell. In this case Ar was filtered by a 0.1 μm pore, 25.4 mm diameter membrane unit (Micron Separations DDN01025) and liquids were filtered by a 0.22 μm filter. Two ESCO Q10063 flat quartz plates, two NaCl plates without a spacer and with a 25 μm Ag wire, Alpha 00307 spacer were used, respectively, for 0.56 μm, 2.2 μm, and 27.7 μm cells. A 0.1 μm pore filter was used for Ar and a 0.22 μm filter was used for liquids in the last three cases.

The Perkin Elmer 1330 Infrared Spectrophotometer and Perkin Elmer Lambda 3B UV/Vis Spectrophotometer were used for spectral measurements. Chemicals were of analytical quality and were used as received from commercial sources. Distilled deionized water was used in solution preparation.

C. Submicrometer Path Length Absorption Spectroscopy

The spatial confinement of planar radiation is diffraction limited by the direction orthogonal to the direction of propagation [35], as summarized in Fig. 22. Hence even in the domain in which the cell path length, l, is smaller than the wavelength of incident radiation, λ, the retention of spectral resolution is consistent with a large orthogonal cell width, **w**. In each absorption cell that was utilized, the orthogonal dimension **w** is more than 1000-fold greater than the wavelength of incident radiation. Nevertheless, diffraction is expected at the sharp variation in index of refraction occurring at the interfaces from the (empty) dry cell. This effect is commonly used to calibrate IR cells. Alternately, in smaller dimension (micrometer and submicrometer) cells, if UV/Vis radiation is employed, this diffraction

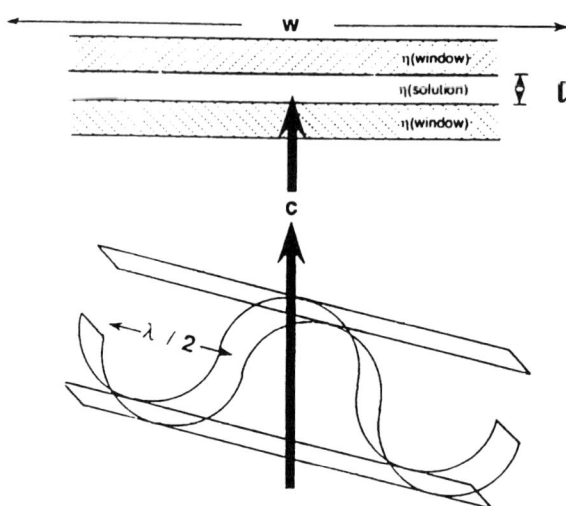

FIG. 22. Schematic representation of subwavelength absorption spectroscopy in which the pathway for light absorption, l, is an order of magnitude shorter than the wavelength of incident monochromatic radiation, λ. The cell dimension transverse to the direction of incident radiation is **w**, **c** is the speed of light, η(window) and η (solution) are respective indices of refraction of the cell windows and contained sample.

can be used to quantify these small path lengths. The measurement of absorption within the dry cell results in a broad interference fringe spectrum. Interference in a cell of pathlength, l, results in **n** fringes over a wavelength range from λ_1 to λ_2 [45]:

$$l = \frac{n\lambda_1\lambda_2}{2|\lambda_2 - \lambda_1|} \tag{28}$$

Two effects suggest that Eq. (28) interference within submicrometer cells will not affect measured absorption spectrum in the solution phase. The amplitude of this interference diminishes to zero as the indices of refraction of window and solution components of the cell approach a single value. A maximum amplitude for an interference fringe spectrum occurs in the dry cell with a difference of indices of refraction = (window) − 1. In each of the investigated liquid phase cells, η (window) − η (solution) < η (window) − 1. As expected, no superimposed interference spectrum is observed in the cells containing solution samples. Second, in accordance with Eq. (28) the interference pattern broadens with decreasing cell width, and the number of interference cycles is no longer significant for pathlength < 100 nm in the UV/Vis/IR region.

Micrometer and submicrometer path length solution phase spectroscopy is sufficiently unusual to require independent measurements to confirm path length calibration. A second approach, in addition to that offered by interpretation of the observed interference in dry cells, may be derived from a consideration of the limiting parameters in very thin cells. Submicrometer path length thin cells can consist of sandwiched optically flat quartz disks in which particulate impurities have been reduced in the ambient environment, and can permit direct observation of solutions substantially more concentrated than with conventional cells. The cell path length is restricted by the specific flatness, f, of both of the two sandwiched optical windows and also the maximum size of particulate impurities present in the cell, p. The specific flatness is limited by either the roughness or the curvature of the optical window. Therefore, from the microscopic perspective, l represents only an average distance over plates separated by surface morphology, curvature, and particulate impurities. The cell path length is limited by:

$$l = p + 2f \tag{29}$$

Finally, a third independent confirmation of path length calibration may be determined by comparison of the Lambert-Beer relationship [Eq. (27)] for two cells i and j of different path lengths. If cell j has a calibrated pathlength l_j, such as found in a conventional (longer path length) cell, then the unknown pathlength, l_i, is simply proportional to the concentration normalized absorption measured in each cell:

$$l_i = \frac{l_j C_j A_i}{C_i A_j} \tag{30}$$

Equations (28) to (30) can provide independent measurements of path length. Equation (29) recognizes that the path length arises from the extremes in microscopic surface morphology of the optical plates and impurities that separate these plates. Equation (30) assumes a constant molar absorptivity over the concentration range studied.

D. Submicrometer Path Length Examples

This section demonstrates near-UV/Vis and IR absorption spectroscopy systematically reduced from the domain of conventional pathlength cells to the submicrometer domain. In this domain spectral resolution can be retained, and the applicability of Eq. (27) to path lengths as short as 77 nm is applied to the absorption of light of up to 3500 nm wavelength. Direct spectroscopic observation of concentrated high extinction coefficient species and in situ analysis of the equilibrium distribution of species in highly concentrated media are also demonstrated. Due to the unusually small path lengths, cell volumes ($\sim 10^{-8}$L) are small compared to the significant exposure area (~ 1 cm^2).

Figure 23 presents the visible absorption spectrum of a high extinction compound ($\varepsilon > 10^5$ M^{-1} cm^{-1}), β carotene, in cells of conventional and

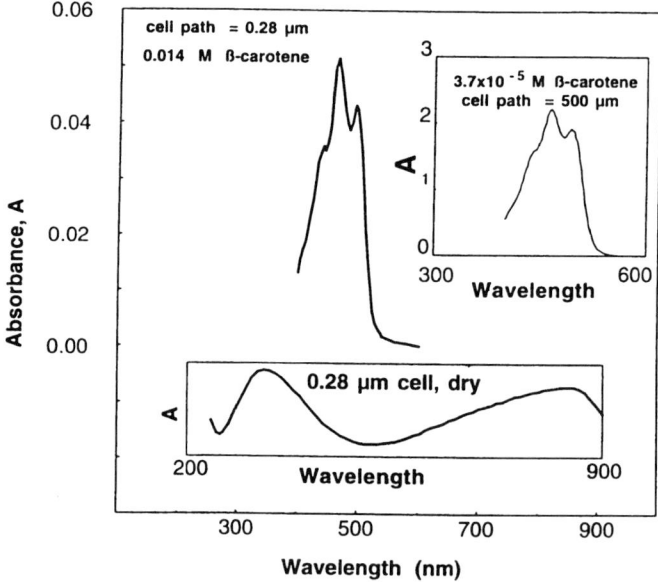

FIG. 23. Absorption spectrum of benzene analytical grade solutions of β-carotene in 0.28 μm and conventional 0.5 cm (top inset) cells. Lower inset: Absorption spectrum (interference pattern) for an unfilled 0.28 μm submicrometer cell.

submicrometer path lengths. As seen in the inset of Fig. 23, and as a result of the high molar absorpitivity of β-carotene, spectra may be observed in a 0.5 cm path length quartz cell for β-carotene in solution of concentrations only below 5×10^{-5} M. Alternatively submicrometer path length cells will allow direct observation of substantially more concentrated solutions of β-carotene. As seen in Fig. 23, the spectrum of a saturated (0.014 M) β-carotene in benzene is obtained in a cell in which the path length is reduced by four orders of magnitude to 0.3 μm. The cell path length is determined by interference fringe [Eq. (28)] as 0.28 μm, by particulate limitation [Eq. (29)] (f = 0.05 μm and p = 0.22 μm) as 0.32 μm, and by the Lambert-Beer relation [Eq. (30)] as 0.31 μm. Application of Eq. (30) to N-N dimethyl-p-toluidine (ε_{254nm} = 12,500 M^{-1}cm^{-1}) and benzaldehyde (ε_{243nm} = 14,000 M^{-1}cm^{-1}) solutions in the same macroscopic and microscopic cells described in Fig. 23 determines the path length to be 0.28 and 0.21 μm, respectively.

The fused silica transmission window is highly transparent from 2.8 to 3.6 μm radiation. This transmission window may be used to study submicrometer path length IR spectroscopy and retention of spectral resolution in this path length domain. Benzene has several closely spaced C-H stretch bands in this region; Fig. 24 presents the infrared transmission spectrum of benzene in conventional and submicrometer pathlength cells. In each cell, including the 0.28 μm path length cell, it is seen that the C-H absorption peaks are resolved to within 6 cm^{-1} approaching the instrumental resolution. Shifts in peak position have been calculated to be significant for cell window materials of various indices of refraction, but are generally small (<5 cm^{-1}) [46].

Benzene absorptivity of 3 μm light is of insufficient magnitude to discern absorption spectra in cells of path length <0.2 μm. Absorption spectra are investigated in cells systematically reduced to even shorter path lengths using cyclohexane with an order of magnitude higher absorptivity in the available transmission window ($\varepsilon_{3.41\mu m}$ = 300 M^{-1}cm^{-1}) and are presented in Fig. 25. In accordance with Eq. (28), the 0.08 (0.077) μm path length cells evidence no discernible interference spectrum; the optical path length was determined from Eq. (29) for these cells (using p = 15 nm and f = 31 nm). Figures 24 and 25 present data demonstrating the continuity of spectral shape, resolution, and position for cell path lengths substantially less than the wavelength of radiation incident upon the cell.

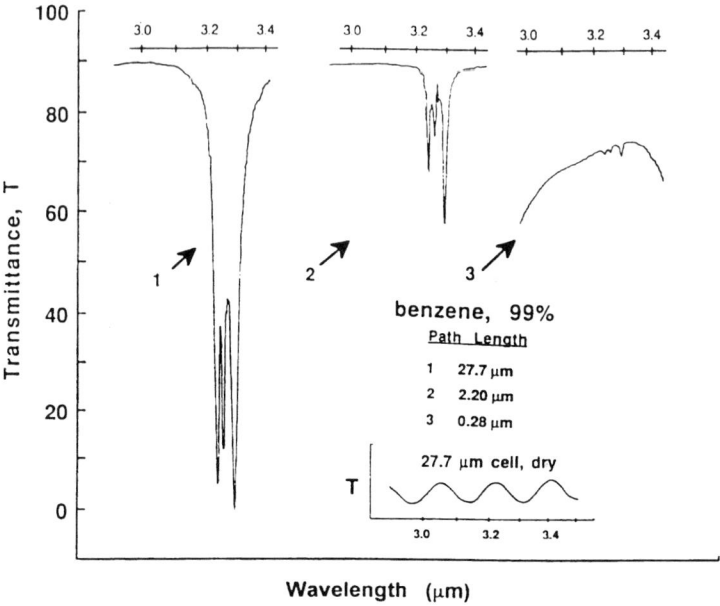

FIG. 24. IR transmission spectrum of benzene in 27.7(±1) μm, 2.2(±0.3) μm, and 0.28(±0.03) μm path length cells. Path lengths were determined from Eq. (3) and the average of several interference patterns as exemplified in the inset for 27.7 μm cell.

Figure 25 illustrates the validity of Lambert-Bouguer relations in submicrometer path length domain. The 0.077 μm path length cell has a path length a factor of 44 shorter than the 3.41 μm wavelength of incident radiation used to measure the cyclohexane spectrum in Fig. 25. No evidence of chemisorption was observed in these cells. The inset of Fig. 25 presents the variation of the measured absorption from 28 μm down to 0.08 μm cell path lengths. The data presented in the Fig. 25 inset extends the Lambert-Bouguer linearity of absorption amplitude with cell path length [Eq. (27)] down to the cell path length domain of less than 10^{-7} m.

The demonstrated linearity of absorption with path length variation and no evidence of chemisorption excludes the possibility that adsorption, chromophore orientation, or applied electric fields have been used to en-

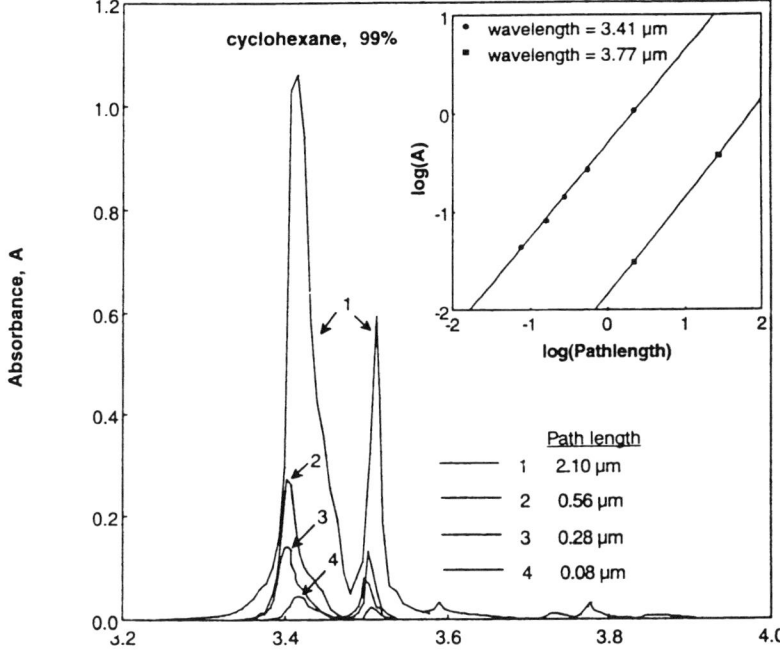

FIG. 25. IR absorption spectra of cyclohexane in 2.2(±0.3), 0.56(±0.06), 0.28(±0.03), or 0.08(–0.01) μm path length cells. Inset: Variation of the indicated cyclohexane absorption maxima from path lengths of 27.7 μm down to 0.077 μm. Inset contains 0.162 μm path length cell data, configured as in the 0.077 μm cell, but utilizing a larger 0.10 μm pore, 25 mm diameter membrane unit. 27.7 μm path length data exceeds the absorption range at shorter wavelengths and is not included in the main figure.

hance effective light absorption. Therefore it can be noted that the results summarized in Figs. 24 and 25 relate solely to bulk chromophore, and not perturbed or oriented chromophore, properties.

Shorter path length cells also permit absorption spectroscopy of increasingly of concentrated aqueous solutions. For example, as shown in Fig. 26, a 13 μm path length cell permits detection of the near-UV spectrum of 0.3 **M** K_2S with or without 13 **M** added KOH. As shown in Fig. 27, an 0.28 μm path length cell permits near UV absorption spectroscopy of over 6 **M** K_2S or KHS solutions [3].

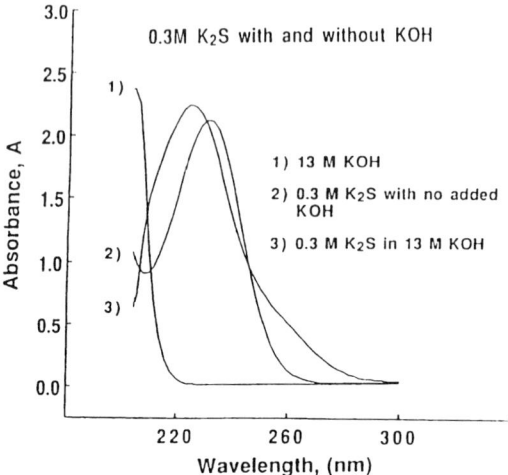

FIG. 26. Near-UV spectroscopy of concentrated KOH and KHS solutions in a 13 μm path length cell.

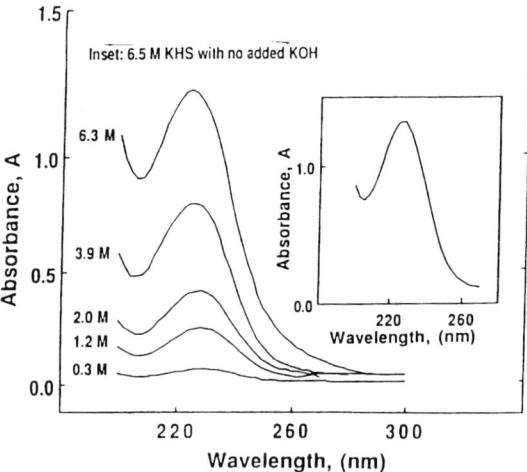

FIG. 27. Near UV spectroscopy of solutions containing equimolar aqueous KHS and KOH in a 0.28 μm path length cell. Inset: Concentration variation of peak absorption in KHS (figure inset) or equimolar KHS/KOH solutions.

VII. FINAL COMMENTS

This chapter has discussed the theory and experimental methodology of several techniques applicable to the analysis of concentrated electrolytes. The analysis of concentrated electrolytes poses fundamental and practical challenges. Yet successful methodologies to analyze these solutions are of importance in understanding chemical processes occurring in the concentrated solution phase and are applicable to on-line analysis of industrial processes, as well as environmental analysis of concentrated toxic effluents. Successful new analytical techniques for analysis of these solutions will be based on physical chemical properties that tend to be accentuated, rather than obscured, with increasing salt concentrations.

ACKNOWLEDGMENTS

Several of the analytical methodologies presented in this chapter were developed together with research students and fellows in my laboratory, including F. Forouzan, D. Peramunage, and K. Longo. The author is grateful for partial support for this project by the Rashi Foundations's Gusttella Award, and this research was supported by the Israel Science Foundation, the Technion V.P.R. Fund, and the Fund for the Promotion of Research at the Technion.

REFERENCES

1. S. Licht, F. Forouzan, and K. Longo, Anal. Chem. *62*:1356 (1990).
2. S. Licht, D. Peramunage, F. Forouzan, and K. Longo, in Proc. of the Symposia on Electrochemical Engineering (C. W. Walton, ed.), Electrochem. Soc. Proc. Vol. *90–10*:241 (1990).
3. D. Peramunage, F. Forouzan, and S. Licht, Anal. Chem. *66*:378 (1994).
4. R. G. Bates, *Determination of pH*, Wiley, New York, 1964.
5. M. W. Lovell, Anal. Chem. *55*:963 (1983).
6. J. R. Jones, Chem. Brit. *7*:336 (1971).
7. G. Eisenman, ed., *Glass Electrodes for pH and Other Cations*, Marcel Decker, New York, 1967.
8. S. Licht, Anal. Chem. *57*:415 (1985).
9. J. F. Wojclk, J. Phys. Chem. *71*:145 (1982).
10. G. Ackerlof and P. Bender, J. Am. Chem. *70*:2366 (1948).
11. W. J. Hammer, Trans. Electrochem. Soc. *72*:45 (1937).

12. R. G. Picknett, Trans. Faraday Soc. *64*:1059 (1968).
13. G. Mattlock, *Advances in Analytical Chemistry and Instrumentation*, Vol. 2, Wiley, New York, 1963.
14. L. Kratz, Glastech. Ber. *28*:35 (1950).
15. G. Mattlock, *pH Measurement and Titration*, Heywood, London, 1961.
16. D. Hubbard and G. F. Rynders, Natl. Bur Stand *39*:561 (1947).
17. S. Licht and J. Manassen, J. Electrochem. Soc. *134*:918 (1987).
18. S. Licht, J. Electrochem. Soc. *135*:2971 (1988).
19. A. N. Kirgintsev and A. V. Luk'yanov, Zhurnal Neorg. Khim. *12*:2032 (1967).
20. J. A. Dean, ed., *Lange's Handbook of Chemistry*, 13th ed., McGraw-Hill, Inc., New York, 1985.
21. (a) R. H. Stokes, in *Activity Coefficients in Electrolyte Solutions*, Vol. 1 (R. M. Pytkowicz, ed), CRC Press, Boca Raton, FL, 1981; (b) R. F. Platford, Chapter 3; (c) J. N. Butler, Chapter 4; (d) J. E. Desnoyers, Chapter 6 (e) Vol. 2.
22. W. R. Bousfield, Trans. Faraday Soc. (1917).
23. D. A. Sinclair, J. Phys. Chem. *37*:495 (1933).
24. R. H. Stokes, J. Amer. Chem. Soc. *69*:1291 (1947).
25. A. T. Williamson, Proc. R. Soc. Edinburgh Sect. A. *195*:97 (1948).
26. R. A. Robinson, J. Phys. Chem. *73*:3165 (1969).
27. H. Braunstein and J. Braunstein, J. Chem. Thermodyn. *3*:419 (1971).
28. F. Forouzan and S. Licht, Anal. Chem. *66*:2003 (1992).
29. T. S. Light, in *Electrochemistry, Past and Present*, ACS Symp. Series 390, 429 (1989).
30. T. S. Light, E. S. Atwood, J. Driscoll, and S. Licht, Anal. Chem. *65*:181 (1993).
31. T. S. Light and S. Licht, Anal. Chem. *59*:2327 (1987).
32. D. Peramunage and S. Licht, Solar Energy *52*:197 (1994).
33. S. Licht, K. Longo, D. Peramunage, and F. Forouzan, J. Electroanal. Interf. Electrochem. *318*:111 (1991).
34. M. J. Bedzyk, G. M. Bommarito, M. Caffrey, and T. L. Penner, Science *248*:52 (1990).
35. F. Oznam and J.-N. Chazalviel, Rev. Sci. Instrum. *59*:242 (1988).
36. R. S. McDonald, Anal. Chem. *58*:1906 (1986).
37. C. S. Yoo and Y. M. Gupta, J. Phys. Chem. *94*:2857 (1990).
38. W. Giggenbach, Inorg. Chem. *10*:1333 (1971).
39. T. Inagaki, J. Chem. Phys. *57*:2526 (1972).
40. K. Lieberman, S. Harush, A. Lewis, and R. Kopelman, Science *247*:59 (1990).
41. E. Betzig, A. Lewis, A. Harootunian, M. Isaacson, and E. Kratschmer, Biophys. J. *49*:269 (1986).

42. A. Harootunian, E. Betzig, M. Isaacson, and A. Lewis, Appl. Phys. Lett. *49*:674 (1986).
43. F. G. Puppels, F. F. M. de Mul, C. Otto, J. Greve, M. Robert-Nicoud, D. J. Arndt-Jovin, and T. M. Jovin, Nature *347*:301 (1990).
44. S. Licht, V. Cammarata, and M. S. Wrighton, Science *243*:1176 (1989).
45. D. C. Smith and E. C. Miller, J. Opt. Soc. Am. *34*:130 (1944).
46. H. J. K. Koeser, Fresenius' Z. Anal. Chem. *317*:845 (1984).

SURFACE PLASMON RESONANCE MEASUREMENTS OF ULTRATHIN ORGANIC FILMS AT ELECTRODE SURFACES

Dennis G. Hanken, Claire E. Jordan, Brian L. Frey, and Robert M. Corn

University of Wisconsin–Madison,
Madison, Wisconsin

I. Introduction   142
II. Background   143
III. SPR Measurements of Monolayer Thickness on Gold Surfaces   148
   A Introduction   148
   B SPR metal and dielectric film thickness parameters   150
   C SPR experimental apparatus   160
   D Examples of SPR thickness measurements   162
IV. SPR Imaging Experiments   179
   A Introduction   179
   B Ex situ SPR imaging experiments of biopolymer adsorption   181
   C In situ SPR imaging experiments of biopolymer adsorption   186
V. SPR Electric Field Measurements   195
   A Introduction   195
   B EM-SPR theory   196
   C Noncentrosymmetric zirconium phosphonate multilayer films   198
   D Modulated SPR measurements on air-gap capacitors   199
   E EM-SPR measurements on multilayer films at electrode surfaces   203
VI. Future Directions   217
   References   218

I. INTRODUCTION

A variety of surface-sensitive spectroscopic methods are currently employed in the characterization of organic thin films and monolayers at electrochemical interfaces. For example, the techniques of polarization modulation Fourier transform infrared reflection absorption spectroscopy (PM-FTIRRAS) [1–6] electrochemically modulated FTIRRAS [7–9], optical second harmonic generation (SHG) [10–12], IR-visible sum frequency generation [13–16], electroreflectance [17,18], and surface-enhanced Raman scattering [19–21] all possess a degree of surface selectivity or specificity that enhances the spectroscopic signal from the interfacial region. Surface plasmon resonance (SPR) methods are surface-sensitive spectroscopic techniques that employ the enhancement of the optical fields that occur at metal (Au, Ag, Cu, Al) surfaces when surface plasmon polaritons (SPPs) are created at the metal/dielectric interface [22–24].

SPPs are coupled photon-plasmon surface electromagnetic waves that propagate parallel to a metal/dielectric interface. SPPs have been used to enhance the surface sensitivity of a variety of spectroscopic measurements (e.g., one- and two-photon surface fluorescence [25–27], Raman scattering [25,26,28–43], and SHG [44–50]). The intensity of the optical fields associated with an SPP decays exponentially away from the metal surface with a decay length on the order of 200 nm [22,24–26,29,51]. In their simplest form, SPPs can be used to probe the index of refraction and thickness of thin films adsorbed to a metal surface using SPR reflectivity measurements. Ellipsometry is another analytical technique that is sensitive to the thickness and index of refraction of thin films. However, ellipsometry is a less sensitive technique than SPR for ultrathin films (<40 nm, depending on the substrate) [52–54].

SPR reflectivity measurements of film thickness have been used at metal surfaces to study the self-assembly of organic monolayers [55,56], the adsorption of biological molecules [6,57–61], and the formation of ultrathin multilayer films [62–64]. SPR measurements have been used to study electrochemical processes such as surface oxidation and underpotential deposition, and to monitor the potential distribution at an electrode surface [17,65–69]. SPR measurements have also been used in conjunction with scanning force [70], atomic force [71–73], and photon scanning tunneling microscopies [51,74] to provide kinetic and spatial information on interfacial processes.

In this chapter, we present a detailed description of the application of SPR methods to the study of molecular adsorption and ultrathin film structure at gold electrodes. Section II is a brief background section that describes the SPR method in more depth and its application to the study of ultrathin organic films. Section III outlines the SPR experimental parameters, including a description of the apparatus used in our work, and provides several examples of SPR thickness measurements of self-assembled monolayer and multilayer films on gold surfaces. The optical technique of SPR imaging is described in Sec. IV, and some examples of its applications to the study of specific and nonspecific adsorption onto charged gold surfaces are presented. Finally, Sec. V describes a novel extension of the SPR technique, denoted electrochemically modulated surface plasmon resonance (EM-SPR), that can be used to determine the electric field strength and electric field profile within ultrathin organic films at charged gold electrode surfaces.

II. BACKGROUND

SPPs are electromagnetic waves that propagate along the interface between a metal and a dielectric medium. They are created by coupling the energy from photons into oscillating modes of electron density at the metal/dielectric interface [22–24]. SPPs are formed from p-polarized light waves (i.e., light with its electric field vector oriented parallel to the plane of incidence) and have propagation vectors or wave vectors, k_{sp}, that lie in the plane of the metal surface. The field amplitudes associated with SPPs are maximum at the metal-dielectric interface and decay exponentially away from the interface [22,24–26,29,51].

The dispersion relation of the surface plasmons at a metal-dielectric interface is given by [24,75]:

$$\omega = ck_{sp}\sqrt{\left(\frac{1}{\varepsilon_m}\right) + \left(\frac{1}{\varepsilon_a}\right)} \tag{1}$$

where ω is the frequency of the light, c is the speed of light, k_{sp} is the wave vector of the photon, and ε_m and ε_a are the dielectric constants of the metal and dielectric (often air or water). The dispersion relationship for a gold/air interface is plotted in Fig. 1 as the solid line. This curve is calculated from Eq. (1) and is based on the free electron gas model for gold, which approximates the dielectric function of the metal $\varepsilon_m(\omega)$ from the

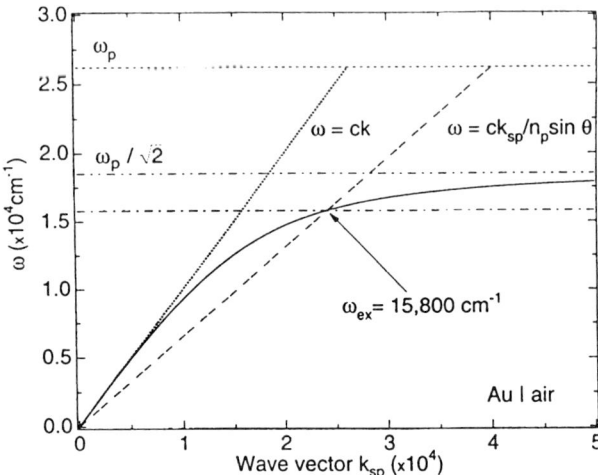

FIG. 1. The dispersion relationship for surface plasmon polaritons (SPPs) at a gold/air interface (solid line). This curve is based on the free electron gas model for gold, which approximates the dielectric function of the metal $\varepsilon_m(\omega)$ from the bulk gold plasmon frequency ω_p ($\cong$26,000 cm^{-1}). The curve approaches an asymptotic frequency called the surface plasmon cut-off frequency $\omega_p/\sqrt{2}$ at $\cong$18,500 cm^{-1}. Above this frequency surface plasmons cannot be created for this particular interface. The dispersion relation for light propagating in air ($\omega = ck$) and in a prism coupler ($\omega = ck_{sp}/n_p \sin\theta$) are also shown in the figure as the dotted and dashed lines, respectively. Only the prism modified light line intersects with the SPP dispersion curve and can lead to the formation of surface plasmon modes at the gold/air interface (see text).

bulk gold plasmon frequency ω_p (~26,000 cm^{-1}) [75]. The curve approaches an asymptotic frequency denoted as the surface plasmon cut-off frequency $\omega_p/\sqrt{2}$ at ~18,500 cm^{-1}. Above this frequency surface plasmons cannot be created for this particular interface. The dispersion relation for light propagating in air ($\omega = ck$) is also shown in Fig. 1 as the dotted line. The SPP dispersion curve always lies to the right of the light line so that direct excitation by external reflection methods on a smooth metal surface will not couple into the surface plasmon modes. Therefore, it is necessary to use a grating or prism coupling arrangement to excite SPPs [22–24]. For prism coupling, either the Otto or Kretschmann configuration, shown in Fig. 2, can be used [76,77]. The Otto setup requires an air or electrolyte

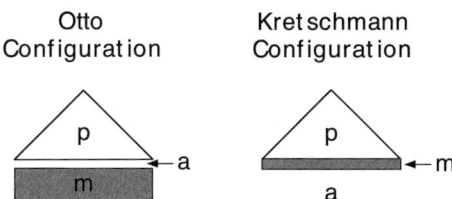

FIG. 2. The two possible coupling configurations to excite surface plasmon polaritons at a metal-dielectric interface. The labels **p**, **m**, and **a** represent a prism, a metal, and a dielectric medium (usually air or water). In the Otto configuration, a thin dielectic medium with a thickness on the order of the wavelength of light is required between the prism and the metal. In the Kretschmann configuration, a thin metal layer with a thickness on the order of the wavelength of light is place in direct contact with the prism. (From Ref. 204.)

gap on the order of the wavelength of light between the prism and the metal. In contrast, the Kretschmann configuration requires a thin metal layer ~50 nm thick that is in direct contact with the prism. In these arrangements, the evanescent light wave produced at the prism/air or prism/metal interface during total internal reflection is coupled into surface plasmon modes at the metal surface due to the increase in the parallel component of the photon's wave vector k_{para} given in the following:

$$k_{para} = n_p \left(\frac{\omega}{c} \right) \sin \theta \qquad (2)$$

where n_p is the index of refraction of the prism and θ is the angle of incidence in the prism. This modified light line is plotted as the dashed line in Fig. 1. This line crosses the SPP dispersion curve at the point labeled ω_{ex}, which is denoted as the excitation frequency and in our experiments corresponds to 15,800 cm^{-1} (632.8 nm). At this point, the momentum matching condition is satisfied ($k_{para} = k_{sp}$), and surface plasmons are created at the gold/air interface. The formation of SPPs is most easily observed as a minimum in a plot of the reflectivity versus incident angle as pictured in Fig. 3. The formation of thin organic films on the metal surface shifts the SPP dispersion curve to larger k_{sp} values, which in turn requires higher incident angles to satisfy the resonance condition of $k_{para} = k_{sp}$. Although most of the work discussed in this review monitors the reflectivity of light at a single wavelength as a function of incident angle in the Kretschmann configura-

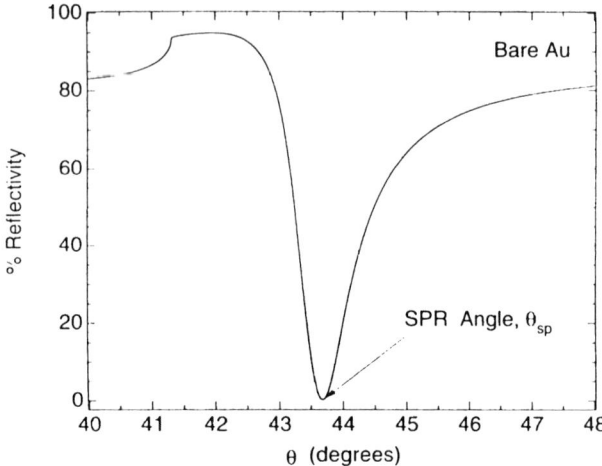

FIG. 3. The formation of surface plasmon polaritons at a gold/air interface. These surface light waves are observed as a minimum in the reflectivity versus incident angle scan. The angular position of this reflectivity minimum is denoted as the SPR angle or θ_{sp}. This curve was obtained from three phase complex Fresnel calculations (see text). (From Ref. 204.)

tion, surface plasmon experiments can also be performed by measuring the reflectivity at a fixed angle as a function of wavelength [78,79]. In addition, several groups have incorporated white light sources or multiple laser wavelengths to monitor reflectivity versus θ scans as a function of wavelength [61,79–86]. This type of arrangement has considerable advantages over single-wavelength scans including the ability to determine both the thickness and index of refraction of a thin film, as long as the film does not absorb over the wavelengths used.

The majority of the initial work using the SPR technique was in the determination of the optical properties of the bare metal and the formation of metal oxide surfaces [22,87]. The application of SPR to the analysis of thin organic films was pioneered in the mid-1970s by Gordon and Swalen [62,88]. This work was expanded to include monolayer films containing highly conjugated dye molecules [80,89,90]. SPR has also been used in electrochemical and other in situ environments to study chemical and physical processes at metal surfaces [17,65–69]. More recently, SPR has been applied to the analysis of thickness and optical properties of LB films

[86,91,92], thin polymer and chromophore-modified polymer films [93–96], and self-assembled monolayer and multilayer films [6,63,64,82,97]. These experimental and theoretical studies have examined the relationship of the optical constants and film thicknesses to the observed SPR reflectivity curves. In addition to measuring thicknesses and optical constants of thin films, the SPR technique has been used to monitor the assembly and reactivity of organic monolayers, photochemical processes, and degradation of thin films [61,72,73,82,95]. Variations of the SPR technique have also been developed to measure nonlinear optical properties of noncentrosymmetric Langmuir-Blodgett multilayers, chromophore-doped polymer films, and self-assembled monolayers [64,91–94,96,98–106].

SPR has also been used extensively to monitor the adsorption of biological molecules onto chemically modified metal surfaces. These studies have been performed on instruments that are generally designed to measure the shift in the SPR minimum to determine an average change in thickness or index of refraction or to produce a spatially resolved image by measuring the percent reflectivity across the surface. These measurements have been done in both ex situ and in situ environments. The ease with which SPR can be performed in situ is extremely advantageous for examining biological samples, which may otherwise denature, and for the investigation of adsorption kinetics [107–111].

The most common SPR instrument measures the average change in thickness or index of refraction caused by adsorption onto a noble metal surface. The metal surface is generally modified to investigate a specific interaction between a molecule immobilized on the surface and some species from solution. Examples of specific interactions that have been studied extensively by SPR include biotin-streptavidin (or avidin) binding [58,97,112,113], antibody-antigen interactions [59,71,114–117], DNA hybridization [61,118,119], and protein adsorption onto silane and alkanethiol self-assembled monolayers [60,120,121]. Supported lipid layers and their interactions with proteins have also been investigated using SPR techniques [122–125].

The in situ SPR reflectivity measurement is the basis of the BIA-CORE SPR adsorption instrument recently introduced and manufactured by Pharmacia [57]. The sensor chip in this instrument is a gold film coated with a thick (100–200 nm) layer of carboxymethylated dextran. This instrument is sensitive to changes in the index of refraction of the dextran matrix caused by the adsorption of species from solution into the matrix. This has been used for the quantitative detection of numerous biological species that interact with the dextran or modified-dextran layer [126]. A review of various applications of the BIACORE instrument has been published previously [127].

The technique of SPR imaging has been used to investigate the nonuniform adsorption of a variety of biological molecules onto surfaces [128]. In some cases, the intentional patterning of a sample surface is used to quickly and accurately investigate the adsorption of a molecule onto multiple surface functionalities; for example, the reflectivity from patterned DNA [118] and polypeptide surfaces has been examined using SPR imaging techniques [129]. Alternatively, nonuniform adsorption of molecules onto a surface may be inherent to the sample; for example, SPR imaging has been used to characterize domain formation in supported phospholipid layers [130,131] and antibody-antigen binding within these domains [132].

The popularity of the optical technique of surface plasmon resonance to monitor adsorption processes is rapidly increasing due to its submonolayer sensitivity and experimental simplicity. This section has provided a brief introduction to the various uses of SPR reflectivity measurements; a further description of the work in the literature may be obtained by referring to the cited reviews. A detailed description of three particular instrumental methods that utilize surface plasmon polaritons is provided in the following sections.

III. SPR MEASUREMENTS OF MONOLAYER THICKNESS ON GOLD SURFACES

A. Introduction

Surface plasmon resonance is a very sensitive method of measuring monolayer and multilayer thickness at gold surfaces. The important experimental and theoretical considerations when making such thickness measurements are described in the following section. First, the necessary parameters required to predict and interpret the formation and changes in the SPR reflectivity curves are obtained from theoretical Fresnel calculations. This is followed by an in-depth description of the SPR experimental apparatus for researchers interested in assembling an SPR instrument. The last part of this section provides some recent examples of the use of SPR to monitor self-assembled films on gold surfaces in both ex situ (air) and in situ (aqueous) environments.

Surface Plasmon Resonance Methods

In an SPR experiment, the percentage of laser light reflected from a prism/gold film/sample multilayer is monitored as a function of incident angle θ as shown in Fig. 4. The resulting reflectivity curves are modeled with theoretical Fresnel calculations to obtain the thicknesses of ultrathin organic films assembled onto the gold surface. These Fresnel calculations are based on reflection and transmission of light from a one-dimensional multilayer dielectric stack that consists of multiple planar phases. The reflected and transmitted light intensities can be described in terms of the optical constants and thicknesses of the various phases (the first (incident) and last phases are taken to have semi-infinite thicknesses). A number of papers have derived three, four, and N phase reflectivity equations to analyze SPR reflectivity curves [56,79,133–135]. These Fresnel calculations are vital to the analysis and interpretation of the experimentally measured SPR reflectivity data and are particularly useful for predicting changes in

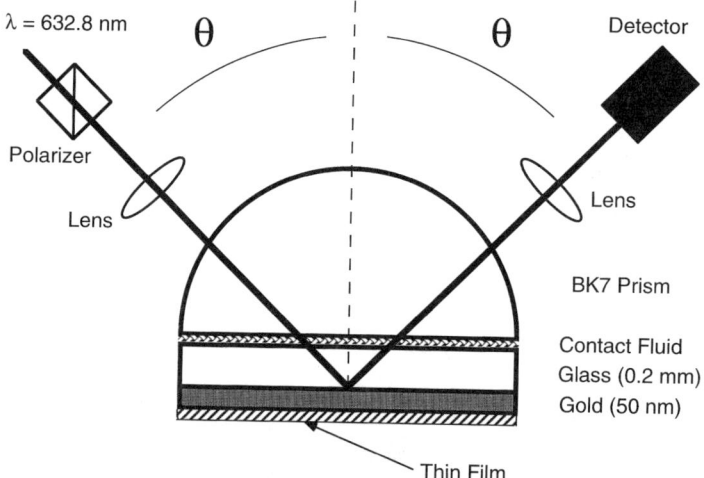

FIG. 4. A view of the prism-gold sample assembly for an SPR instrument. A 50 nm gold film evaporated onto a glass microscope slide cover serves as the substrate for the formation of a thin film. This sample is brought into optical contact with a hemispherical prism in order to couple the laser light into the surface plasmon modes at the gold surface. Rotating the prism-gold sample assembly allows the reflectivity to be recorded as a function of the incident angle, θ. (From Ref. 6.)

SPR reflectivity curves upon ultrathin film formation. In addition, these theoretical calculations are critically important to the design of other SPR experiments.

B. SPR Metal and Dielectric Film Thickness Parameters

Optimization of the optical parameters affecting SPR thickness measurements can be performed through the use of theoretical Fresnel calculations. The SPR experiments described in this review utilize the Kretschmann configuration in which a thin gold film is in direct contact with a glass prism. The thickness of the gold layer is a key parameter for obtaining highly sensitive SPR measurements. Figure 5 displays a series of theoretical Fresnel reflectivity curves calculated for a BK7 glass prism/gold interface as the gold layer thicknesses is varied. The 0 nm curve corresponds to the reflectivity expected for a simple BK7 prism/air

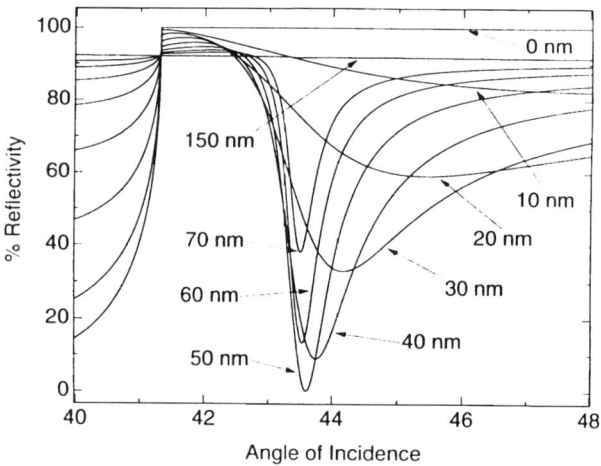

FIG. 5. SPR reflectivity curves for various gold thicknesses determined from complex Fresnel calculations. The curve corresponding to 50 nm of gold on the prism gives the largest drop in reflectivity at the SPR angle, and therefore is the optimal thickness for SPR measurements under these conditions. The following values were used in the calculations and are similar to those determined experimentally: $n_{prism} = 1.515$, $n_{Au} = 0.154 + 3.55i$, $n_{air} = 1$, and the wavelength of light was 632.8 nm. (From Ref. 204.)

Surface Plasmon Resonance Methods

interface; notice that total internal reflection (100% reflectivity) is observed at all angles greater than the critical angle (41.3°). As the gold thickness is increased, the reflectivity at angles *less than* the critical angle increases due to some reflection by the gold. At angles *greater than* the critical angle, the reflectivity decreases due to the creation of surface plasmons at the gold/air interface. For thicknesses up to 50 nm, the SPR reflectivity curves get sharper and steeper and eventually have a minimum of 0.1% reflectivity or less. The angle at which this minimum occurs is denoted as the surface plasmon angle θ_{sp}. Above 50 nm, the curves remain sharp, but show less of a decrease in reflectivity at θ_{sp}. Gold films of 150 nm or greater no longer allow coupling of the light into surface plasmon modes and simply show a constant reflectivity of about 92% regardless of the angle of incidence. To obtain the greatest sensitivity from the SPR measurements, a steep drop to nearly 0% reflectivity is desired; therefore, a gold thickness of approximately 50 nm is used. The optimal thickness depends on the optical constants of the gold film, and these can vary depending on the exact deposition conditions of the gold thin film. Fitting the experimental SPR reflectivity curves for each sample to theoretical curves is necssary to obtain accurate thickness results. This need arises from the sample-to-sample variations in the gold layer thickness that occur during the metal deposition step. These gold thickness variations result in different SPR angles for the different clean gold surfaces. For this reason, all of the SPR data are fit to theory curves for each clean gold surface and each successive monolayer.

An example of this fitting process is shown in Fig. 6. This figure graphs the experimental SPR reflectivity curve obtained for a clean gold surface (open circles) and the theoretical fit (solid line) as determined from three-phase (BK7, Au, air) Fresnel calculations. In the fitting routine, three parameters are kept constant (n_{BK7} = 1.515 for the prism, n_{air} = 1.0, and λ = 632.8 nm) and three parameters are varied: the gold layer thickness (d_{Au}) and the real (n_{Au}) and imaginary (k_{Au}) components of the index of refraction. The bulk values for the gold index of refraction (n_{Au} = 0.166, k_{Au} = 3.15)[136] may be used as an initial guess for the calculation of the SPR reflectivity curve. First, d_{Au} is adjusted to fit the critical angle region since the percent reflectivity at angles less than the critical angle (near 40°) is determined by the gold thickness rather than total internal reflection of the prism. The second step deals with changing k_{Au} in order to match the theoretical and experimental SPR angles. The third parameter, n_{Au}, is adjusted to give the correct reflectivity past the SPR angle in the 46–48° region. This three-step procedure frequently requires more than one iteration since

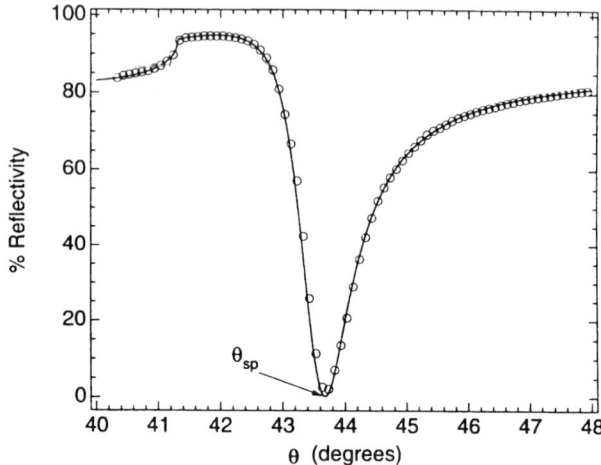

FIG. 6. SPR reflectivity curve for a clean gold surface. The percent of reflected laser light is plotted versus the incident angle, θ. The circles give the experimentally measured values and the solid line is the result of three-phase (BK7, Au, air) complex Fresnel calculations and yield the gold thickness $d_{Au} = 47.0$ nm and index of refraction $n_{Au} + ik_{Au} = 0.163 + i3.52$. The angle of minimum reflection is defined as the SPR angle, θ_{sp}. (From Ref. 204.)

d_{Au}, k_{Au}, and n_{Au} do not act independently on the critical angle region, the SPR angle, and the 46–48° region, respectively. However, these are the regions most affected by each of those parameters. A more complete list of the effects of d_{Au}, k_{Au}, and n_{Au} on the SPR reflectivity curve follows:

> Increase in d_{Au}: Increases reflectivity from 40–41°, shifts the SPR angle to a smaller angle, and narrows the SPR curve.
> Increase in k_{Au}: Moves SPR angle to smaller angles and significantly narrows the SPR curve.
> Increase in n_{Au}: Decreases reflectivity in the 46–48° region and broadens the SPR curve.

Despite the interrelated nature of these parameters, excellent fits to the data result after only two to three iterations. In this example, the theoretical fit corresponds to $d_{Au} = 47.0$ nm and $n_{Au} + ik_{Au} = 0.163 + i3.52$.

The electric field, E, associated with the formation of surface plas-

Surface Plasmon Resonance Methods

mons at the gold/air interface, decays exponentially away from the gold surface as shown in Fig. 7. This plot gives E^2/E_o^2 as a function of the distance from the gold surface where E_o^2 is the square of the electric field of the incident radiation. These E^2/E_o^2 values were calculated at the SPR minimum angle of 43.5°. At the gold surface, E^2/E_o^2 is 74, which corresponds to an electric field enhancement of $\sqrt{74} = 8.6$. This enhancement decays exponentially away from the surface and reaches the 1/e value at 164 nm, which gives an estimate of the distance scale probed by surface plasmons. A number of surface spectroscopic techniques have made use of this enhancement including Raman, fluorescence, and second harmonic generation (see Sec. I).

Material adsorbed to the gold surface affects the creation of surface plasmons, which changes the SPR reflectivity curves and allows measurement of an effective thickness of the adsorbed material. Figure 8 plots the SPR reflectivity curves for various thicknesses of adsorbed thin films. The 0 nm curve is simply for a clean gold surface, and the other curves give the

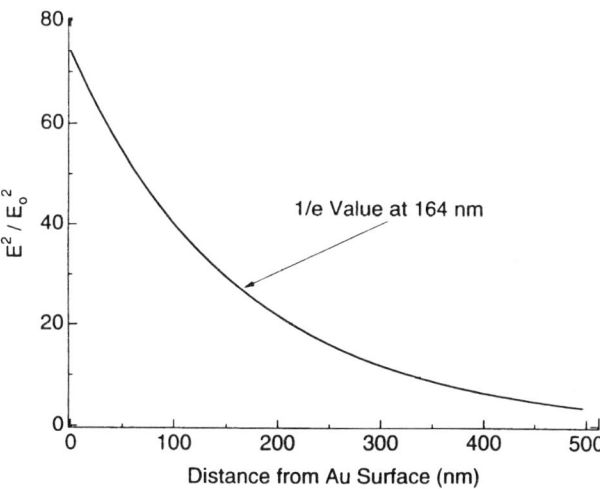

FIG. 7. Electric field strength of the surface plasmon waves as a function of distance from the gold surface. The electric field of the surface plasmons, E, is enhanced over the electric field of the incident radiation. E_o, as shown by the ratio E^2/E_o^2. The electric field strength decays exponentially with distance from the gold surface. (From Ref. 204.)

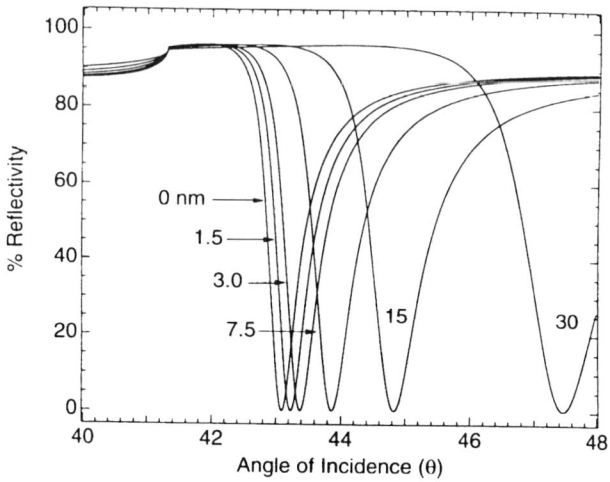

FIG. 8. SPR reflectivity curves for thin films of varying thickness. The 0 nm curve corresponds to that of a clean gold surface. As the film thickness increases, the SPR angle of minimum reflectivity shifts to higher angles and the curve shape becomes significantly broader. These theoretical plots demonstrate that experimental shifts in SPR angle, $\Delta\theta$, can be related to the thickness of material adsorbed to the gold surface, provided the index of refraction is known (in these plots $n_{thin\ film}$ = 1.5). (From Ref. 204.)

thicknesses in nanometers for thin films with an index of refraction of 1.5. The SPR angle shifts to higher values as more material adsorbs to the gold surface. As the film thickness increases, other changes in the reflectivity curves occur as well. The curves get much broader as seen in Fig. 8. In addition, the percent reflectivity at the minimum increases, but this effect is not easily distinguished in the figure. A plot of $\Delta\theta$ versus the adsorbed layer thickness is shown in Fig. 9 where $\Delta\theta$ is obtained from Fig. 8 by taking the difference in angular position of the SPR minimum reflectivity for the various thickness curves relative to the 0 nm curve. Clearly, the shift in SPR angle as a function of adsorbed layer thickness is nonlinear, and the shift per nanometer is larger for the thicker films.

The shift in the SPR angle not only depends on the thickness of the adsorbed layer, but also upon its index of refraction, n. Figure 10 displays the theoretical reflectivity curves for several 7.5 nm films with different indices of refraction (n = 1.3–1.7). Films with higher values of n

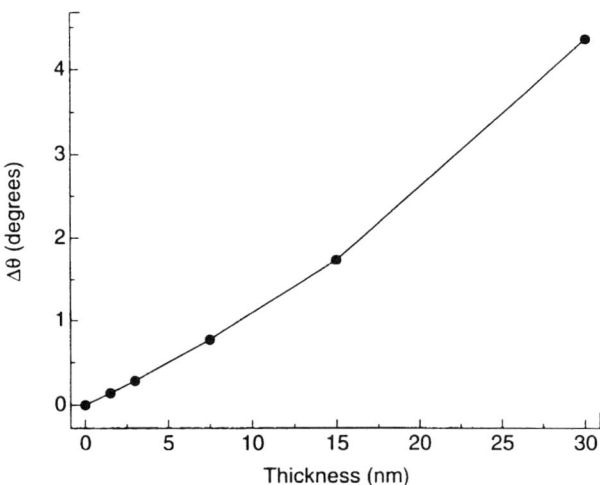

FIG. 9. The shift in SPR angle versus thin film thickness. The values for the shift in SPR angle, $\Delta\theta$, for various thicknesses were obtained from the graph in Fig. 8. Note that a 10 nm thin film gives a shift of over one degree, and that the thicker films give even larger shifts per nanometer. This plot can serve as an approximate conversion of $\Delta\theta$ to thickness for experimental data. (From Ref. 204.)

have larger shifts in the SPR angle, despite having the same film thickness. Note that the indices of refraction used for the film layer in these calculations have only a real component. SPR measurements can be performed on thin films of *absorbing* materials that have an imaginary component to their index of refraction. The conversion of $\Delta\theta$ to thickness then requires a complex index of refraction value. These absorbing thin films also result in broad SPR reflectivity curves that have much higher percent reflectivities at their minima. None of the ultrathin organic films examined in the next sections absorbed light at the experimental wavelength of 632.8 nm, and therefore only the real index of refraction values were required.

In principle, at a single wavelength both the index of refraction and thickness of the monolayer can be extracted from the SPR reflectivity curve by fitting the entire curve rather than just considering the shift in SPR angle. Figure 11 shows two theoretical reflectivity curves that have the same minimum angle but different shapes: the narrower curve is due to

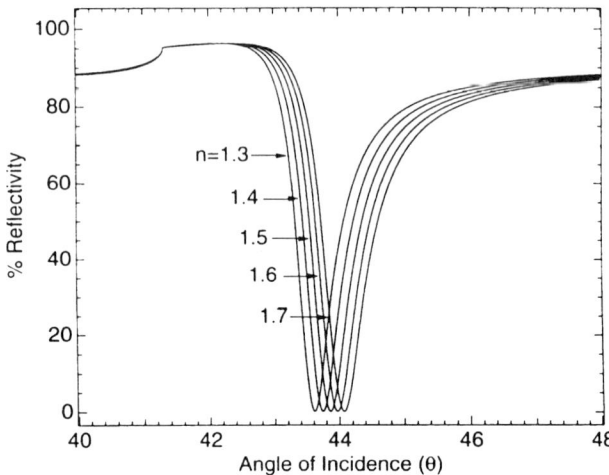

FIG. 10. SPR reflectivity curves for thin films with various indices of refraction. All of these curves correspond to films with a thickness of 7.5 nm, but with different index of refraction values (n = 1.3, 1.4, 1.5, 1.6, and 1.7). Higher index of refraction values produce larger angular shifts. This is similar to that observed for increased film thickness. Therefore, thickness and index of refraction for a thin film cannot be uniquely determined from a $\Delta\theta$ value measured at a single wavelength. (From Ref. 204.)

a film 19.2 nm thick with n = 1.45, whereas the wider curve has a thickness of 15.6 nm and n = 1.60. The inset is an expanded view demonstrating that these films have the same SPR angles. Despite the significant difference in their indices of refraction, these two films give nearly identical SPR reflectivity curves. One could attempt a regression analysis in order to extract both thickness and index of refraction from the experimental data; however, experimental uncertainties such as sample-to-sample variations in the flatness of the gold surface or the uniformity of the thin films have much larger effects on curve shape than the small theoretical difference seen in Fig. 11. Therefore, curve shape alone cannot realistically provide both the thickness and index of refraction for monolayer films [133,134]. However, other researchers have obtained both of these parameters for a monolayer film from SPR experiments made at multiple wavelengths or in different solvents [61,81–86,134,135].

Surface Plasmon Resonance Methods

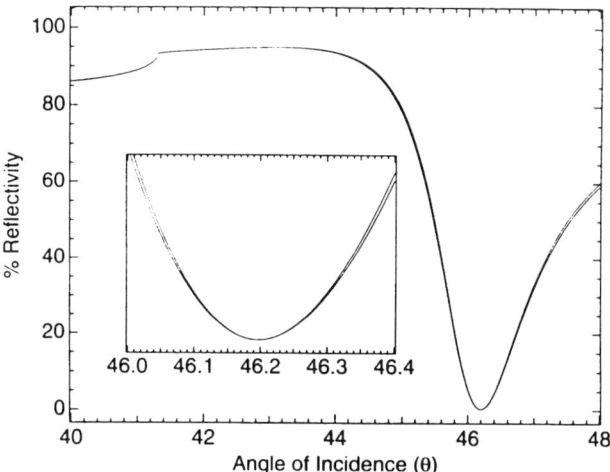

FIG. 11. SPR curves for two films with different thicknesses and indices of refraction, but the same SPR angle. The narrower curve corresponds to a thickness of 19.2 nm and n = 1.45, whereas the broader curve is 15.6 nm thick with n = 1.60. The inset clearly shows that these two films have the same SPR angle of approximately 46.2°. Despite their identical SPR angles, these films have slightly different curve shapes. The experimental curve shape could conceivably provide values for both the thickness and index of refraction for the same film; however, this subtle difference in curve shape is much less than that produced by experimental uncertainties. (From Ref. 204.)

An example of the shift in the surface plasmon angle upon film formation is shown in Fig. 12. This figure plots two experimental surface plasmon reflectivity curves: one is the bare gold curve from Fig. 6 (circles) and the second curve is of the same sample after self-assembly of a monolayer of 11-mercaptoundecanoic acid (MUA) (squares). As mentioned above, a fit to the bare gold film from three-phase Fresnel calculations was used to determine the optical constants and thickness of the gold layer. These values are not changed while determining the thickness of a monolayer film adsorbed to that gold surface using four-phase Fresnel calculations; the only parameters adjusted are the monolayer film thickness (d_{ML}) and index of refraction (n_{ML}). Since it is not possible to determine both d_{ML} and n_{ML} from the same SPR curve at a single wavelength, n_{ML} is usually es-

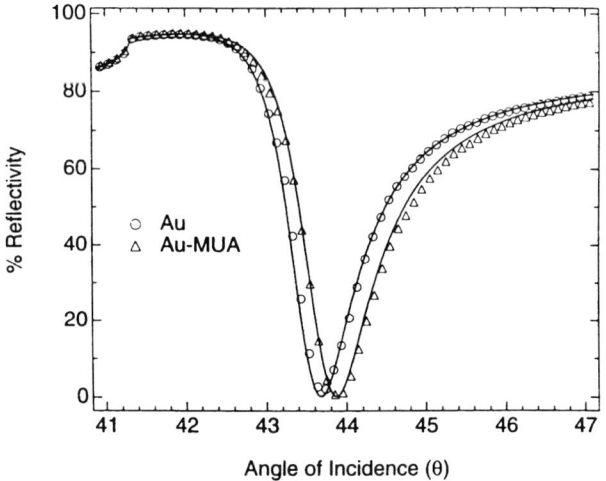

FIG. 12. SPR reflectivity curves for a clean gold surface (circles) and following self-assembly of a monolayer of 11 mercaptoundecanol (MUA) (triangles). The percent of reflected laser light is plotted versus the incident angle, θ. The solid lines are the result of three-phase (BK7, Au, air) and four-phase (BK7, Au, MUA, air) complex Fresnel calculations using the thickness and index of refraction values of gold determined in Fig. 6 of $d_{Au} = 47.0$ nm and $n_{Au} + ik_{Au} = 0.163 + i3.52$ and a MUA index of refraction of 1.45. The fit to the MUA SPR curve corresponds to a monolayer thickness of 1.7 nm. (From Ref. 204)

timated from bulk values (which for MUA is about 1.45) or is obtained from another analytical method and then kept constant. The remaining parameter d_{ML} is varied until the theory curve SPR angle exactly matches the experimental SPR angle. The resulting value of d_{ML} is the thickness of the monolayer film, which in the case of the MUA monolayer in Fig. 6 corresponds to 1.7 nm. This process is repeated for each successive monolayer adsorbed to the surface.

The SPR experiments described here focus only on the shift in the SPR angle and its conversion to a thickness value based on an index of refraction, which is estimated, or determined by another analytical method. An error in the thickness calculation is introduced by assuming a value for n rather than measuring it. Similar uncertainties are present in the determination of monolayer and multilayer film thicknesses on metal surfaces

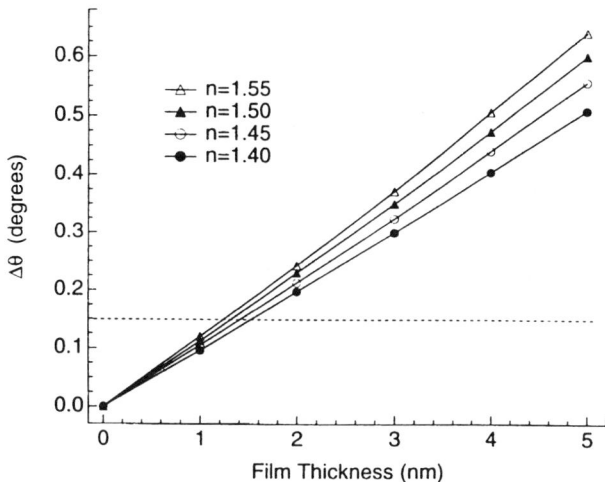

FIG. 13. Calculations of shift in SPR angle versus film thickness for various indices of refraction. The results of a four phase (BK7, Au, film, air) complex Fresnel calculation giving the expected change in SPR angle, $\Delta\theta$, for thin films with thicknesses up to 5.0 nm and different refractive indices: n = 1.40, 1.45, 1.50, and 1.55. The sensitivity of the SPR technique is given by the slope of approximately 0.12°/nm. The dashed line represents a shift of 0.15°, which corresponds to 1.4 nm at n = 1.45. An uncertainly of ±0.05 in the thin film index of refraction gives an error of roughly 0.1 nm for this 1.4 nm film as indicated by the intersections of this dashed line with the solid lines. (From Ref. 6.)

from ellipsometric measurements and have been discussed in the literature [133,137]. To examine the uncertainty caused by estimating n, a four-phase (BK7, Au, film, air) complex Fresnel calculation was performed for a range of values of n. Figure 13 plots the results of these calculations and shows the shift in the SPR angle as a function of film thickness for thin films with n = 1.40, 1.45, 1.50, and 1.55. A variation of ±0.05 in the monolayer index of refraction leads to an error of approximately ±7% (i.e., 0.1 nm for every 1.5 nm of film thickness). Although n may vary somewhat from the bulk index of refraction, it is unlikely to change by more than ±0.05. If the bulk value of n is not known for a given compound, it can usually be estimated within these same limits from the indices of refraction for related compounds.

C. SPR Experimental Apparatus

This section presents a detailed description of the *ex situ* scanning surface plasmon resonance apparatus used in the SPR thickness measurements. The SPR instrument is shown schematically in Fig. 14. The HeNe laser (Newport Corp.) that has a power output of 1 mW, a wavelength of 632.8 nm, and a polarization ratio of 500:1 is used in this setup. To increase the sensitivity, the laser beam is mechanically chopped at a frequency of 930 Hz by a Stanford Research Systems SR540 chopper. Lenses L1 and L2 are employed to collimate the laser beam, which would otherwise diverge and produce too large a spot on the sample. The mirrors M1 and M2 are used

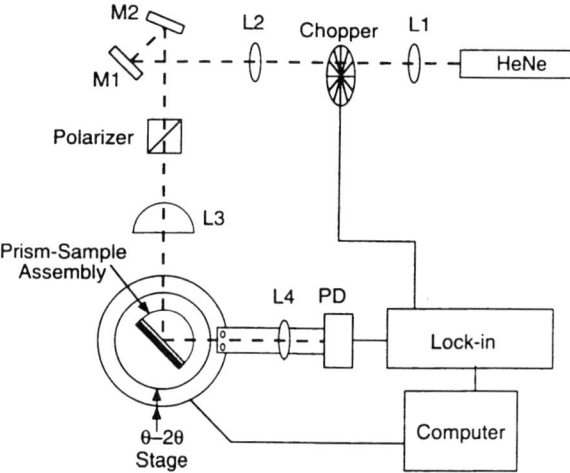

FIG. 14. Overall experimental arrangement for the scanning SPR instrument. Lenses L1 and L2 collimate the HeNe laser beam, and mirrors M1 and M2 are used to align the beam onto the center of the prism-sample assembly. The polarizer is set to p-polarization. The combination of lens L3 (a hemispherical BK7 prism) and the BK7 hemispherical sample prism produces collimated light on the gold sample surface. The incident angle, θ, is changed by rotation of the prism-sample assembly mounted on a θ-2θ stage controlled by a computer with stepper motors. The reflected light is collected using a detector arm, connected to the 2θ part of the stage, which consists of a lens L4 and a photodiode PD. A lock-in amplifier demodulates the chopped signal, and a computer records the reflectivity versus incident angle. (From Ref. 204.)

Surface Plasmon Resonance Methods 161

to align the beam onto the prism-gold sample assembly so that it strikes the gold layer at the center of rotation of the θ–2θ stage in order to avoid walking the beam across the sample during the experiment. A calcite polarizer (Newport Corp) with an extinction ratio of 10^{-5} polarizes the light parallel to the plane of incidence (p-polarization) since s-polarized light does not create surface plasmons. The combination of lens L3 and the hemispherical prism is used to provide a collimated beam impinging on the gold sample. In fact, in order to compensate for the strongly focusing prism, the lens L3 is an identical hemispherical prism placed approximately 2 inches in front of the sample prism. Thus, the laser beam is tightly focused by the prism L3, and then as it diverges the sample hemispherical prism recollimates it to give a narrow range of incident angles on the gold film.

Figure 4 presents an expanded view of the prism-gold sample assembly. The sample consists of a BK7 glass microscope slide cover coated with a thin layer of gold that is subsequently modified with self-assembled monolayer films. The glass substrates prior to gold deposition are rigorously cleaned in concentrated sulfuric acid, rinsed copiously with Millipore water, and then silanized with (3 mercaptopropyl)trimethoxysilane using conditions optimized previously [138].

Gold is vapor deposited onto these substrates using a Balzers vapor deposition apparatus equipped with a turbo molecular pump. The deposition occurs at a pressure of about 1×10^{-6} mbar with around 3.6 amperes running through the tungsten boat containing gold pellets (99.99%) obtained from Canadian gold coins. Gold films of 49 ± 2 nm are deposited at approximately 0.1–0.2 nm/sec by monitoring the thickness with a quartz crystal microbalance (QCM) inside the vacuum chamber. The gold is deposited at room temperature, and the gold/glass samples are subsequently annealed at approximately 300°C for one hour. This procedure results in surfaces with atomically flat Au(111) terraces of about 120 nm in diameter, as determined by STM measurements [6].

The gold-coated samples are brought into optical contact with a BK7 hemispherical prism using an appropriate index matching fluid [ethylene glycol (EG) (n = 1.43), glycerol (n = 1.47), or methyl salicylate (n = 1.53)]. The hemispherical prism ensures that the laser beam is normal to the prism regardless of the angle of incidence. By entering the prism perpendicularly, the beam is not refracted and thus calibration of the incident angle is simplified. Furthermore, the laser spot on the gold surface does not move as the prism-gold sample assembly is rotated.

Therefore, the same spot is sampled throughout the angular scan. Scanning of the incident angle is accomplished by rotating the prism-gold sample combination using a θ-2θ rotation stage assembly with computer-controlled stepper motor drivers that result in an angular resolution of 0.004°.

The entire laser beam that is reflected from the prism-gold sample assembly is collected by lens L4 and focused onto the photodiode PD (Hamamatsu S1336-5BK) (see Fig. 14). The response from the photodiode detector is sent to the lock-in amplifier (Stanford Research Systems SR510) to demodulate the chopped signal. This reflectivity signal is then recorded as a function of incident angle, θ, by a computer, which also controls the stepper motor rotation of the sample and detector.

D. Examples of SPR Thickness Measurements

Three sets of exemplary scanning SPR measurements that monitor monolayer and multilayer adsorption onto a gold surface are presented: (1) Ex situ SPR measurements are used in conjunction with vibrational spectroscopy to monitor electrostatic adsorption of a biopolymer onto a self-assembled monolayer-modified gold surface. These modified surfaces are then used to control protein immobilization. (2) The same combination of techniques is used to monitor the structure and assembly of multilayer zirconium phosphonate (ZP) films having variable indices of refraction. (3) in situ SPR thickness measurements for both biopolymer films and mixed ZP multilayer films are examined.

1. Biopolymer/Protein Adsorption Studies

Through a combination of polarization-modulation Fourier transform infrared reflection-absorption spectroscopy (PM-FTIRRAS) and ex situ SPR measurements, it has been demonstrated that the multiply charged polypeptide poly-L-lysine (PL) will strongly adsorb onto gold surfaces modified with a self-assembled monolayer of 11 mercaptoundecanoic acid (MUA) creating a robust MUA/PL bilayer on the gold surface (see Fig. 15).[6] The PM-FTIRRAS spectroscopic measurements indicate that the PL adsorbs onto the MUA monolayer via the formation of multiple carboxylate-ammonium ion pairs, and that the PL can be released from this surface by rinsing with a buffer solution adjusted to a pH that destroys the ion pairing interactions. In these initial studies it was determined that only a fraction of the PL amino groups participate in the

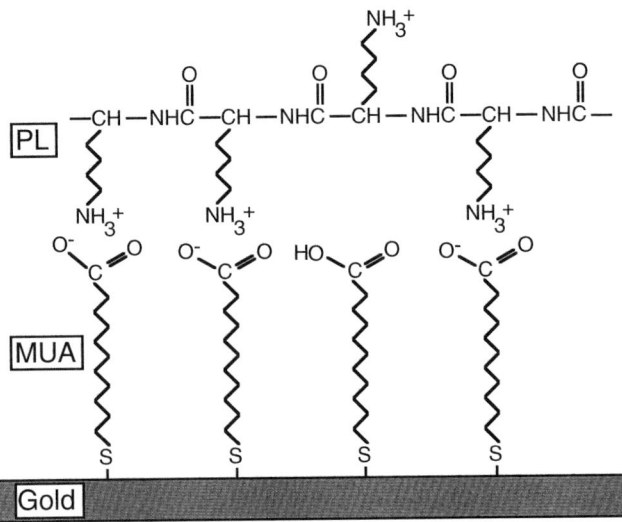

FIG. 15. Schematic diagram of PL and MUA monolayers adsorbed onto a gold substrate. A self-assembled alkanethiol monolayer of 11-mercatoundecanoic acid (MUA) is formed on the gold surface. A monolayer of poly-L-lysine (PL) adsorbs to the MUA coated gold surface via ion pair formation between the ammonium and carboxylate groups. (From Ref. 6.)

electrostatic binding. This leaves a large fraction of the ε-amino groups available to interact with proteins either by direct chemical cross-linking or by specific interaction with chemical functionalities incorporated into the adsorbed PL monolayer.

As an example of controlling protein immobilization onto gold surfaces with PL monolayers, the specific adsorption of the protein avidin was examined in a previous paper [97]. Avidin (MW = 67,000) is a tetrameric glycoprotein that contains four specific binding sites for biotin ($K_{aff} = 10^{15}$ M^{-1}), and the presence of multiple binding sites on the protein has led to its use in numerous biosensor applications [71,139–148]. Avidin and the related protein streptavidin have been absorbed previously onto gold surfaces by specific interactions with surface-immobilized biotin and by simple nonspecific adsorption [59,113,143,149–151].

The approach for the specific adsorption of avidin using PL monolayers is shown schematically in Fig. 16. As discussed above, a self-assem-

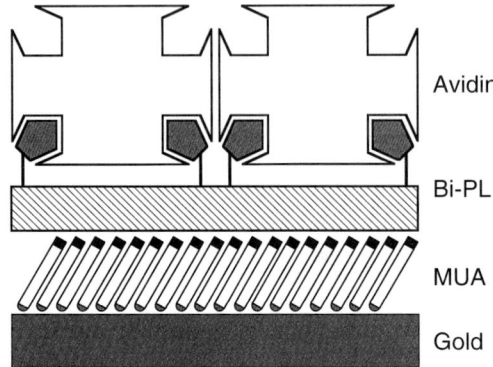

FIG. 16. Schematic diagram showing the specific adsorption of the protein avidin onto a gold surface coated with a biotinylated poly-L-lysine (Bi-PL) monolayer. The Bi-PL is immobilized via ion pair formation with a self-assembled monolayer of 11-mercaptoundecanoic acid (MUA). This diagram is not intended to imply that every avidin is bound to the surface with two biotin moieties. (From Ref. 97.)

bled monolayer of MUA is first prepared on the surface of a vapor-deposited thin gold film. Next, a monolayer of PL in which some of the lysine ε-amino groups have been modified with biotin functionalities (Bi-PL) is electrostatically adsorbed onto the surface. Finally, exposure of this surface to an avidin solution results in the specific adsorption of the protein. To control the surface coverage of avidin, the biotin moieties are attached to the PL prior to its adsorption. This strategy provides for the controlled variation of the percentage of lysine residues that are modified with biotin.

Surfaces with varying numbers of specific adsorption sites for avidin were prepared by chemically modifying PL with different amounts of biotin (0–22% of the lysine residues). These various Bi-PL conjugates were adsorbed electrostatically onto MUA-coated gold surfaces and then exposed to avidin solutions. In this manner, the percent biotinylation of lysine residues needed to produce complete avidin monolayers was determined to be about 20%. The experimental SPR reflectivity curves for the adsorption of monolayers of MUA, 22% Bi-PL, and avidin onto a gold surface are depicted in Fig. 17. The experimentally measured shifts in the SPR angle and the calculated film thicknesses for the various monolayer

Surface Plasmon Resonance Methods

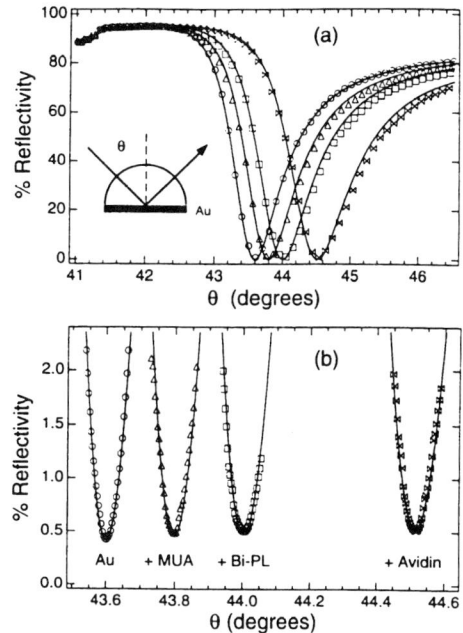

FIG. 17. (a) The SPR reflectivity curves for a clean gold surface (○), and the same surface after the sequential adsorption of a MUA monolayer (△), a 22% Bi-PL monolayer (□), and a layer of the protein avidin (■). (b) An expanded view of the SPR curves for the same sample as in (a), but with data points taken every 0.004° to precisely determine the reflectivity minimum also referred to as the SPR angle. The shift in the SPR angle upon adsorption is used to determine the thickness of the adsorbed layer via complex Fresnel calculations (solid lines). (From Ref. 97.)

films are listed in Table 1 (part A). A shift of $\Delta\theta = 0.180 \pm 0.005°$ from the bare surface was observed after the adsorption of the self-assembled MUA monolayer. Using a four-phase (BK7, Au, MUA, air) complex Fresnel calculation, $\Delta\theta$ was converted to a thickness of 1.7 ± 0.1 nm by assuming a MUA index of refraction of $n_{MUA} = 1.45$. This film thickness is in excellent agreement with previous ellipsometric film thickness measurements of 1.6–1.9 nm for the MUA monolayer and confirms the accuracy of the SPR technique [152–154]. A theoretical thickness of about 1.8–1.9 nm is ex-

TABLE 1

SPR Angle Shifts and Calculated Layer Thicknesses

	Layer	Total $\Delta\theta^a$ (degrees)	Additional $\Delta\theta$ (degrees)	Index of refraction[b]	Layer thickness (nm)	Surface coverage[c] (cm^{-2})
Part A	MUA	0.180 ± 0.005	0.180	1.45	1.7 ± 0.1	
	22% Bi-PL	0.385	0.205	1.52	1.7	
	Avidin	0.870	0.485	1.45	4.1	3×10^{12}
Part B	MUA	0.180	0.180	1.45	1.7	
	PL lysines	0.310	0.130	1.52	1.05	4×10^{14}
	Avidin	0.310	0.00	1.45	0	

[a]Total shift in SPR angle from that of the bare gold surface.
[b]Index of refraction estimated from bulk values.
[c]Absolute surface coverage determined from fluorescence measurements.
Source: Ref. 94.

pected for a fully extended MUA monolayer oriented normal to the surface [152,153,155], and so the measured value of 1.7 nm is in agreement with a picture of the MUA molecules forming a close-packed self-assembled monolayer where the methylene chains are oriented nearly perpendicular (i.e., with an average tilt angle of ≤30°) to the surface. The 22% Bi-PL monolayer was adsorbed onto the MUA monolayer from a 5 mM NaHCO$_3$ buffered solution (pH = 8.5). The observed SPR angle shift of 0.205° corresponds to an additiona thickness of 1.7 nm for the 22% Bi-PL monolayer, which is somewhat thicker than a monolayer of unmodified PL (1.05 nm). SPR measurements on monolayers of 0.3–15% Bi-PL gave intermediate thicknesses between the 1.05 and 1.7 nm as expected.

The biotin moieties attached to the surface via the PL acted as specific adsorption sites for avidin. Thus, exposure of the 22% Bi-PL–coated surface to avidin in the same NaHCO$_3$ buffer solution produced an additional shift in SPR angle of nearly 0.5° corresponding to an avidin layer thickness of 4.1 nm, which is in good agreement with the expected thickness for a single monolayer of avidin [59,140,141,143,145,147,149]. Thus, the monolayer thickness of 4.1 nm obtained from the SPR results demonstrates that a complete monolayer of avidin has adsorbed onto the 22% Bi-PL–coated surface. In fact, full monolayers of avidin approximately 4.0 nm thick were also obtained for Bi-PL monolayers with as little as 11% bi-

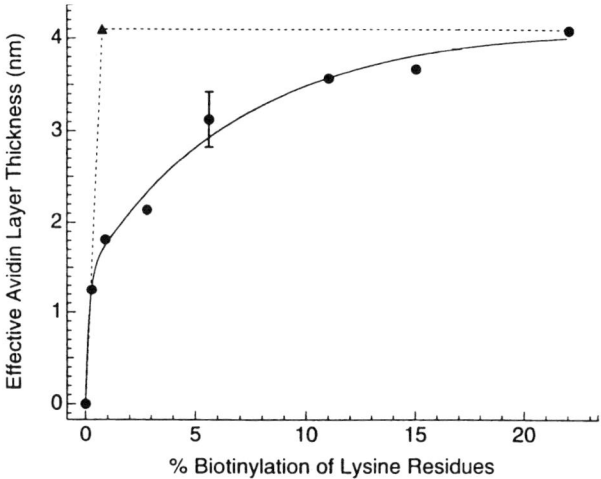

FIG. 18. A plot of effective avidin layer thickness versus the percent biotinylation of the lysine residues of the adsorbed Bi PL. The experimental data points (circles) were determined by SPR thickness measurements and fit with a double exponential (solid line). The calculated curve (dashed line) was obtained via the fluorescence measurements—see discussion in the text. Comparison of the experimental and calculated curves suggests sterically hindered binding of avidin, which could be overcome by increasing the number of specific adsorption sites (i.e., the %Bi-PL). (From Ref. 97.)

otinylation, as shown in Fig. 18. This result agrees with that obtained by Spinke et al., who found that 10% of a biotin-thiol mixed with 90% 11-mercaptoundecanol led to the adsorption of a full streptavidin monolayer [59]. Complete 100% biotin thiol monolayers, however, were observed to hinder streptavidin adsorption and yield lower surface coverages [59,151].

Two control experiments were performed to verify that the avidin was specifically adsorbed onto the biotin sites present on the gold surface. First, SPR measurements determined that no avidin adsorption occurred onto a PL monolayer that had not been biotinylated (Table 1, part B). Second, no avidin adsorption was observed when the Bi-PL–coated surface was exposed to a solution of avidin that had been first saturated with d-biotin. The prevention of nonspecific adsorption of the protein avidin onto the gold surface by the PL (or Bi-PL) monolayers can be ascribed to the fact that, at the pH of the buffered deposition solutions, both PL and avidin are positively charged. The adsorption of avidin onto charged surfaces was investigated further in the SPR imaging experiments described in Sec. IV.

In addition to the full monolayers of avidin obtained from 11–22% Bi-PL monolayers, partial avidin monolayers could be formed by decreasing the amount of biotin on the PL. The effective avidin layer thickness, which is proportional to surface coverage, was determined with SPR measurements. This avidin layer thickness is plotted in Fig. 18 versus the %Bi-PL. Notice that partial monolayers of avidin resulted from Bi-PL monolayers with less than 11% biotinylation. The %Bi-PL was fixed by the stoichiometry of the one-step solution phase reaction between biotin and the lysine residues of PL *prior* to adsorption of the Bi-PL onto the surface. This method of controlling the spacing of biotin groups on the surface with Bi-PL avoids problems associated with mixed alkanethiol monolayers such as phase segregation [156] and solution composition versus surface composition mismatches [157,158]. This method is also preferable to a surface reaction between biotin and PL already adsorbed to the gold surface, which was found to offer little control, knowledge, or reproducibility of the biotinylation percentage of the PL. The solution phase reaction prior to adsorption, however, provided precise control over the amount of biotin on the gold surface and the resulting avidin coverage.

As seen in Fig. 18, the avidin surface coverage is not a linear function of the %Bi-PL. If all of the biotin moieties in the Bi-PL monolayer were available for binding with avidin, the avidin layer thickness should have increased linearly with the %Bi-PL. Once monolayer (saturation) coverage is reached, no further increase in avidin coverage should be observed. The dashed line in Fig. 18 depicts this "ideal" behavior. The triangle marks the point of saturation (1:1 ratio of biotin and avidin on the surface), which should be 0.75% Bi-PL as determined from fluorescence measurements. The solid line in Fig. 18 is a double exponential fit to the data; several fits were tried, and all gave nearly the same initial slope. This initial slope at low %Bi-PL closely matches the slope expected for "ideal" behavior (dashed line). This agreement indicates that at low surface coverages an avidin molecule binds to virtually every biotin moiety in the Bi-PL monolayer.

At higher %Bi-PL, however, the surface coverage of avidin no longer increases as quickly as expected. This discrepancy is likely due to the sterically hindered binding of avidin; similar steric effects have been

Surface Plasmon Resonance Methods 169

observed previously for biotin-lipid doped Langmuir-Blodgett films on quartz [144]. Once the Bi-PL surface is covered with a significant amount of avidin, avidin molecules from solution have a difficult time finding an adsorption site that is not partially blocked. Nonetheless, at higher %Bi PL (ca. 20%), a large excess of specific adsorption sites is provided, the steric hindrance problem is overcome, and a complete monolayer of avidin is formed.

2. Zirconium Phosphonate Multilayer Studies

The combination of PM-FTIRRAS and ex situ SPR measurements has also been used to characterize the structure and assembly of ultrathin multilayer films on gold substrates formed by the zirconium phosphonate (ZP) ligand coupling chemistry first reported by Mallouk and co-workers [63,144,159,160]. These films consist of alternating layers of organic (alkyl or aryl) and inorganic (zirconium phosphonate) structures and are thermally robust and highly reproducible. In this example, ZP multilayers were used to construct ultrathin organic films with both a specific thickness and an index of refraction that could be tuned within a given range ($1.51 \leq n \leq 1.64$ at $\lambda = 632.8$ nm). To accomplish this, ZP multilayer films were formed which contained two different bis(phosphonate) ions: 1,10-decanediylbis(phosphonate) (DBP) and 4,4' azo-bis[(p-phenylene)methylene]bis(phosphonate) (AZO). As depicted in Fig. 19, monolayers of either DBP or AZO were deposited sequentially onto a substrate primed with a monolayer of phosphorylated 11 mercaptoundecanol (MUD) and one self-assembled monolayer of DBP. For example, a 1:1 AZO/DBP film consisted of alternating self-assembled monolayers of AZO and DBP. By varying the AZO/DBP ratio in the ZP multilayer, the overall index of refraction of the thin film can be controlled.

ZP multilayer films formed from the sequential self-assembly of 20 DBP monolayers onto a primer layer on a gold substrate were examined with a combination of PM-FTIRRAS and ex situ SPR measurements. The positions and relative intensities of the infrared bands did not change with the number of DBP multilayers, indicating that each self-assembled DBP monolayer had an equivalent molecular structure, although the alkyl chains show more disorder compared to long-chain alkanethiol monolayers on gold. The thickness of the DBP multilayer film was determined by a series of ex situ SPR measurements that were performed during the film deposition process. These effects are shown in Fig. 20 during the forma-

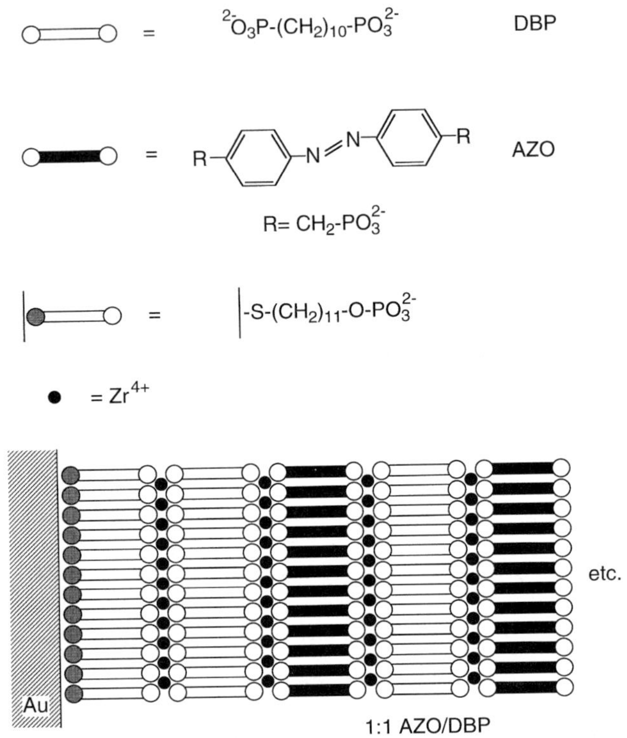

$^{2-}O_3P\text{-}(CH_2)_{10}\text{-}PO_3^{2-}$ DBP

$R=CH_2\text{-}PO_3^{2-}$ AZO

$|\text{-}S\text{-}(CH_2)_{11}\text{-}O\text{-}PO_3^{2-}$

● = Zr^{4+}

1:1 AZO/DBP

FIG. 19. Schematic diagram for the construction of mixed zirconium phosphonate (ZP) multilayer films. The ZP films are created on vapor-deposited gold substrates that have been first primed with a self-assembled monolayer of phosphorylated 11 mercaptoundecanol (MUD) and one self-assembled monolayer of 1,10 decanediylbis(phosphonate) (DBP). Mixed multilayers are formed by the sequential self-assembly of either 1,10-decanediylbis(phosphonate) (DBP) or 4,4'-azobis [(p phenylene)methylene]bis(phosphonate) (AZO) molecules onto the surface. For example, the 1:1 AZO/DBP multilayer film will have alternating self-assembled monolayers of AZO and DBP. (From Ref. 63.)

tion of a multilayer DBP film. Figure 21 plots the shift ($\Delta\theta$) in the SPR angle observed after the addition of each self-assembled monolayer during the formation of a DBP multilayer film relative to the MUD-DBP primer layer.

The shift in the SPR angle observed during the formation of the DBP

Surface Plasmon Resonance Methods

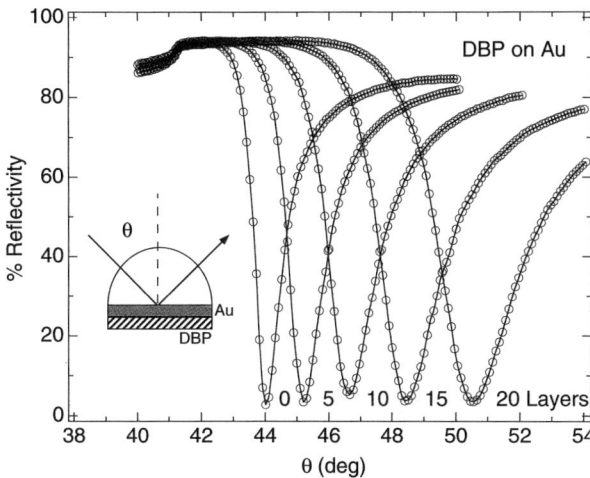

FIG. 20. SPR reflectivity curves obtained during the formation of a 100% DBP multilayer film. These SPR reflectivity curves were taken after the self-assembly of 0, 5, 10, 15, and 20 DBP monolayers. The shifts observed in the SPR angle can be quantitatively related to the thickness and index of refraction of the ZP multilayer. (From Ref. 63.)

multilayer can be used to determine the film thickness (d) if the index of refraction of the film (n) is known. One possibility is to use the bulk index of refraction of $n = 1.54$; this approach was employed by Mallouk et al. in order to obtain a monolayer thickness of 1.64 nm from ellipsometric data [161]. In principle, ellipsometric measurements of the DBP multilayer on the gold substrate could be used to determine n and d simultaneously; in practice, however, the small shifts in the ellipsometric parameters Δ (the phase difference upon reflection) and Ψ (the arctangent of the amplitude ratio for the s and p components of light upon reflection) observed from monolayers on metal surfaces make this determination difficult [160,162]. An alternate approach employed in this work was to determine n for the various films by ellipsometric measurements on ZP multilayers that had been deposited onto a transparent silica substrate. The primer layer used for the transparent substrates (a bifunctional silane monolayer) was not as well defined as the alkanethiol monolayer employed on the gold substrates and resulted in larger sample to sample variations in film thickness for a

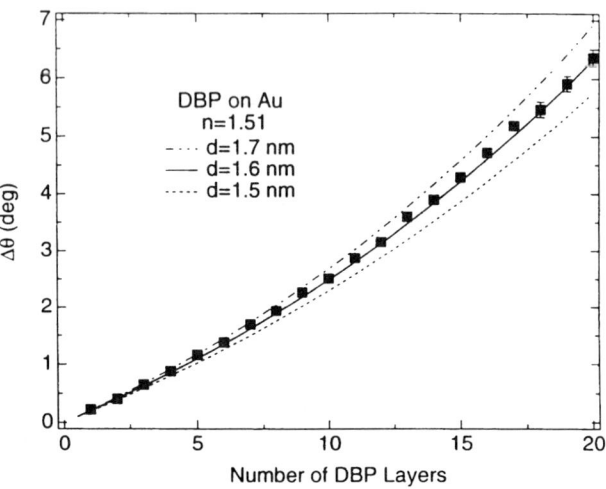

FIG. 21. Shifts in the SPR angle ($\Delta\theta$) measured during the formation of a 20 monolayer 100% DBP film. The shifts $\Delta\theta$ were determined from SPR reflectivity curves such as those shown in Fig. 20 with the SPR angle of the 0 Layer surface defined as $\Delta\theta = 0$. The three lines in the figure are the results of complex Fresnel calculations using an index of refraction of 1.51 (as determined from ellipsometric experiments) for the ZP multilayer and a DBP monolayer thickness of either 1.5, 1.6, or 1.7 nm. From the experimental data a DBP monolayer thickness of 1.6 ± 0.05 nm is determined. (From Ref. 63.)

given number of monolayers. However, for films of 10–30 monolayers, the changes in the ellipsometric parameters Δ and Ψ were large enough to determine the index of refraction to ±0.01. For example, the Δ and Ψ values for 10, 15, 20, and 25 DBP multilayers deposited onto a silica substrate are plotted in Fig. 22. Also plotted in the figure (the solid and dashed lines) are the theoretical (Δ, Ψ) values expected as a function of film thickness with n fixed at either 1.50, 1.51, or 1.52. The experimental data clearly show that n for the DBP multilayer film is determined to be equal to 1.51 ± 0.01. Similar ellipsometric measurements were used to determine n for the other ZP multilayers described in this paper and are listed along with the value for the 100% DBP film in Table 2.

After determining that the index of refraction for the DBP multilayers is equal to 1.51, the SPR angle shifts in Fig. 21 can be analyzed to as-

Surface Plasmon Resonance Methods

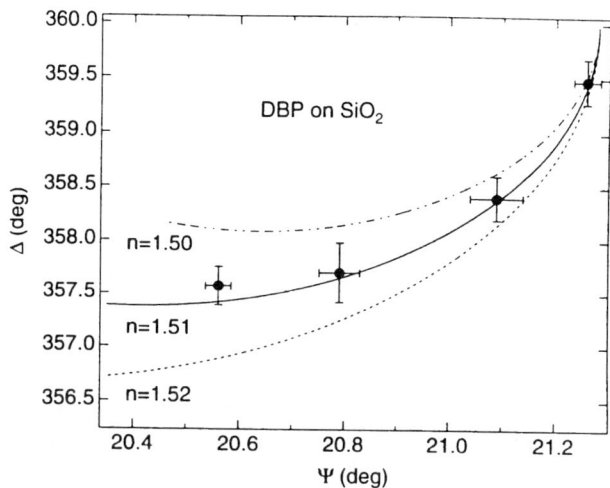

FIG. 22. Ellipsometric measurements for 100% DBP multilayers formed on a fused silica substrate. Values of Δ and ψ after the self-assembly of 10, 15, 20, and 25 DBP monolayers onto the silica surface are plotted as the solid circles in the figure. The three curves in the figure are the results of Fresnel calculations for a three-phase system (SiO_2, film, air) as the film thickness is varied and its index of refraction is set to either 1.50, 1.51, or 1.52. The film thickness for the three curves increases from right to left. From the experimental data, an index of refraction of 1.51 ± 0.01 is obtained for the 100% DBP multilayer. The results of ellipsometric measurements on other ZP multilayers are listed in Table 2. (From Ref. 63.)

TABLE 2

Monolayer Thicknesses and Indices of Refraction for the ZP Multilayer Films

Film	n^a	d (nm/layer)[b]	$\Delta\theta_{20}$ (deg)	d_{20} (nm)
DBP	1.51 ± 0.01	1.6 ± 0.05	6.36 ± 0.15^c	32.0 ± 0.5
AZO	1.64	1.6	8.33	32.0
1:1	1.57	1.9	9.65	38.0

[a]Determined by ellipsometry on fused silica substrates.
[b]Determined by SPR on vapor-deposited gold slides.
[c]Reported uncertainty is the standard deviation of three gold-ZP film samples.
Source: Ref. 60.

certain the thickness of the self-assembled DBP monolayers. Using a six-phase Fresnel calculation, the theoretical curves for $\Delta\theta$ as a function of the number of ZP monolayers are plotted in Fig. 21 for monolayer thicknesses of 1.5, 1.6, and 1.7 nm. A comparison of the experimental data points and the theoretical $\Delta\theta$ curves demonstrate that a DBP monolayer thickness of 1.6 ± 0.05 nm can be obtained from the SPR measurements. This thickness agrees favorably with values reported previously [161]. The accuracy of better than ± 0.1 nm is typical for the SPR technique and is remarkable for an optical measurement at $\lambda = 632.8$ nm. Using a space-filling model, the inorganic portion of the DBP monolayer from the center of the first carbon above to the first carbon below the inorganic layer is about 0.65 nm. In an all-*trans* configuration that is oriented perpendicular to the surface, the additional nine methylene groups contribute about 1.14 nm for a total thickness of approximately 1.7–1.8 nm [163,164]. An experimentally determined average monolayer thickness of 1.6 nm is plausible given the possibility of tilting and the presence of alkyl chain disorder in the DBP monolayer as observed in the PM-FTIRRAS spectrum (not shown).

For comparison with the 100% DBP monolayer, a second ZP multilayer was created by the sequential self-assembly of 20 AZO monolayers onto a primer layer on a gold substrate. Again, the positions and relative intensities of the various infrared bands in the PM-FTIRRAS spectrum for the 100% AZO multilayer did not change with film thickness, indicating that each self-assembled AZO monolayer possessed an equivalent molecular structure. As with the DBP film, SPR measurements were performed during the deposition of the AZO multilayer film, and the shift ($\Delta\theta$) in the SPR angle observed after the addition of each self-assembled monolayer is plotted as the squares in Fig. 23. The SPR data for the DBP film from Fig. 21 is also plotted in the figure as the triangles. A larger $\Delta\theta$ per monolayer was observed for the AZO multilayer film; this larger shift is attributed to a higher index of refraction of the AZO multilayer. As listed in Table 2, ellipsometric measurements of AZO multilayer films on a transparent substrate yielded an index of refraction of $n = 1.64 \pm 0.01$. This value is within the range observed for azobenzene-derivatized polymers (1.623–1.758) [165] and is significantly higher than the value of 1.51 observed for the DBP film. Using an index of refraction of 1.64, the six-phase Fresnel calculations can be used to analyze the SPR shifts and determine an average AZO monolayer thickness of 1.6 ± 0.05 nm, which is the same value obtained for the DBP monolayer. Using a space-filling model with an inorganic section thickness of 0.65 nm, the AZO monolayer in which the

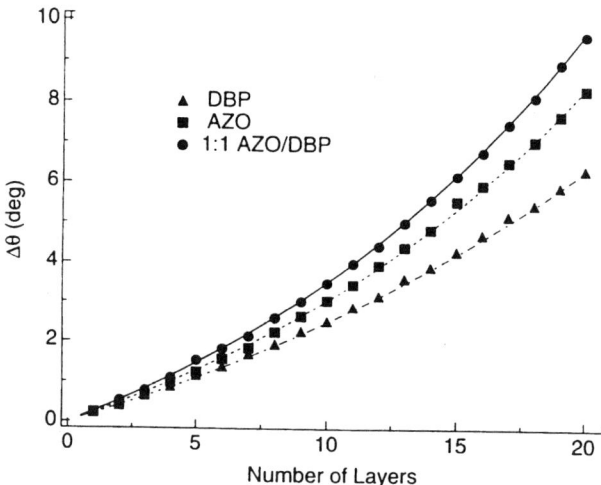

FIG. 23. Shifts in the SPR angle ($\Delta\theta$) measured during the formation of 20 monolayer ZP films of 100% DBP (triangles), 100% AZO (squares), and 1:1 AZO/DBP (circles). The shifts in $\Delta\theta$ were determined from SPR reflectivity curves with $\Delta\theta = 0$ defined as the SPR angle for the primed gold surface (see text for details). The solid and dashed lines are the results of complex Fresnel calculations using the indices of refraction and average monolayer thicknesses for DBP, AZO, and 1:1 AZO/DBP films listed in Table 2. An unexpectedly large average monolayer thickness of 1.9 nm is observed for the 1:1 AZO/DBP film. (From Ref. 63.)

azobenzene is oriented perpendicular to the surface (0.9–1.0 nm) should have a total thickness of approximately 1.7–1.8 nm [163,164]. The smaller experimentally determined value of 1.6 nm again suggests tilting and disorder in the AZO film.

The similarity of the DBP and AZO thicknesses suggests that mixed monolayers of these two molecules may yield stable ZP multilayer films possessing an intermediate index of refraction. Having determined that a DBP multilayer film has an index of refraction n = 1.51 and that an AZO multilayer film has an index of refraction n = 1.64, one would expect that a ZP multilayer film formed from alternating self-assembled monolayers of AZO and DBP would possess an index of refraction of n = (1.64 + 1.51)/2 = 1.575. Indeed, the ellipsometric measurements of a 1:1 AZO/DBP multilayer film on a transparent substrate yield a value of n = 1.57 (see Table 2).

However, SPR measurements on the 1:1 AZO/DBP multilayer suggest that even though n = 1.57 as expected, the structure and thickness of the ZP film has changed. As with the previous two ZP films, SPR measurements were performed during the deposition of the 1:1 AZO/DBP multilayer film in order to determine the average monolayer thickness. The shift $\Delta\theta$ observed after the addition of each self-assembled monolayer is plotted as the circles in Fig. 23 along with the data from the DBP and AZO multilayers. Given that n = 1.57 for this film, one would expect that the $\Delta\theta$ curve for this film would fall in between the curves for the AZO and DBP multilayers. Unexpectedly, the data for the 1:1 AZO/DBP multilayer systematically shows a *larger* $\Delta\theta$ shift than either of the other two films. Using n = 1.57, this $\Delta\theta$ curve corresponds to an average monolayer thickness of 1.9 nm, which is a 19% increase in the average thickness as compared to the 1.6 nm obtained for both the AZO and DBP monolayers. To demonstrate how unusual this number is, if the monolayer thickness is fixed at 1.6 nm, the $\Delta\theta$ curve for the 1:1 AZO/DBP multilayer can only be fit using an index of refraction of 1.70, which is clearly an unreasonable number given n = 1.64 for the 100% AZO multilayer. The uniform shift in the SPR angle for each additional monolayer and PM-FTIRRAS spectra indicate that this 19% increase in thickness results from the incorporation of 19% more DBP in the 1:1 AZO/DBP film than the amount suggested in the simple picture in Fig. 19. The additional DBP could be incorporated into the 1:1 AZO/DBP multilayer film by increasing the packing density within the mixed monolayers and/or by intercalation into the preceding AZO regions of the film.

3. *In Situ SPR Studies*

The two sets of SPR thickness measurements discussed in Sec. III.D.1 and III.D.2 were made in an ex situ environment (air). A major advantage of the SPR technique is that it can be easily modified to monitor monolayer and multilayer film assembly and thickness in an in situ or an electrochemical environment (usually aqueous, but other organic solvents can also be used). This allows for the real-time monitoring of the self-assembly, adsorption, and reaction kinetics of ultrathin films. Since direct optical access to the metal/liquid interface is not required to follow the formation of surface plasmons, in situ SPR measurements require only minor modification of the prism/gold sample assembly to incorporate an in situ flow cell or electrochemical cell.

In addition to an in situ cell, a small number of experimental changes

Surface Plasmon Resonance Methods

are also required in moving to an in situ environment. Since the bulk medium of water has a larger optical index of refraction as compared to air ($n_{H_2O} = 1.33$ at 632.8 nm), the surface plasmon angle shifts to a higher incident angle (>70°) for a BK7 prism/gold sample combination. One method of compensating for this increase in the bulk medium's index of refraction is to use a higher index of refraction prism/gold sample system such as sapphire (n = 1.75) or SF10 glass (n = 1.72). For the in situ SPR measurements shown in this chapter, either SF10 hemispherical prisms or 60° prisms are coupled to gold-coated SF10 microscope slide covers using an index matching fluid (1-iodonapthalene or Cargill n = 1.725 immersion fluid). The in situ surface plasmon angle for the SF10 prism/gold sample system occurs at significantly lower incident angles (~57°) than for the BK7/gold/water case.

Figure 24 plots the in situ SPR curves for a bare gold surface (open triangles) and after modification with a self-assembled monolayer of MUA and

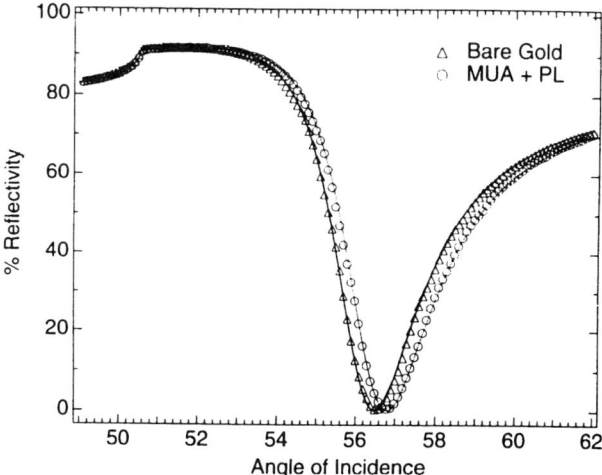

FIG. 24. In situ scanning SPR reflectivity curves obtained for a bare gold substrate (open squares) and the same substrate following self-assembly of a bilayer of MUA and poly-L-lysine (open circles). The curve is the measured reflectivity (%R) of p-polarized light at 632.8 nm from the hemispherical SF10 prism/Au/film/water interface as a function of incident angle, θ. The solid lines in the figure are a five-phase Fresnel fit to the SPR data (see text).

an electrostatically adsorbed poly-L-lysine monolayer (open circles). The SPR angles for these curves are shifted to higher incident angles relative to the corresponding ex situ curves. In addition, the SPR reflectivity curves are broader than the ex situ data. The solid lines are five-phase Fresnel calculations which are fit to the SPR reflectivity data using the previously determined indices of refraction for MUA and poly L-lysine. The thicknesses determined for the MUA and poly-L lysine layers from these Fresnel calculations are 1.6 and 1.3 nm, respectively. These in situ results are in good agreement with those determined ex situ in Sec. III.D.1 (see Table 1). The slight increase in the PL monolayer thickness measured in situ may possibly be due to hydration of the PL monolayer in the aqueous environment.

In situ SPR measurements have also been used to determine the thicknesses of mixed ZP films of DBP and the nonlinear optical molecule HAPA ([5-[4-[[4-[(6-hydroxyhexyl)sulfonyl]phenyl]azo]phenyl] pentoxy]-phosphonic acid). These mixed ZP films of DBP and HAPA are used in Sec. V to measure the electric field profile within multilayer films at charged gold electrodes. In the in situ SPR measurements shown in Fig. 25, the reflectivity (%R) of a p-polarized HeNe laser was monitored as a function of the incident angle (θ) for a 60° SF10 prism/gold/ZP film/electrolyte interface. The ZP film in this experiment consisted of a phosphorylated MUD primer monolayer, one HAPA ZP monolayer, and one DHP (dihexadecylphosphate) monolayer (see Sec. V.C for details). The solid line in Fig. 25 is a five-phase Fresnel calculation using index of refraction n_{DBP} and n_{HAPA} values determined previously [64] and an overall thickness d of 6.7 nm as measured from ex situ scanning SPR experiments. The inset graph in Fig. 25 plots the angular shift in θ_{sp} relative to a single HAPA monolayer film with increasing DBP overlayers. The solid line in the inset is a seven phase Fresnel fit to the data, again using previously determined values n_{DBP} and n_{HAPA}, and corresponds to an increase in the ZP film thickness of 1.6 ± 0.2 nm per DBP monolayer. This DBP layer spacing is equivalent to that measured for pure multilayer DBP films in the ex situ experiments descried in Sec. III.D.2.

The SPR data described in this section and PM-FTIRRAS experiments presented previously [6,63] demonstrate that the optical techniques of surface plasmon resonance and PM-FTIRRAS are a powerful combination of methods for the determination of monolayer and multilayer film thickness and structure. The scanning SPR technique has the additional advantage that it can be easily adapted to monitor film formation and adsorption processes in electrochemical, biological, and other in situ environments. Additional in

Surface Plasmon Resonance Methods

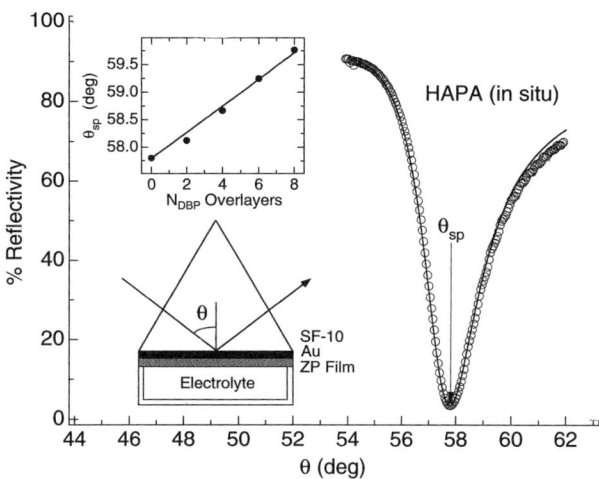

FIG. 25. In situ scanning SPR reflectivity curve obtained for a ZP film that consisted of a phosphorylated MUD primer layer, one HAPA monolayer, and one DHP capping monolayer on a gold substrate (open circles). The curve is the measured reflectivity (%R) of p-polarized light at 632.8 nm from the 60° SF10 prism/Au/ZP film/electrolyte interface as a function of incident angle, θ. The solid line in the figure is a five-phase Fresnel fit to the SPR data (see text). The inset graph follows the shift in θ_{sp} upon the formation of N DBP monolayers where N varied from 0 to 8. The solid line in the inset graph is a seven phase Fresnel fit to the data and yields the average ZP film thickness (see text). (From Ref. 105.)

situ SPR methods for studying multilayer organic film formation on gold surfaces are detailed in the next two sections.

IV. SPR IMAGING EXPERIMENTS

A. Introduction

Along with the scanning SPR technique described in the previous section, fixed angle SPR imaging can also be employed to measure adsorption onto patterned alkanethiol self-assembled monolayers. SPR imaging and microscopy have been used previously by various researchers to study adsorption onto patterned surfaces [128,131,132]. The ability of the SPR imaging technique to observe patterned surfaces is useful when it is desirable to monitor adsorption onto many surface functionalities simultaneously, as required in many biosensor applications. For example, SPR imaging allows differential biopolymer adsorption studies to be performed very accurately because two functionalized surfaces can be placed on the same substrate and exposed to identical conditions.

A schematic diagram of the SPR imaging apparatus is shown in Fig. 26. In this experiment, an expanded HeNe laser beam is incident on the prism/thin gold film sample assembly near the SPR angle, and the reflected light is detected with an inexpensive CCD camera to produce images such

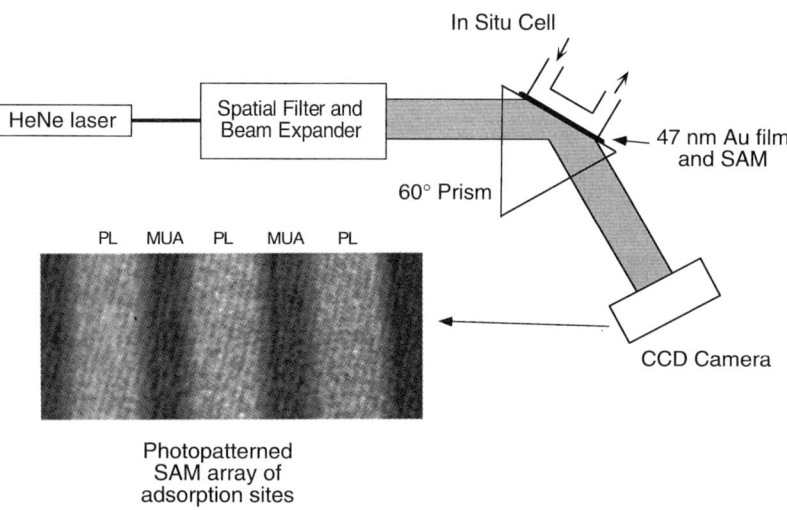

FIG. 26. Optical layout for the SPR imaging instrument. A HeNe laser is sent through a spatial filter and beam expander, which are required so that the entire sample surface can be illuminated. This expanded beam is then directed at the prism sample assembly. A 47 nm thick gold film vapor deposited onto a glass slide is used as the sample substrate for the formation of patterned self-assembled monolayer (SAM) films. This is in contact with a 60° prism which is required to couple the incident p-polarized light into the surface plasmon modes at the interface. In the imaging setup the reflected intensity, at some fixed angle, is then measured across the beam using a CCD camera. For in situ experiments a flow cell is attached to the back of the prism sample assembly so that the gold sample is in contact with solution. An image of a photopatterned SAM array obtained with this instrument is also shown in the figure. (From Ref. 129.)

Surface Plasmon Resonance Methods 181

as those shown in the figure. Any change in the index of refraction or the thickness of a film adsorbed to the gold surface results in a change in the intensity of the reflected light observed at a fixed angle. By creating two different alkanethiol self-assembled monolayers (SAMs) with UV photopatterning on a single thin gold film, a differential adsorption measurement can be performed by monitoring changes in the SPR image as the sample is exposed to a solution of interest. These differential adsorption measurements are equal in sensitivity to the SPR angle shift measurements of biopolymer adsorption performed previously with the scanning SPR instrument. The SPR imaging technique can be used to monitor the adsorption of submonolayer amounts of material in both ex situ and in situ configurations. The following examples demonstrate the use of SPR imaging to characterize the adsorption of proteins, polypeptides, and oligonucleotides onto modified gold surfaces.

B. Ex Situ SPR Imaging Experiments of Biopolymer Adsorption

SPR imaging experiments are performed in our laboratory at a single fixed angle to monitor differences in biopolymer adsorption onto photopatterned chemically modified gold surfaces. For example, a negatively charged MUA monolayer can be used to electrostatically adsorb the polycation poly-L-lysine or any positively charged protein such as avidin (see Fig. 27). One method of determining the difference in thickness between the PL monolayers and the avidin monolayers formed by this electrostatic adsorption process is to measure the different SPR angle shifts obtained from ex situ scanning SPR experiments, as described in the previous section. Figure 28a plots the experimental and theoretical SPR curves for a PL monolayer and an avidin monolayer formed on a negatively charged MUA surface. The circles in the figure correspond to the experimental SPR curve obtained from a PL monolayer, and the squares are the experimental data from an avidin monolayer. In addition to the SPR angle shift, the change in %R at a fixed angle near the SPR minimum can be used to quantitate the difference in thickness between the two monolayers. Figure 28b plots the differential percent reflectivity ($\Delta\%R$) curve, which is simply the difference in the two observed percent reflectivities (%R) as a function of incident angle. A maximum in the differential reflectivity is observed just below the SPR angle; SPR imaging experiments are typically performed at this angle to obtain the highest sensitivity to changes in the thickness or index of refraction of the adsorbed material.

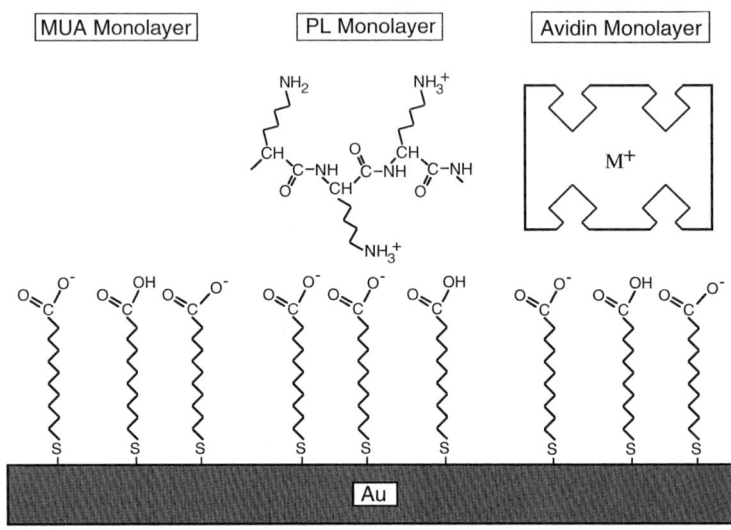

FIG. 27. A schematic diagram of three monolayers adsorbed onto a gold substrate: a negatively charged 11 mercaptoundecanoic acid (MUA) monolayer, a positively charged poly L-lysine (PL) monolayer electrostatically adsorbed to MUA, and the protein avidin electrostatically adsorbed to MUA. (From Ref. 129.)

In order to make differential adsorption measurements with the SPR imaging apparatus, two-component surfaces are created that consist of alternating stripes of two different surface functional groups. These surfaces are prepared by a series of adsorption/self-assembly, photochemical desorption, and rinsing steps. For example, the procedure required for the formation of surfaces with stripes of both PL and MUA monolayers (MUA-PL surface), used for all protein differential adsorption experiments, is shown in Fig. 29. First, a SAM of MUA was formed on a gold surface after which PL was electrostatically adsorbed onto the MUA as described previously [6]. After formation of the PL/MUA bilayer, the surface was placed behind a mask and irradiated with UV light. The UV irradiation causes the gold thiolate bond to be oxidized [166–168] so that both the MUA and PL layers are removed from the exposed stripes by rinsing with ethanol and water. This photopatterned surface is then re-exposed to an ethanolic MUA solution, resulting in a surface with alternating regions

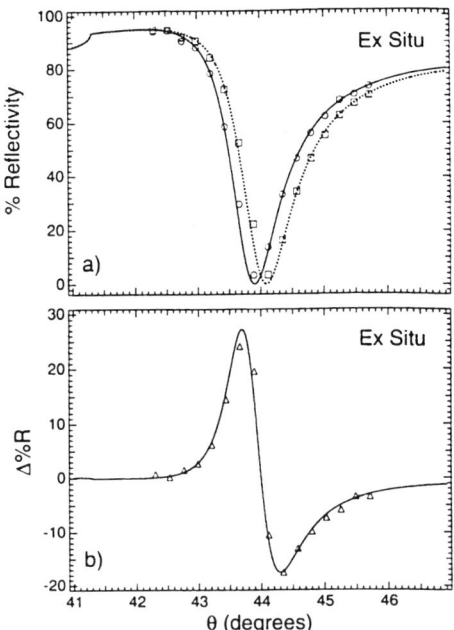

FIG. 28. SPR reflectivity and differential reflectivity curves for PL and avidin monolayers measured ex situ. (a) Experimental and theoretical SPR reflectivity curves for PL and avidin surfaces. The circles and squares show, respectively, the experimental % reflectivities for PL and avidin monolayers measured as a function of incident angle, θ. The solid and dotted lines are the result of four phase complex Fresnel calculations for PL and avidin films and the shift in the angle of minimum %R is due to the difference in thickness between the two monolayers. (b) The experimental and theoretical differential reflectivity curves obtained from the difference in the SPR curves in (a) for PL and avidin monolayers. (From Ref. 129.)

of MUA and PL monolayers. At a neutral or slightly basic pH, the areas of PL will be positively charged and those of MUA will be negatively charged; this surface can thus be used to examine the electrostatic adsorption of proteins. When the reflectivity from such a surface is measured in an SPR imaging experiment at a fixed angle, images of the type shown in Fig. 26 are obtained. The %R values for images are obtained by taking a ratio of the intensity of images observed using p- and s-polarized incident

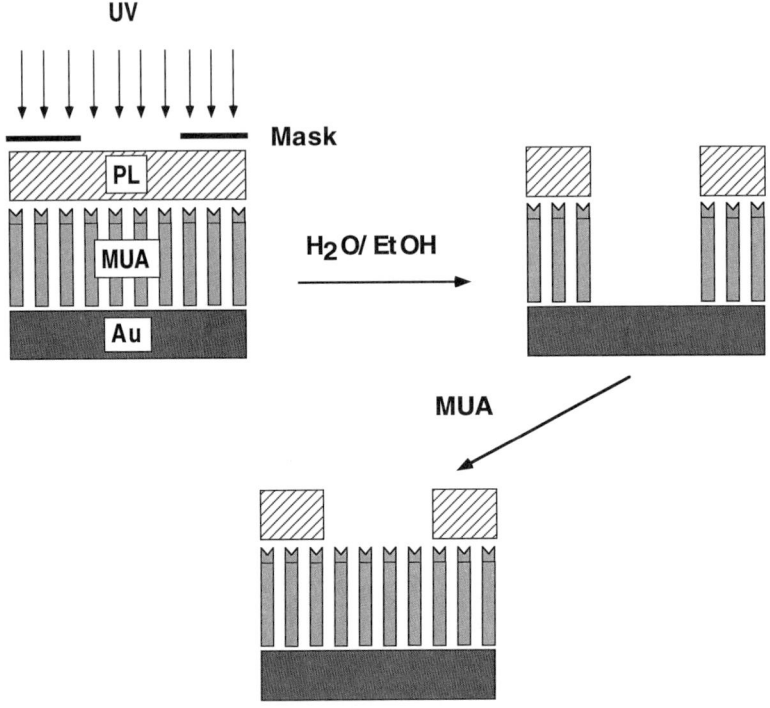

FIG. 29. Schematic diagram outlining the process for making patterned MUA and PL surfaces. Initially MUA is adsorbed onto the gold surface from an ethanolic solution and PL is electrostatically bound to the MUA. This PL monolayer is then placed behind a mask and exposed to UV light. The UV light oxidizes the gold sulfur bond so that both the MUA and the PL can be removed in the exposed areas by rinsing with water and ethanol. MUA can then be readsorbed to the exposed gold creating a surface containing areas of MUA and areas of PL, which is used to study differential protein adsorption. (From Ref. 129.)

light. These images are represented quantitatively by averaging the %R values measured at each pixel of the CCD camera along the stripes and generating a "line profile" across the image.

For example, line profiles generated from SPR images of two MUA-PL surfaces are plotted as the solid lines in Fig. 30. The SPR images were obtained at an incident angle below the SPR angle, and the line profiles

Surface Plasmon Resonance Methods

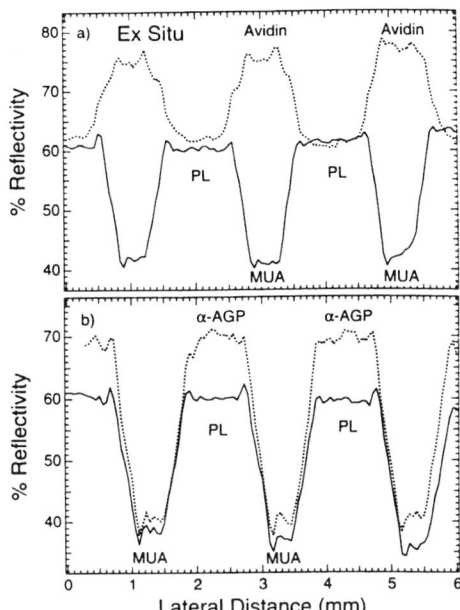

FIG. 30. Line profiles showing the adsorption of avidin and α-acid glycoprotein onto patterned MUA PL surfaces. The solid lines in both (a) and (b) are the % reflectivities measured for gold samples photopatterned with stripes of PL and MUA. The dotted line in (a) is the %R measured after exposing the sample to a pH 8.5 avidin solution and subsequent rinsing with water. The adsorption of α-AGP at pH 8.5 onto a MUA-PL sample is shown by the dotted line in (b). (From Ref. 129.)

from these images show that the %R from the PL monolayer is higher than the MUA monolayer as expected since it is thicker by 10.5 Å [6]. This surface should contain both negatively charged MUA regions and positively charged PL regions when exposed to an aqueous solution of pH 8. The dotted lines in Fig. 30 are the line profiles for two such surfaces that have been exposed to solutions of either avidin (isoelectric point (pI) = 10) or α-acid glycoprotein (α AGP, pI = 3). The surfaces have been removed from the adsorption solution and rinsed with a buffer solution before being imaged. From these ex situ measurements, it is evident that the avidin nonspecifically binds to the MUA regions of the surface, and the α-acid glycoprotein nonspecifically binds to the PL regions of the surface. Thus, the nonspecific adsorption of these two proteins appears to be dominated by the electrostatic interactions between the proteins and the surface functional groups.

While in principle this information could have been obtained from a series of scanning SPR measurements, the SPR imaging technique is a rapid and very sensitive method for studying protein adsorption. The speed and sensitivity of these experiments arise from the fact that adsorption onto a single surface containing multiple areas with different functional groups can be measured simultaneously and that each functional group is exposed to identical adsorption conditions. Very small changes in %R corresponding to submonolayer amounts of biopolymer adsorption can be observed with this measurement, and if the data are normalized (as in Fig. 30) to obtain quantitative %R values, an effective average thickness can be determined.

C. In Situ SPR Imaging Experiments of Biopolymer Adsorption

1. Avidin Adsorption onto Patterned MUA/PL Surfaces

The same SPR imaging experiments that were performed ex situ on the adsorption of avidin onto MUA-PL surfaces can also be performed in situ with the photopatterned monolayers in contact with an avidin adsorption solution. The in situ SPR imaging differential adsorption measurements are employed to examine how the isoelectric point of a protein, the solution pH, and the electrolyte concentration can affect the electrostatic adsorption of proteins onto a chemically modified gold surface. As in the previous section, a comparison of the SPR curves obtained from in situ scanning SPR measurements for PL and avidin monolayers can help to quantitate the changes in %R expected from an in situ SPR imaging experiment. Figure 31 plots the SPR reflectivity curves and differential reflectivity curves for these two monolayers obtained from an in situ cell utilizing an SF10 coupling prism. As shown in the figure, the SPR minima have shifted to higher angles as compared to the ex situ data, and the SPR reflectivity curves have broadened, due to the index of refraction of the monolayers being closer to that of aqueous solution than air. This broadening leads to a smaller differential reflectivity for a given change in thickness than that observed ex situ. However, the qualitative results are the same: if an SPR imaging experiment is performed at a fixed angle just below the SPR minimum, then an increase in the %R is observed as avidin adsorbs to the surface.

Figure 32 plots the line profiles from a series of in situ SPR imaging

Surface Plasmon Resonance Methods

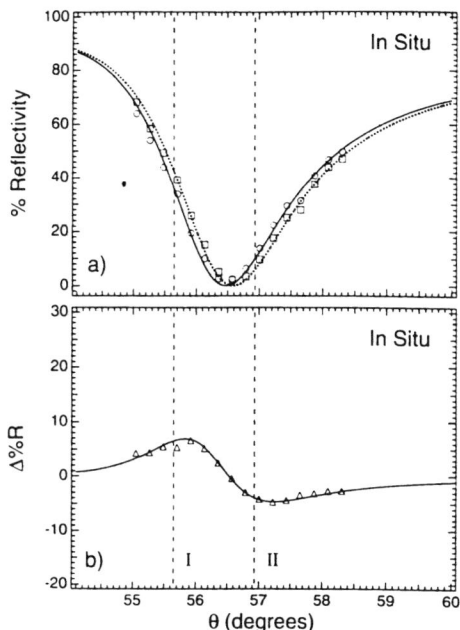

FIG. 31. SPR reflectivity and differential reflectivity curves for a PL monolayer and an avidin monolayer measured in situ. (a) The circles and the solid line show the experimental and theoretical SPR curves for a PL monolayer and the squares and dotted line show the same for an avidin monolayer. Notice that when measured in situ the SPR reflectivity curves are shifted to higher angles and are considerably broadened compared to the ex situ SPR curves shown in Fig. 28a for the same monolayers. (b) The differential reflectivity resulting from the SPR curves in (a); notice here that the differential reflectivities are much less than those seen in Fig. 28b. The dashed lines labeled I and II correspond to the two angles at which the calculated differential reflectivity vs. thickness is considered in Sec. IV.C.3. (From Ref. 129.)

measurements on MUA-PL surfaces identical to those used in the ex situ experiments. Notice that the two areas on the MUA-PL surface show a smaller difference in %R in this in situ experiment than in the corresponding ex situ experiment (Fig. 30), as predicted from the scanning SPR curves in Fig. 31. As in the ex situ measurements, exposure of this surface to an avidin solution results in the strong adsorption of the protein avidin onto the MUA monolayer portions of the surface (dashed line). Using the

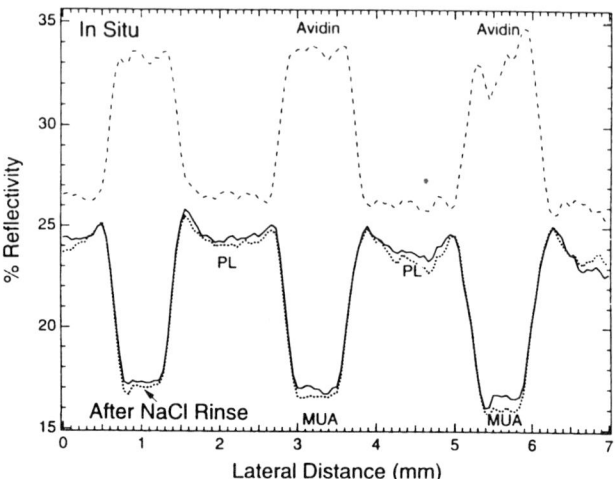

FIG. 32. Line profiles measured in situ for avidin adsorption onto a patterned MUA-PL surface and the subsequent removal of avidin by high electrolyte concentration. The solid line is the %R from a MUA-PL surface imaged in buffer and the dashed line is taken after exposure of the sample to a pH 8.5 avidin solution and rinsing with buffer. The dotted line is the %R measured in the buffer after exposure of the sample to 2 M NaCl (pH 8.5). The dotted line shows that a high concentration of electrolyte will remove all of the avidin from a MUA-PL surface, as expected for electrostatically bound avidin. (From Ref. 129.)

in situ measurements, the avidin adsorption is found to reach a constant level in about 5 minutes; this level of adsorption does not change when the avidin solution is replaced with pH 8 buffer.

Also shown in Fig. 32 is the line profile of the photopatterned SPR image after rinsing with a 2 M NaCl solution (dotted line). The line profile shows that the avidin monolayer has been completely removed upon exposure to a solution with high electrolyte concentration due to the screening of the PL and MUA charges. This observation agrees with our previous assertion that the avidin is adsorbed electrostatically to the MUA monolayer. The rate of desorption of the avidin and the residual amount of the protein on the surface varied with the salt concentration; complete desorption of the monolayer occurred within 20 minutes for all NaCl solutions above 1.0 M.

Interestingly, it is shown in the figure that the electrostatically adsorbed PL layer on the patterned sample does not desorb upon exposure to

Surface Plasmon Resonance Methods

the high-salt solution. However, if an electrostatically adsorbed PL monolayer is exposed to a high-salt solution directly after deposition, it will completely desorb as expected. It has been observed that if the PL/MUA bilayer is heated at 100°C for one hour before exposure to a 2 M NaCl solution, little or no PL desorption is detected. We hypothesize that this irreversible binding of the PL/MUA bilayer is due to hydrogen bond formation caused by dehydration and/or conformational changes which occur during heating. In the process of making the patterned MUA-PL samples (shown in Fig. 29) the PL/MUA bilayers were heated, and for this reason there was no loss of PL during the 2 M NaCl rinse shown in Fig. 32.

Further in situ SPR imaging experiments were performed in order to examine the effect of solution pH upon avidin adsorption to the MUA monolayer. As the pH is lowered, the carboxylic acid groups of the MUA monolayer should become protonated and the electrostatically adsorbed avidin molecules should desorb. The %R measured for an avidin mono-

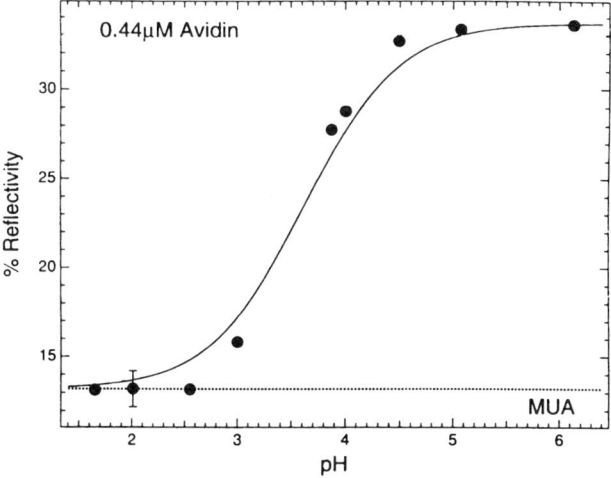

FIG. 33. The % reflectivity measured from a MUA surface in equilibrium with a 0.44 μM avidin solution as a function of solution pH. At a pH below 2.5 no avidin adsorbs to MUA. This can be seen by the overlap of the circles below pH 2.5 with the dotted line, where the dotted line shows the %R from MUA before exposure to avidin. At a pH greater than 4.5 the amount of avidin adsorption plateaus at a %R which corresponds to a full monolayer of avidin. This titration curve was fit to a Langmuir adsorption isotherm for $K_{ads} = 550$ (solid line). (From Ref. 129.)

layer adsorbed to a MUA SAM from a 0.44 μM solution is shown as a function of pH in Fig. 33; the data in this figure were obtained from a series of SPR imaging experiments of a MUA-PL surface. Above a pH of 4.5, a full avidin monolayer is adsorbed onto the MUA monolayer. However, below this pH only partial avidin monolayers are observed, and by pH 2.5 the avidin is completely removed from the surface. Further experiments (not shown) also indicate that avidin will completely desorb in a pH 13 solution. Since this pH is significantly above the pI of avidin, this observation also supports the assertion that avidin is electrostatically adsorbed to the MUA SAM. The pH at which 50% of an avidin monolayer is adsorbed to the MUA surface, as observed in Fig. 33, is significantly less than the measured pK_a for a MUA monolayer. This difference results from the fact that both the pK_a for the MUA surface and the ion pairing constant between avidin and the MUA monolayer are important in determining the surface coverage of electrostatically adsorbed avidin [129].

2. Molecular Weight Dependence of Poly-L-lysine Adsorption

Another factor that affects the electrostatic adsorption of polypeptides onto surfaces is the number of monomers in the polypeptide. In a second set of in situ SPR imaging experiments, this effect is examined by measuring the amount of PL adsorbed to MUA as a function of PL molecular weight. We have previously observed in ex situ measurements that PL molecules with an average of 67 lysine residues will form a full monolayer on a MUA surface when deposited from a pH 8.5 solution, but that lysine monomers will not adsorb at this pH [6]. Figure 34 shows the results of in situ SPR imaging experiments in which the amount of PL adsorbed onto a MUA monolayer was determined for a series of solutions with an average PL length that varied from 1 to 270 lysine residues. In these experiments, a constant lysine residue concentration of 0.7 mM was maintained in each solution. Each point in Fig. 34 was obtained by measuring the amount of PL adsorbed to the MUA portions of a gold thin film that was photopatterned with areas of MUA and 11 mercaptoundecanol (MUD). A striped MUA-MUD surface was used in these experiments since PL does not adsorb to a MUD SAM, and therefore the percent reflectivity of the MUD surface could be used as an internal standard. With these MUA-MUD surfaces, we were able to monitor the very small changes in the percent reflectivity observed for the partial monolayers formed on the MUA monolayer from low molecular weight PL. The MUA-MUD surface was regenerated in between exposures to the various PL solutions by rinsing with an acidic (pH

Surface Plasmon Resonance Methods 191

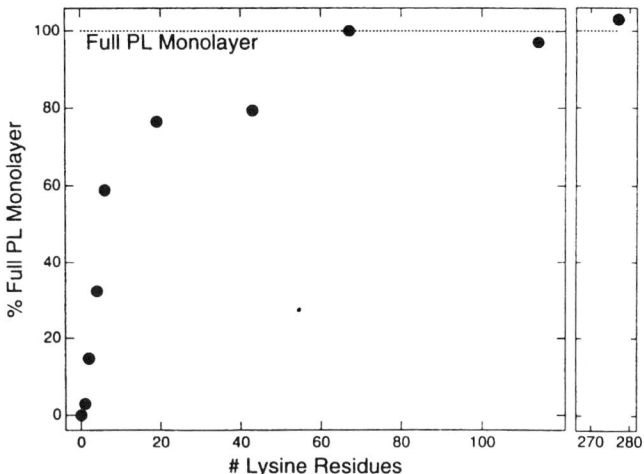

FIG. 34. The % of a full PL monolayer electrostatically adsorbed onto MUA as a function of the molecular weight measured in pH 8.5 solutions each with a constant lysine residue concentration of 0.7 mM. Percent reflectivities were obtained from a surface patterned with stripes of MUA and mercaptoundecanol (MUD). The percentage of a full PL monolayer was obtained by normalizing the difference in %R seen for each length PL on MUA to the maximum PL coverage. No adsorption is seen for L-lysine but the percent of adsorption increases rapidly with polymer chain length until a full monolayer of PL is adsorbed for chain lengths longer than 67 monomer units. (From Ref. 129.)

2) buffer. The percentage of a full monolayer (shown in Fig. 34) was obtained by normalizing the %R measured for each length PL on MUA to the maximum PL coverage. The amount of adsorbed PL increases quickly from no observable adsorption for lysine monomers to 60% of a monolayer for PL with 6 residues and then levels off until a plateau is reached at full monolayer coverages for chain lengths above 67 lysine residues. Since the lysine residue concentration is held constant during these experiments, the increased PL adsorption with chain length is due solely to an increased affinity between the longer PL molecules and the MUA surface. It was also determined that monolayers formed from PL solutions with average chain lengths greater than 19 residues showed no desorption when rinsed with a pH 8.5 buffer solution, but some loss of PL was observed for the shorter

PL chains upon rinsing. This indicates that, in addition to the adsorption coefficient, the adsorption kinetics are also affected by the polypeptide molecular weight.

3. Multiple Oligonucleotide Hybridizations Onto Immobilized DNA

In a final example, in situ SPR imaging has been used to detect the attachment of DNA oligonucleotide probes onto chemically modified gold surfaces and to characterize the subsequent in situ hybridization of oligonucleotide targets to these probes. As seen previously, SPR is sensitive to the thickness and index of refraction of an adsorbed layer. However, if the surface coverage of a particular analyte is low, it may be necessary to use some type of amplification strategy to improve the SPR response. In this section, an in situ chemical amplification strategy based on multiple DNA hybridizations is demonstrated as a means of improving the detection limit of DNA hybridization adsorption by SPR methods [169].

In this experiment, a DNA oligonucleotide containing 30 bases which is modified at the 5′ end with an amine group has been attached to a gold surface coated with MUA and electrostatically bound PL using the bifunctional linker 1,4-phenylene diisothiocyanate (PDITC) [170]. DNA which is covalently bound to the surface is referred to as "Probe" DNA in the inset of Fig. 35. Once the probe is bound to the surface, it is possible to hybridize a complementary (target) piece of DNA to it, shown as target A in the inset of Fig. 35. Target A is designed so that only half of its 30-base sequence is involved in hybridization to the surface probe and the remaining free bases can be used to hybridize a layer of target B to the surface. Target B can then hybridize a third strand of target DNA to the surface; in particular, its sequence is such that it will hybridize a second layer of target A. By sequentially exposing the surface to solutions of first target A and then target B, it is possible to perform multiple hybridizations to a single strand of probe DNA using only two different target oligonucleotides. Figure 35 shows the plot profiles from an in situ SPR imaging experiment where multiple hybridizations are used to amplify the SPR signal produced by the initial covalently bound probe DNA. In this experiment two spots containing amine-modified DNA with different sequences were created on a PDITC-modified gold surface and imaged in situ. The curve labeled "Probes" in Fig. 35 is a line profile taken through these two spots. A solution containing the complement to one of these DNA probes (target A) was injected into the in situ cell. It can be seen from line A1 in the figure that

Surface Plasmon Resonance Methods 193

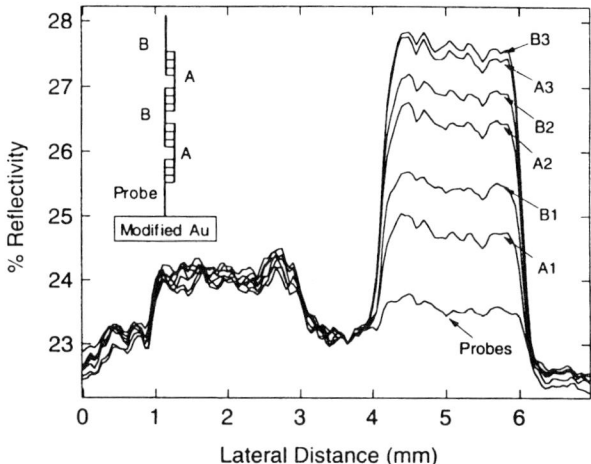

FIG. 35. Line profiles showing the amplification of the SPR response from DNA hybridization at a modified gold surface. The inset shows the strategy used to amplify the SPR response by hybridizing two different DNA oligonucleotides (A and B) to a single probe DNA molecule attached to the surface. The SPR line profile from a surface with two different DNA oligonucleotide probes immobilized on it is labeled "Probes" in the figure. Line A1 shows the response after exposing the surface to the complement to one of the DNA molecules. All of the bases in oligonucleotide A are not involved in hybridization to the probe and the free bases are able to hybridize to oligonucleotide B (line B1). A second exposure of the surface to oligonucleotide A results in further hybridization shown as line A2. More hybridization results from subsequent exposures to oligonucleotides B and A. (From Ref. 170.)

this complement adsorbs only to one of the spots, which clearly demonstrates the specificity of this surface hybridization. Then the sample was exposed to target B, and further adsorption to the spot was observed (line B1). This sequential hybridization will occur many times as shown by the increase in SPR response of each of the first five line profiles in Fig. 35. Notice that this increase is not linear, but levels out after five hybridization steps. There are two possible reasons for this behavior: the SPR response is not linear with increasing thickness, or hybridization efficiencies of each adsorption step are less than 100%.

The nonlinearity in the SPR response is quantified in Fig. 36, which

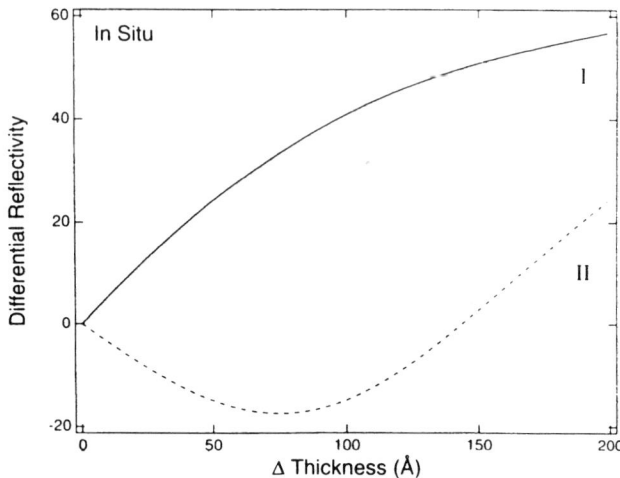

FIG. 36. Differential reflectivity vs. change in thickness of a thin film, with refractive index of 1.45, calculated in situ at two fixed angles shown as I and II in Fig. 31. The solid line is the differential reflectivity at angle I and shows a region of about 50 Å for which the differential reflectivity changes linearly with thickness before it starts to level out. The dashed line is the differential reflectivity at angle II and is an example of an angle where the differential reflectivity first decreases with increasing thickness.

plots the theoretical change in %R expected for an in situ SPR imaging experiment as a thin film with index of refraction of 1.45 adsorbs to the gold surface. For example, if an SPR imaging experiment is performed at the fixed angle labeled "I" in Fig. 31, the differential reflectivity would increase linearly over the first 50 Å and then slowly level out. In contrast, if an angle on the other side of the SPR minimum were chosen ("II" in Fig. 31), an initial decrease in the %R would be observed upon adsorption of the thin film. These two examples demonstrate that the change in differential reflectivity is highly dependent on the incident angle and so it cannot be assumed that the SPR response is linear with thickness. This problem can in principle be overcome by measuring the differential %R at multiple fixed angles. If the SPR imaging measurements are performed at multiple fixed angles such as those plotted in Figure 31, then changes in effective film thickness can be calculated more reliably during an adsorption experiment, and the SPR imaging apparatus can be used to examine films over a wider range of thicknesses.

The magnitude of the increase in %R per hybridized layer shown in Fig. 35 certainly does drop off, but is this decrease due to the inherent nonlinearity in the SPR response or to a low hybridization efficiency? After the six hybridization steps were performed, the incident angle at which the images were taken was increased. A seventh hybridization step was subsequently performed and the SPR response increased by approximately 1 %R (data not shown). The fact that this increase is larger than that seen at the original angle for the sixth hybridization indicates that the continuously smaller changes in SPR signal must be at least partially due to the nonlinear increase in the differential reflectivity with added thickness. This experiment indicates that multiple DNA hybridizations can best be characterized by measuring the SPR imaging response at multiple fixed angles.

As these examples show, the use of SPR imaging experiments for differential adsorption measurements at patterned surfaces is a natural extension of the SPR scanning measurements. SPR imaging experiments have been used to investigate the adsorption of proteins, polypeptides, and oligonucleotides onto chemically modified gold surfaces. SPR imaging is well suited for these studies because (1) adsorption onto a patterned array of several chemically different surfaces can be observed simultaneously with this technique, (2) any sample-to-sample variations in both ex situ and in situ differential adsorption measurements can be eliminated by using these patterned surfaces, and (3) quantitative adsorption information can be readily extracted from the resulting SPR images and line profiles. In the next section, in situ SPR experiments are applied to an electrochemical environment in order to measure the electrostatic fields within self-assembled multilayer films on a gold electrode surface.

V. SPR ELECTRIC FIELD MEASUREMENTS

A. Introduction

In an electrochemical environment, the optical technique of SPR has been used to probe a variety of chemical and physical processes at metal surfaces. We have recently demonstrated that a novel extension of the surface plasmon resonance technique called electrochemically modulated surface plasmon resonance (EM-SPR) can be used to monitor electrostatic fields inside organic monolayer and multilayer films at electrode surfaces [64]. The strength of the electrostatic fields within the interfacial region of a chemically modified electrode is a fundamental parameter that controls the reactivity and electrochemistry of any chemical species incorporated into the thin film [171–176]. To date, there have been a limited number of studies that have determined field strengths within monolayer films at electrode surfaces [177–182], and only a few measurements of the variations in the electric fields as a function of position in the ultrathin film have been reported [183–185]. In this section, we describe the EM-SPR technique and demonstrate how it can be used to determine the electric field strength and field profile within noncentrosymmetric self-assembled zirconium phosphonate (ZP) multilayer films at electrode surfaces.

B. EM-SPR Theory

The EM-SPR technique utilizes surface plasmons to measure the minute changes in the index of refraction of a noncentrosymmetric thin organic film that occur upon application of an external electrostatic field. The change in index of refraction (Δn) of a film upon application of an external electric field is referred to as the linear electro-optical effect [186]; this response requires a noncentrosymmetric material and is described by the electro-optic coefficient, r. The electro optic coefficient is a tensor quantity, and for a non-birefringent material the tensor element r_{33} has the general form of [96]:

$$r_{33} = \frac{2\Delta n_z}{n^3 \Delta E_z} \tag{3}$$

where n is the isotropic index of refraction of the material and Δn_z is the change in n in the z-direction created by a change in an applied field along the z-axis, ΔE_z. The value of r_{33} for ultrathin noncentrosymmetric ZP films of the molecules HAPA ([5-[4-[[4-[(6 hydroxyhexyl)sulfonyl]phenyl]azo]phenyl]pentoxy]phosphonic acid) and PY AZO ([1-[4-[4-[(N-(2-hydroxyethyl)-N methyl)amino]phenyl]azo](5-phosphonopentyl)] pyridinium bromide) (see Fig. 37a) has been determined from modulated SPR experiments on air-gap capacitors; these r_{33} measurements are described in more detail in Sec. V.D.

As in the scanning SPR measurements discussed in Sec. III, surface plasmons are created at a gold/electrolyte interface in an EM-SPR experiment by the prism coupling of a HeNe laser to a thin gold film at the SPR

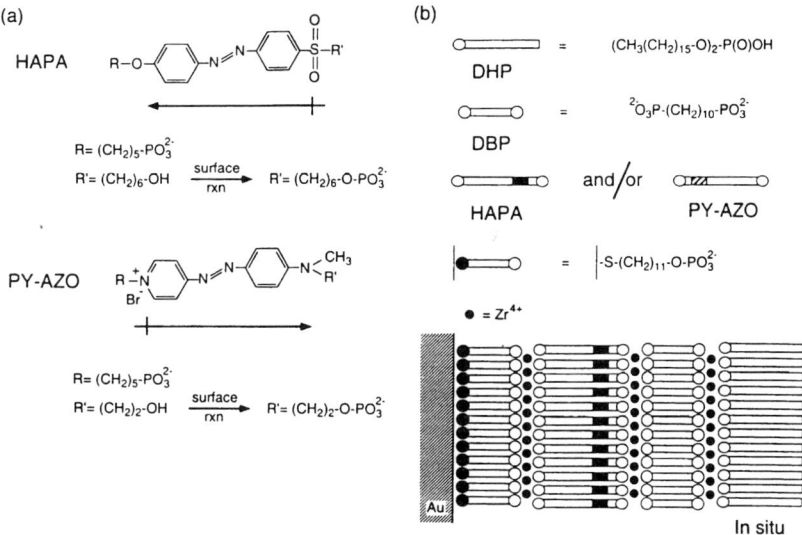

FIG. 37. (a) The structures and associated dipole moments (black arrows) for the nonlinear optical chromophores HAPA ([5 4-[[4-[(6-hydroxyhexyl)sulfonyl]phenyl]azo]phenyl]pentoxy]phosphonic acid) and PYAZO ([1-[4-[4-[(N-(2-hydroxyethyl)-Nmethyl)amino]phenyl]azo](5-phosphonopentyl)]pyridinium bromide) used in these experiments to form noncentrosymmetric zirconium phosphonate (ZP) multilayer films on gold surfaces. (b) Schematic diagram for the construction of noncentrosymmetric mixed zirconium phosphonate multilayer films. The ZP films are created on vapor-deposited gold substrates that have been primed with a self-assembled monolayer of phosphorylated MUD. Exposure of the zirconated surface to solutions of the nonlinear optical compounds HAPA and/or PY-AZO following the procedure of Katz and co-workers led to noncentrosymmetric multilayer films. For electric field profile experiments, N DBP (1,10-decane diylbis(phosphonate)) layers where N = 0–8 are self-assembled onto the HAPA monolayer. The films were then capped with a self assembled monolayer of DHP (dihexadecyl phosphate), and characterized in an electrochemical environment with EM-SPR. (From Ref. 105.)

angle (θ_{sp}). This SPR angle depends upon the thickness and index of refraction of the dielectric medium in contact with the metal surface. The EM-SPR experiment measures shifts in θ_{sp} by the changes in reflectivity ($\Delta\%R$) that occur upon potential modulation ($\Delta\phi_m$) as a function of incident angle θ. These experimental EM-SPR "differential reflectivity" or "$\Delta\%R$" curves are modeled with complex Fresnel calculations in order to relate the measured Δ%R to the change in the film's index of refraction Δn. From this Δn value, the change in electric field strength (ΔE) within the film due to potential modulation at a charged electrode can be determined by [96,187]:

$$\Delta E = \frac{2\Delta n}{n^3 r_{33}} \quad (4)$$

where n is the optical index of refraction of the film and r_{33} is the electro-optic coefficient for the noncentrosymmetric ZP film determined previously. Note that in Eq. (4) we have dropped the z subscripts on ΔE and Δn since the electric fields are always normal to the electrode surface (defined as the z-direction).

The linear electro-optical effect is a nonlinear optical mixing process similar to optical second harmonic generation (SHG) where one of the electric field components is at zero frequency [186]. The electro-optic coefficient r_{33} for the noncentrosymmetric ZP films can be related to the complex second-order surface nonlinear susceptibility $\chi_{zzz}^{(2)}(-\omega; \omega, 0)$ by [96,188–190]:

$$\chi_{zzz}^{(2)}(-\omega; \omega, 0) = -n^4 r_{33} \quad (5)$$

This surface nonlinear susceptibility has both a magnitude and a phase component associated with it. In Sec. V.E.3 the phase of the electro optical signal is used to determine the relative orientation of different nonlinear optical chromophores in mixed ZP multilayer films.

C. Noncentrosymmetric Zirconium Phosphonate Multilayer Films

As discussed in Sec. V.B, the linear electro-optical effect is nonzero only for noncentrosymmetric thin films. This requirement can be satisfied by the creation of self-assembled multilayer films based on the zirconium phosphonate (ZP) ligand chemistry first reported by Mallouk and co-workers for centrosymmetric films [159,160] and subsequently modified by Katz et al. to create noncentrosymmetric multilayers [162,191,192]. For the EM-SPR experiments described here, ultrathin (<50 nm) noncentrosymmetric organic films were formed on gold substrates by the sequential deposition of self-assembled zirconium phosphonate monolayers of the molecules HAPA or PY-AZO (see Fig. 37a). The HAPA molecule contains an azobenzene nonlinear optical (NLO) chromophore with sulfonyl electron-accepting and phenyl ether electron-donating groups, and is terminated with a phosphonic acid on one end and an alcohol functional group on the other. The PY-AZO molecule consists of a pyridinium electron-accepting group and an amino electron-donating group separated by an azodye Π-conjugated ring system. HAPA and PY-AZO were synthesized because of their large molecular hyperpolarizabilities and dipole moments. Note that the dipole moment of PY-AZO points in the opposite direction of the HAPA molecule; i.e., in PY AZO, the dipole moment points away from the phosphonate group as compared to the dipole moment in HAPA, which points towards the phosphonate end of the molecule.

Noncentrosymmetric films of HAPA or PY-AZO were formed by the method outlined by Katz et al. (see Fig. 37b), in which the NLO chromophore adsorbs as a self-assembled monolayer on a Zr^{4+} ion–coated primer layer composed of phosphorylated MUD (11 mercaptoundecanol) via the phosphonic acid group, and then the terminal alcohol groups on the other end of the molecules are converted to phosphates which can bind a new monolayer of Zr^{4+} ions for the adsorption of additional monolayers [153,162]. The in situ electric field profile measurements in Sec. V.E.2 employed mixed noncentrosymmetric ZP films that contained the centrosymmetric molecule DBP (1,10 decanediylbis(phosphonate)) and were terminated with a self-assembled monolayer of DHP (dihexadecyl phosphate). This DHP capping monolayer produced very hydrophobic gold electrode surfaces that helped decrease electrolyte penetration into the film. These films were all characterized with a combination of scanning SPR and PM-FTIRRAS as detailed in Sec. III.D prior to performing modulated SPR measurements.

D. Modulated SPR Measurements on Air-Gap Capacitors

The determination of the linear electro-optic coefficient r_{33} for a ZP/HAPA monolayer film was accomplished by the use of modulated SPR measurements on air-gap capacitors. This method has been used previously for the determination of the electro-optic coefficients of ultrathin Langmuir-Blodgett NLO organic films and micron thick NLO poled-polymer films [64,91–94,96,98–104]. In these experiments, the ultrathin NLO organic film is deposited onto a thin (47 nm) gold film that serves as one electrode of an air-gap capacitor as shown in Fig. 38. An external AC voltage (on the order of 10–50 V with a frequency of 1–10 kHz) is applied to the capacitor in order to create an oscillating electrostatic field on the order of 10^4 V/cm that modulates the index of refraction of the ultrathin NLO organic film via

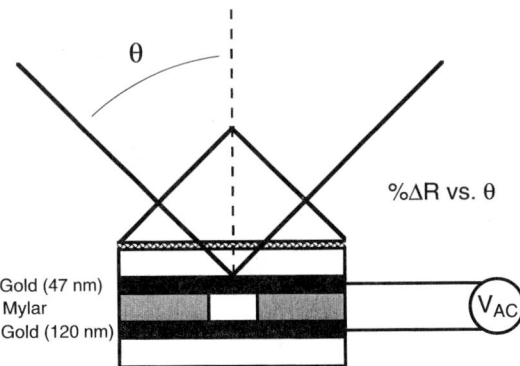

FIG. 38. The modulated surface plasmon resonance apparatus used for the determination of the electro-optical response of noncentrosymmetric ZP films. This apparatus measures the reflectivity (%R) and differential reflectivity (Δ%R) of p-polarized light at 632.8 nm (HeNe laser) from the BK7/Au/ZP film/air interface as a function of incident angle, θ. The samples for these experiments consist of a thin Mylar dielectric film which is pressed between the ZP modified gold samples and a second gold electrode. A small square is cut out of the Mylar so the surface plasmon wave is evanescent in air upon excitation. Electrical contact is made to the two gold surfaces, and a sinusoidal waveform is applied across the gap.

the electro-optical effect. This change in the index of refraction results in a shift of the SPR angle that can be observed by detecting the modulated change in reflectivity as a function of incident angle, θ.

Figure 39 plots the modulated change in reflectivity (Δ%R) observed for a single ZP/HAPA monolayer film as a function of incident angle near the surface plasmon angle ($θ_{sp}$). This signal was obtained by the application of a sinusoidal voltage ΔV = 30 V at a frequency of 10 kHz across the air-gap capacitor. The thickness of the air-gap in the capacitor in this experiment was determined from capacitance measurements to be 15 μm, and a static or zero frequency dielectric constant ε of six was assumed for the ZP/HAPA monolayer films. (This value was chosen to match the experimentally determined dielectric constant for similar NLO azodye multilayer ZP films reported by Katz et al. [193]). The conditions of this experiment corresponded to an electric field strength ΔE = ΔV/εd of 3.3 ×

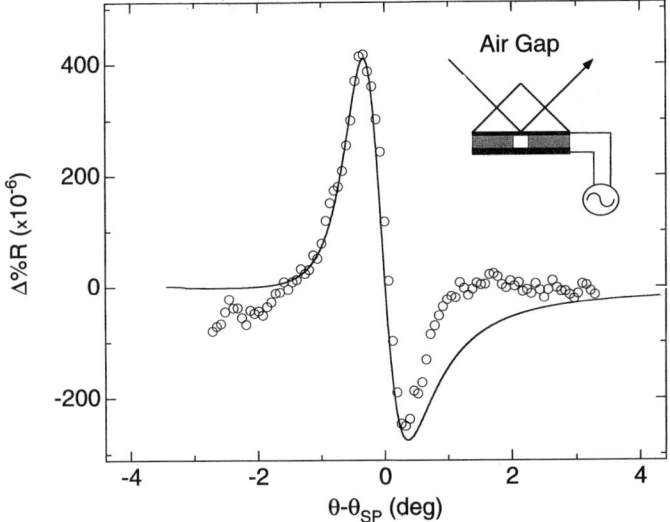

FIG. 39. Differential reflectivity (Δ%R) obtained by modulated surface plasmon resonance measurements in air for a ZP film consisting of a monolayer ZP/HAPA film formed onto a primer layer on a thin gold film. This signal which is attributed to a modulation in the dielectric constants of the HAPA monolayer through the linear electro-optical effect is obtained from application of a sinusoidal waveform (±15 V, 10 kHz) applied across the air gap (15 μm). The differential reflectivity caused by this modulation is recorded as a function of the incident angle, θ. The solid line in the figure is a complex Fresnel fit to the data obtained by varying the real component of the dielectric constant of the HAPA layer. (From Ref. 64.)

10^3 V/cm within the ZP/HAPA monolayer film. The change in reflectivity at 10 kHz was obtained by lock-in amplifier detection of the photodiode current (see Sec. III.C) and has the functional form of the derivative of the standard surface plasmon reflectivity curve. The magnitude of the differential reflectivity (Δ%R) increased linearly with the applied AC voltage for the HAPA sample as expected and was constant over the frequency range of 1–10 kHz. In addition, no signal was observed in the absence of the ZP/HAPA monolayer.

These experimental modulated SPR differential reflectivity curves can be modeled with complex Fresnel calculations in order to relate the measured Δ%R to the change in the film's index of refraction Δn. An example of how Fresnel calculations can be used to relate Δ%R and Δn is depicted in Fig. 40. Figure 40a plots two surface plasmon theory curves for two thin films that differ in index of refraction. The difference between these two curves produces the differential reflectivity curve shown in Fig. 40b. This theoretical Δ%R curve is used to fit the experimental air-gap SPR data and results in the determination of Δn for a given ΔE. The solid line in Fig. 39 is a five-phase Fresnel calculation in which the real part of the index of refraction for the HAPA monolayer is varied by $\Delta n = 8.0 \times 10^{-6}$. (Only the real component of the refractive index was considered because the HAPA monolayer does not absorb at the wavelength employed in this ex-

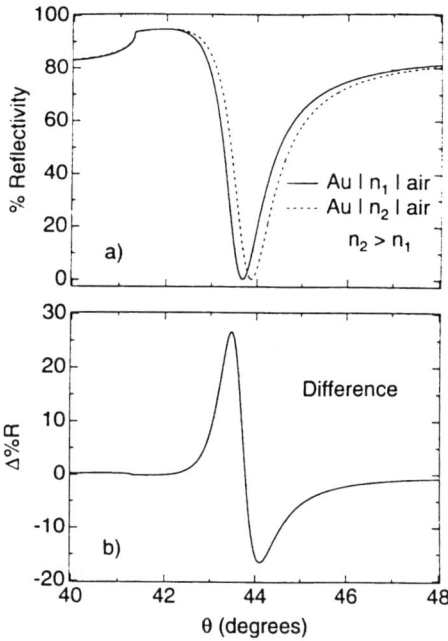

FIG. 40. Method for modeling the measured air-gap modulated SPR differential reflectivity curves. (a) Two surface plasmon curves generated from Fresnel calculations which differ in the film's index of refraction. (b) The differential reflectivity curve produced from the difference between the two curves in (a). This theoretical Δ%R curve is used to fit the experimental modulated SPR data and results in the determination of Δn for a given ΔE. (From Ref. 105.)

periment.) A similar change in the SPR reflectivity can be obtained by a change in the thickness of the film (Δd) with applied voltage but we assume that all changes are due to index of refraction effects via the electro-optic coefficient. Using Eq. (3), a change in Δn of 8.0×10^{-6} corresponds to an electro optic coefficient r_{33} of 11 pm/V at 632.8 nm. This r_{33} is an off-resonance value since the monolayer film does not absorb light at this wavelength. From the value of r_{33}, the surface nonlinear susceptibility $\chi_{zzz}^{(2)}$ ($-\omega; \omega, 0$) for the ZP film is calculated from Eq. (5) to be 41 pm/V (9.7×10^{-8} esu). Using this $\chi_{zzz}^{(2)}$, the HAPA molecular hyperpolarizability b_{zzz} is estimated to be 9.3×10^{-29} esu at 632.8 nm, assuming a chromophore tilt angle of 27° as obtained from second harmonic generation (SHG) experiments and a number density for the HAPA molecules in the ZP film of 10^{21} molecules/cm^3 [194]. The magnitude of the electro-optic coefficient in these ZP films is less than that for LiNbO$_3$ (31 pm/V), but is within the range of the values observed in NLO poled polymer films (1–55 pm/V) [96,188–190,195–198]. Moreover, the noncentrosymmetric character of the HAPA monolayer did not decrease with time, as is often the case for the polymer systems. Similar air-gap capacitor modulated SPR experiments were performed on a monolayer PY-AZO film, and an electro-optic coefficient of –8 pm/V at 632.8 nm was obtained. The negative sign indicates that the relative electro-optical response of the PY-AZO monolayer is opposite to that observed for the HAPA monolayer, as expected from the difference in the direction of the dipole moment for the two chromophores [106].

E. EM-SPR Measurements on Multilayer Films at Electrode Surfaces

1. Electric Field Measurements

The SPR modulation experiments used with the air-gap capacitors can also be applied to in situ electrochemical environments to monitor the electrostatic fields within ultrathin organic films via the electro-optical effect. In this electrochemically modulated surface plasmon resonance (EM-SPR) experiment, the electrochemical potential of the gold electrode is modulated sinusoidally with a magnitude $\Delta\phi_m$. The EM-SPR experiment measures the change in reflectivity that occurs due to this potential modulation as a function of incident angle θ. As in the air-gap SPR experiments, these experimental EM-SPR differential reflectivity ($\Delta\%R$) curves can be modeled with complex Fresnel calculations in order to determine the change in the film's index of refraction (Δn) for a given $\Delta\phi_m$. Using this Δn value and the value of r_{33} obtained from the SPR air-gap experiments, the change in electric field strength (ΔE) within the film due to the potential modulation can be determined from Eq. (4).

A schematic diagram of the electrochemical cell used in the EM-SPR experiments is shown in the inset of Fig. 41. This cell consisted of a simple teflon body pressed against the modified gold electrode with an o-ring seal. A three-electrode potentiostat with the thin gold film as the working electrode was used to apply the potential modulation. A cyclic voltammogram (CV) of a gold electrode coated with a mixed HAPA/DBP multilayer film of thickness 13 ± 1 nm (as determined from in situ SPR scanning measurements) is shown in Fig. 41. The cyclic voltammetry ex-

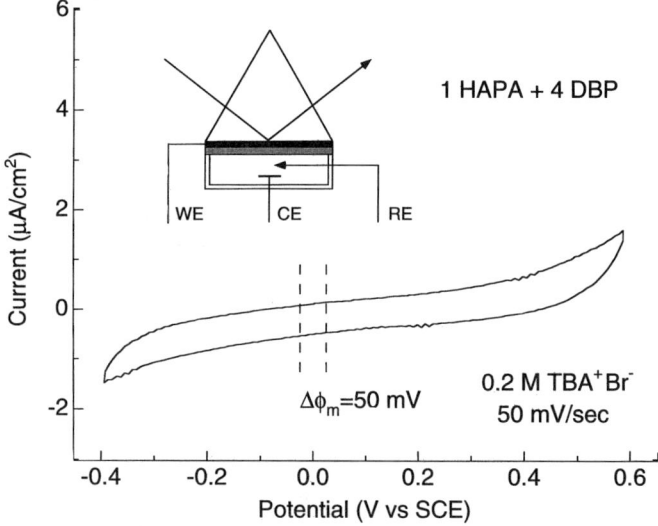

FIG. 41. Cyclic voltammogram for a 13 ± 1 nm thick ZP multilayer film that consisted of a phosphorylated MUD primer layer, one HAPA monolayer, four DBP monolayers, and one DHP capping monolayer on a vapor-deposited gold substrate. This sample is referred to as the "1 HAPA + 4 DBP" ZP film. A three-electrode assembly was formed with the ZP coated Au film as the working electrode, a saturated calomel reference electrode, and a platinum counter electrode. The CV was obtained in the region from –0.4 to 0.6 V vs. SCE in 0.2 M tetrabutylammonium bromide at a scan rate of 50 mV/sec. For EM-SPR, the electrode potential was fixed to 0.0 V vs. SCE and a small sinusoidal modulation ($\Delta\phi_m$ = 50–760 mV, at 1 kHz) was applied to the cell. (From Ref. 105.)

hibited no faradaic processes over a wide potential range from −0.4 to 0.6 V vs. SCE. For the EM-SPR experiments, the electrode potential was set to 0.0 V vs. SCE, and a sinusoidal waveform $\Delta\phi_m$ of 50–760 mV was applied at a frequency of 1 kHz. No changes in the measured EM-SPR response were observed for modulation frequencies from 0.2 to 1.5 kHz.

Typical in situ EM-SPR differential reflectivity curves obtained from a potential modulation of 50 mV are plotted in Fig. 42. Two $\Delta\%R$ curves are shown in the figure: the open squares are data from a 6.7 nm ZP/HAPA

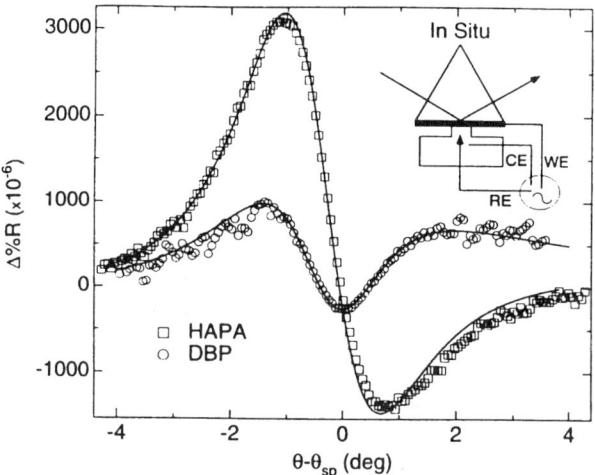

FIG. 42. Differential reflectivity ($\Delta\%R$) obtained by EM-SPR measurements made in situ. The sample for this experiment shown in the inset of the figure consists of a teflon cell which is pressed against the ZP modified gold sample. A three-electrode assembly is made with the ZP modified gold sample as the working electrode. A potentiostat controlled the applied potential relative to SCE. A small sinusoidal waveform (50 mV, 1 kHz) was then applied to the cell at a potential where no electrochemistry took place. The open circles are obtained for a ZP film consisting of two monolayers of DBP, and one monolayer of DHP formed onto a prime layer on a thin gold film. This signal results from a modulation in the gold dielectric constants. The open squares are obtained for a ZP film consisting of one monolayer of HAPA, and one monolayer of DHP formed onto a primer layer on a thin gold film. The solid lines for both sets of data are complex Fresnel fits obtained by varying the real component of the dielectric constant of the HAPA layer and the gold. (From Ref. 64.)

film consisting of a single HAPA monolayer film formed on a phosphorylated MUD primer monolayer and capped with a DHP monolayer (see Fig. 37b), and the open circles are the data obtained for a centrosymmetric ZP/DBP multilayer consisting of a primer monolayer of phosphorylated MUD, followed by two layers of DBP and one capping layer of DHP. The modulation in the SPR signal from the centrosymmetric film results solely from the electro-optical contributions of the gold electrode substrate. This effect was not observed in the air-gap experiment to within the signal-to-noise ratio of the experiment, but changes in the optical constants of metal electrode surfaces have been observed previously in electroreflectance measurements [17,68]. The solid line that fits these data is a seven-phase Fresnel calculation which assumes that the real part of the index of refraction of the first 0.05 nm of the gold electrode surface changes upon modulation of the electrode potential by an amount $\Delta n = -1.6 \times 10^{-3}$. The thickness of this layer is taken as 0.05 nm in order to approximate the Thomas-Fermi screening depth of the metal [17]. Increasing this thickness reduced the size of Δn, but had no effect on the shape of the calculated fit. Note that for these data, the minimum in $\Delta \%R$ occurs at the surface plasmon angle θ_{sp}.

In contrast, the larger differential reflectivity curve in Fig. 42 obtained from the *noncentrosymmetric* HAPA film exhibits a minimum and a maximum in $\Delta \%R$ above and below θ_{sp}, respectively, as observed previously in the HAPA air-gap capacitor experiments. The solid line that fits the data is a seven-phase Fresnel fit taking both the modulation of the index of refraction of the HAPA monolayer and the modulation of the index of refraction of the gold surface into account. This theoretical fit yields a value for the change in index of refraction for the HAPA monolayer of $\Delta n = 3.3 \times 10^{-5}$.

Using the value of r_{33} determined for a HAPA monolayer in Sec. V.D, the experimentally determined Δn for the ZP/HAPA film can be related to the change in electric field strength (ΔE) within the multilayer films at the electrode surface. The value for Δn of the HAPA monolayer film is a factor of four greater than that measured in the air-gap experiment, and therefore corresponds to a change in electric field strength during potential modulation of 1.4×10^4 V/cm for $\Delta \phi_m = 50$ mV (2.8×10^5 V/cm for $\Delta \phi_m = 1$ V) within the ultrathin organic film.

A second set of EM-SPR experiments is shown in Fig. 43, which plots the differential reflectivity curves observed from the "1 HAPA + 4 DBP" mixed multilayer for a series of different modulation potentials. The

Surface Plasmon Resonance Methods

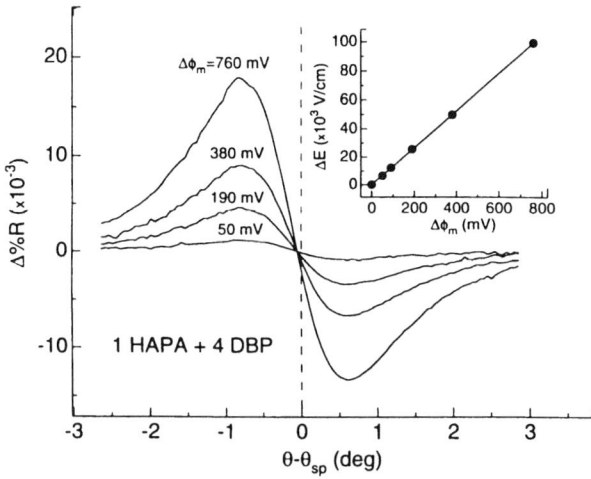

FIG. 43. Differential reflectivity (Δ%R) obtained by EM-SPR measurements in 0.2 M tetrabutylammonium bromide for the 1 HAPA + 4 DBP ZP film. The curves are the overlay of increasing electrode modulation potentials ($\Delta\phi_m$) of 50, 190, 380, 760 mV about 0.0 V vs. SCE, at 1 kHz, respectively. The differential reflectivities (Δ%R) measured in situ are converted to a change in electric field strength (ΔE) within the self-assembled multilayer film by comparison to air-gap electro-optical experiments. The change in the electric field strengths within the multilayer ZP film shown in the inset graph are surprisingly linear with the applied modulation potential in the experimentally measured electrochemical potential window from –0.4 to 0.6 V vs. SCE. (From Ref. 105.)

shape of the Δ%R curves does not change with increasing modulation voltage and is similar to the HAPA monolayer film. The Fresnel calculations predict a zero-crossing point of $\theta-\theta_{sp} = -0.06°$ for all of the Δ%R curves; this prediction agrees with the experimental data in the figure. From the Fresnel fits of the data, values for Δn are obtained and range from Δn = 1.6 × 10^{-5} for $\Delta\phi_m$ = 50 mV to Δn = 2.4 × 10^{-4} for $\Delta\phi_m$ = 760 mV. The ΔE values obtained from these Δn values by Eq. (4) are plotted as a function of $\Delta\phi_m$ in the inset graph of Fig. 43. The change in field strength is found to be surprisingly linear for applied modulation potentials from 0 to 760 mV and corresponds to a ΔE of 6.5 × 10^3 V/cm for a $\Delta\phi_m$ of 50 mV (1.3 × 10^5 V/cm per volt modulation). These values for ΔE within the 13 ± 1 nm multilayer ZP film are consistent with the field strengths determined by Pope and Buttry inside a monolayer film at a silver electrode from fluorescent shift measurements [179,180].

We have also recently examined the EM-SPR response of ZP films incorporating the noncentrosymmetric molecule PY-AZO (pictured in Fig. 37a) [106]. The EM-SPR differential reflectivity ($\Delta\%R$) signal observed from a chemically modified gold sample that consisted of phosphorylated MUD, three PY-AZO monolayers, and one DHP monolayer is shown in Fig. 44. The series of curves labeled a–e are the overlay of increasing modulation potentials ($\Delta\phi_m$ = 50–200 mV). As with the ZP/HAPA films, the electro-optical signals were linear with the applied modulation potential. The shape of the PY-AZO EM-SPR differential reflectivity curves was constant with increasing modulation voltage, and the minimum and maximum in $\Delta\%R$ occur to the left and right of θ_{sp}, respectively. This is the reverse of that measured for the ZP/HAPA samples and is due to the opposite

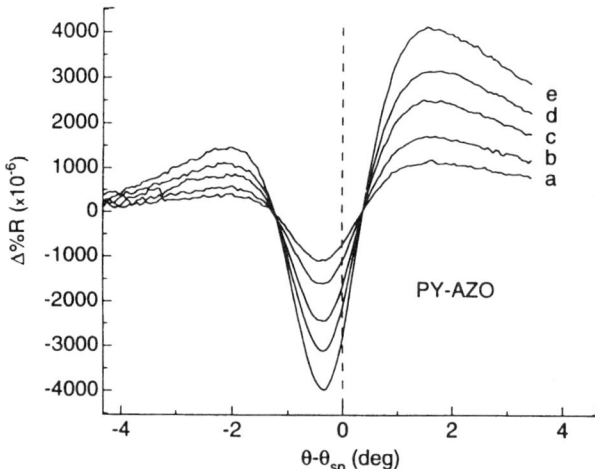

FIG. 44. Differential reflectivity ($\Delta\%R$) obtained by electrochemically modulated surface plasmon resonance measurements in 0.2 M tetrabutylammonium bromide for a three-layer ZP/PY-AZO sample. The curves labeled a–e are the overlay of increasing electrode modulation potentials ($\Delta\phi_m$) of 50, 80, 120, 160, 200 mV about 0.0 V vs. SCE, at 1 kHz, respectively. The measured differential reflectivity is a result of the electro-optical effect. (From Ref. 106.)

sign of r_{33} for these two films. Although the Δ%R curves indicate that the phase of $\chi^{(2)}(-\omega, \omega, 0)$ for the PY-AZO films is opposite that of HAPA films, the shape of the differential reflectivity waveforms is not exactly inverted due to the contributions from the gold substrate. Note also that the curves in Fig. 44 contain two zero-crossing points, one on each side of θ_{sp}. The Δ%R curves for ZP/HAPA films contain only one zero crossing point of the left of θ_{sp}; this difference is due to the substrate contributions to the differential reflectivity.

The EM-SPR response from the ZP/PY-AZO multilayer films was again modeled with theoretical Fresnel calculations. Figure 45 plots a seven phase Fresnel calculation of the differential reflectivity curve for a ZP/PY-AZO monolayer film where the underlying surface gold layer is taken into account. Note that the shape of the theoretical curves very accurately fit the experimental data. From these calculations, the change in index of refraction for the PY-AZO monolayer was determined to be Δn =

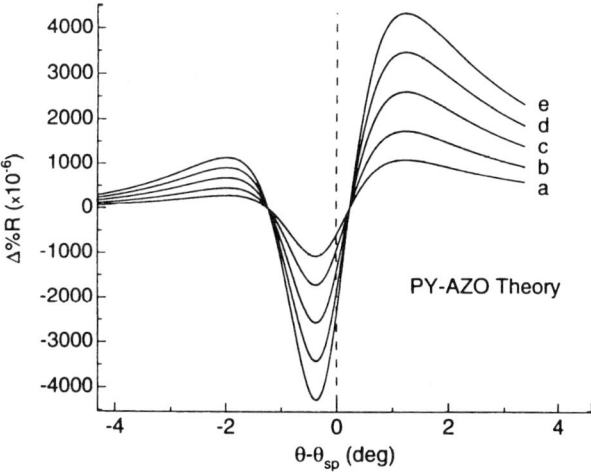

FIG. 45. Theoretical Fresnel differential reflectivity (Δ%R) calculations for a ZP/PY-AZO sample. The curves labeled a–e are the difference of two surface plasmon curves which differed by increasing values of Δn_{Au} = (−1.4, −2.3, −3.4, −4.5, −5.6) × 10^{-3}, respectively, in the real component of the index of refraction of the surface gold layer (0.05 nm) and Δn = (−1.4, −2.3, −3.4, −4.5, −5.6) × 10^{-5}, respectively, for the ZP/PY-AZO film. (From Ref. 106.)

−1.4 × 10^{-5} for $\Delta\phi_m$ = 50 mV. This value of Δn is less than half that measured for a ZP/HAPA monolayer film. Air-gap electro-optical experiments indicate that the signal levels for PY-AZO and HAPA films should be approximately equivalent in magnitude (−8, 11 pm/V, respectively). The lower than expected Δn for the PY AZO sample implies that there is a decrease in the electrostatic fields inside the PY-AZO film as compared to the HAPA film due to more ion and solvent penetration into the multilayer. This result is reasonable given that, unlike HAPA, the PY-AZO species is charged and the film must contain additional counterions.

2. Electric Field Profile Measurements Inside Multilayers

In a series of experiments on mixed HAPA/DBP multilayers, EM-SPR measurements were used to map out the electric field profile within ultrathin films at electrode surfaces [105]. In these experiments, a single HAPA monolayer was used to probe the electric field at a specific point within the ZP film. A first set of EM-SPR experiments was performed on a series of five HAPA/DBP monolayers that consisted of a phosphorylated MUD primer monolayer, one HAPA monolayer, N DBP monolayers (where N varied from 0 to 8), and a DHP capping monolayer. In each of these films, the HAPA monolayer was the first ZP monolayer deposited onto the electrode surface. The mixed multilayer films were characterized prior to the EM-SPR experiments with PM-FTIRRAS and scanning in situ SPR measurements (see Sec. III.D.3). These measurements indicate that the thicknesses of the ZP films increase by 1.6 ± 0.2 nm per DBP monolayer (total film thicknesses from 6.7 to 20 nm for 0–8 DBP overlayers) with no observable changes in HAPA film structure and packing density. This DBP monolayer thickness is equivalent to that measured ex situ for multilayer DBP films in Sec. III.D.2. These results indicate that stable, reproducible mixed multilayers of HAPA and DBP are formed on the gold electrode surface. The change in electric field strength at the HAPA monolayer position for a $\Delta\phi_m$ of 50 mV was determined from EM-SPR differential reflectivity measurements on each of the films; the results of these experiments are plotted in Fig. 46. A decrease in ΔE of up to 65% is observed as the number of centrosymmetric DBP monolayers is increased.

This decrease in the measured EM-SPR signal is attributed to the increase in overall thickness (d) of the ZP film due to the formation of DBP overlayers and has the functional form of $1/d$. A simple Helmholtz model of the electrochemical double layer also predicts that the electric field strength should decrease with film thickness as $1/d$ assuming a fixed static

Surface Plasmon Resonance Methods

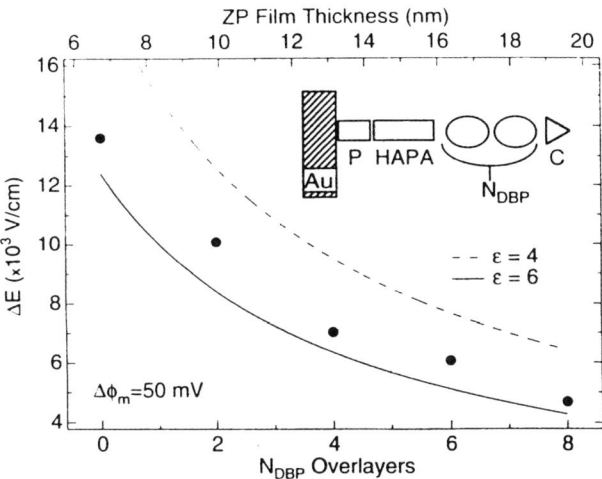

FIG. 46. Change in electric field strength (ΔE) within a noncentrosymmetric ZP/HAPA film for increasing numbers of DBP overlayers (N_{DBP} = 0–8). The ΔE values are determined via the EM-SPR differential reflectivity curves for a HAPA monolayer whose position was fixed in the film (see inset). The positions labeled P and C in the inset correspond to the phosphorylated MUD primer layer and the DHP capping monolayer, respectively. The solid and dashed lines are fits to the experimental data using a simple Helmholtz model. (From Ref. 105.)

dielectric constant ε [171]. For pure ZP films of DBP and HAPA-like chromophores, the static dielectric constants have been determined previously to be four and six, respectively [193,199]. The solid line in Fig. 46 is a theory curve generated from the Helmholtz model using the ZP film thicknesses determined from in situ and ex situ SPR measurements and a static dielectric constant of six. The dashed line in the figure is the same theory curve except using a value of 4 for ε. Although the experimentally determined ΔE values parallel the Helmholtz model using an ε of 6, the effective dielectric constant for these mixed ZP films should vary from 6 to 4 as more DBP overlayers are assembled.

Deviations from a simple Helmholtz picture for the measured ΔE values inside the ZP multilayers are even more apparent in a second set of experiments. In these measurements, a series of mixed HAPA/DBP films were prepared in which the total thickness was held constant and the position of the HAPA monolayer was varied. These films consisted of a phosphorylated MUD prime layer, one HAPA monolayer, six DBP monolayers, and a DHP capping monolayer ("1 HAPA + 6 DBP" ZP films). As depicted in the inset of Fig. 47, the position of the HAPA monolayer was varied from closest to the electrode surface (N = 1) to furthest away from the surface (N = 7). These mixed ZP films were characterized prior to the EM-SPR experiments with PM-FTIRRAS and scanning in situ SPR measurements; no significant changes in film structure, molecular orientation, or overall film thickness (17 ± 1 nm) were observed as the HAPA monolayer position was varied [105]. Figure 47 plots the change observed in the electric field strength (ΔE) obtained from the differential reflectivity curves for a potential modulation $\Delta \phi_m$ of 50 mV as a function of the position of HAPA in a sample (N). As the HAPA layer was placed further away from the electrode surface, a linear decrease in ΔE was ob-

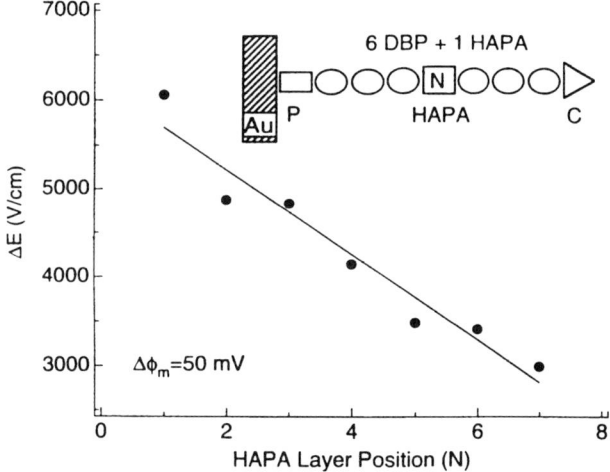

FIG. 47. The variation in the change in electric field strength (ΔE) within a 17 ± 1 nm thick multilayer ZP film that consisted of a phosphorylated MUD primer monolayer, one HAPA monolayer, six DBP monolayers, and one DHP capping monolayer. These ΔE values are obtained via changes in the EM-SPR reflectivity curves by systematically varying the position of the HAPA monolayer within the multilayer ZP film (see inset). The linear decrease in ΔE is attributed to ion and solvent penetration into the multilayer film, and implies a quadratic decrease with thickness of the local potential ϕ inside the ZP film. (From Ref. 105.)

Surface Plasmon Resonance Methods

served with a maximum loss of 50% at the outermost (N = 7) position. This decrease is not expected from a simple parallel-plate capacitor model of the interface, which would predict a constant electric field inside the film [171].

A number of possibilities exist to account for the observed decrease in ΔE: (1) the orientation or tilt angle of the HAPA chromophores is changing with position in the multilayer film; (2) changes in the HAPA layer packing density as a function of position in the mixed ZP films are taking place; (3) loss of noncentrosymmetry in the ZP film is due to HAPA intercalation and randomization into the DBP portions of the multilayer; (4) ion and solvent penetration occurs into the ZP multilayer resulting in a decrease in the electric field strength within the film. The PM-FTIRRAS and in situ SPR characterization measurements made on these mixed ZP films eliminate the first three possible sources to the measured decrease in ΔE; all of the data show that during the ZP film deposition process no significant changes in film uniformity, packing density, molecular orientation, or overall film thickness take place for the various multilayer films. This suggests that the experimentally measured linear decrease in ΔE is a result of ion and solvent penetration into the ZP film in the electrochemical environment. A linear decrease in ΔE implies a quadratic decrease with thickness of the local potential ϕ inside the ZP film; i.e., redox species incorporated into this ZP film at the N = 7 position would only experience 25% of the applied potential ϕ_m. The effect of the local potential on the electrochemical response of ultrathin organic films at electrode surfaces has been discussed previously [171–176,178,200,201], and the EM-SPR measurements described in this chapter provide a method for quantitation of the local electrostatic potential within these films.

3. EM-SPR Interference Effects Within Mixed HAPA/PY-AZO Multilayers

A final set of EM-SPR experiments demonstrates how interference effects between two different noncentrosymmetric ZP monolayers can be measured in order to establish the relative directional order of chromophores in mixed multilayers at electrode surfaces. These experiments incorporate mixed ZP films composed of the nonlinear optical chromophores HAPA and PY-AZO. In previous sections the EM-SPR differential reflectivity curves of ZP multilayers of these two chromophores were shown to be opposite in shape, and the electro-optic coefficients r_{33} for PY-AZO and

HAPA were determined to be opposite in sign. These values reflect the difference in the direction of the dipole moment of the PY-AZO molecule compared to that of the HAPA molecule as discussed in Sec. V.5.C.

As mentioned previously, the electro-optical response r from the interface can also be described as the complex surface nonlinear susceptibility $\chi^{(2)}(-\omega, \omega, 0)$. In addition to a magnitude, $\chi^{(2)}(-\omega, \omega, 0)$ also has a phase component associated with it. The phase of $\chi^{(2)}$ has been used previously in SHG experiments to ascertain the absolute orientation of a monolayer film [10]. For example, interference effects have been observed in the SHG from mixed monolayers of two similar chromophores pointing in opposite directions at a liquid/liquid interface [202]. This same type of cancellation effect should also be observable in EM-SPR experiments on mixed ZP films containing both HAPA and PY-AZO monolayers.

To demonstrate this effect, EM-SPR measurements were performed on a set of three mixed HAPA/PY-AZO multilayer films: (1) a ZP film with two PY-AZO monolayers and one HAPA monolayer, (2) a ZP film with one PY AZO monolayer and two HAPA monolayers, and (3) a ZP film with one PY AZO and one HAPA monolayer. Each of these ZP multilayers contained a phosphorylated MUD primer layer and a DHP capping monolayer; each of these ZP films was characterized with PM-FTIRRAS prior to the EM-SPR measurements. The PM-FTIRRAS spectra indicated that the HAPA and PY AZO monolayers in the mixed films had molecular structures and packing densities equivalent to those observed in pure films of the two chromophores. The EM-SPR differential reflectivity curves for these three films are shown in Figures 48 and 49. The $\Delta\%R$ curve for the "1 PY-AZO + 2 HAPA" ZP film in Fig. 48a has a waveform similar to the $\Delta\%R$ curve for a single HAPA monolayer, whereas the $\Delta\%R$ curve for the "2 PY-AZO + 1 HAPA" sample in Fig. 48b has a waveform similar to the $\Delta\%R$ curve for a single PY-AZO monolayer. The difference in the waveform for these two samples is due to the interference between the electro-optical response of the HAPA and PY AZO monolayers in the ZP films. Figure 49 plots the $\Delta\%R$ curve (solid line) for the "1 PY-AZO + 1 HAPA" ZP film and shows a complete cancellation of the electro optical response of the ZP multilayer. The residual EM-SPR differential reflectivity signal observed from this sample is due to contributions from the metal substrate and is consistent with the $\Delta\%R$ curves obtained from a centrosymmetric DBP ZP film of similar thickness (dashed line in Fig. 49).

These measurements are the first observation of interference effects

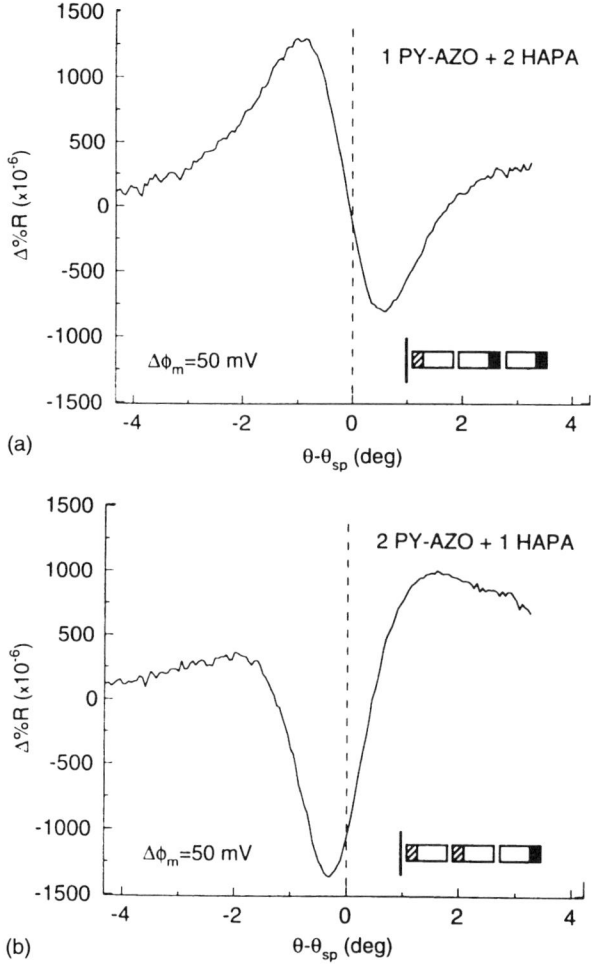

FIG. 48. Differential reflectivity ($\Delta\%R$) obtained by EM-SPR measurements for (a) a ZP film with one PY-AZO monolayer and two HAPA monolayers, 1 PY-AZO + 2 HAPA, and (b) a ZP film with two PY-AZO monolayers and one HAPA monolayer, 2 PY-AZO + 1 HAPA. Each of these ZP multilayer films contained a phosphorylated MUD primer layer and a DHP capping monolayer. The curve in (a) has the waveform of a HAPA sample, while (b) has the waveform of a PY-AZO sample. The difference in the waveform for these two samples is due to the interference between the electro-optical responses of the HAPA and PY-AZO monolayers in the ZP films. (From Ref. 105.)

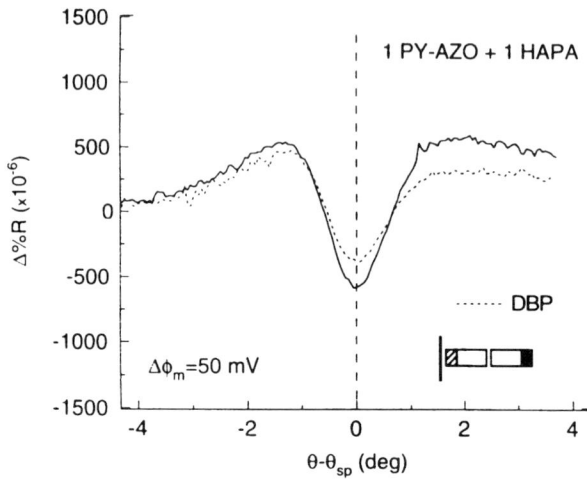

FIG. 49. Differential reflectivity (Δ%R) obtained by EM-SPR measurements for a ZP film with one PY-AZO and one HAPA monolayer, 1 PY-AZO + 1 HAPA (solid line). This ZP multilayer film also contained a phosphorylated MUD primer layer and a DHP capping monolayer. The solid curve shows a complete cancellation of the electro-optical response of the ZP multilayer. The residual EM-SPR differential reflectivity signal observed from this sample is due to contributions from the metal substrate. This substrate response is the same as that observed from a centrosymmetric DBP ZP film of similar thickness (dashed line). (From Ref. 105.)

in EM-SPR measurements on self-assembled noncentrosymmetric films and demonstrate unequivocally that mixed ZP multilayers with different oriented chromophore monolayers can be created. The amount of interference observed from the mixed ZP multilayers also depends upon the electric field profile through the ultrathin film. For example, a less pronounced interference effect was observed from a "1 HAPA + 1 PY-AZO" ZP film as compared to the "1 PY-AZO + 1 HAPA" ZP film shown in Fig. 49. Differences between these two films are due to the fact that the second ZP monolayer in these samples experiences a slightly lower local electric field strength (as expected from the results of Sec. V.E.2), which in the case of HAPA and PY-AZO happens to cancel out the slight difference in the magnitude of r_{33} for the two different chromophores.

VI. FUTURE DIRECTIONS

As demonstrated by the examples in the chapter, the SPR scanning and imaging experiments are two extremely sensitive methods for characterizing the thickness of ultrathin films on metal surfaces. When used in conjunction with other analytical techniques such as polarization modulation Fourier transform infrared reflection absorption spectroscopy, detailed information about the assembly, structure, and reactivity of adsorbed monolayer can be ascertained. For electrochemical systems, the EM-SPR measurement is a very useful tool for the elucidation of the electrostatic fields inside noncentrosymmetric films at electrode surfaces.

In general, the availability of the commercial BIACORE SPR instrument since the early 1990s has led to a large increase in the use of the SPR technique in sensor applications. In the future, SPR sensor systems based on multiple wavelength excitation and fiber-optic probes may extend this technique to the on-line monitoring of catalytic processes and other chemical reactions. In an electrochemical environment, SPR fiber-optic probes combined with techniques such as anodic stripping voltammetry have been used to detect trace metal ions with a suggested eventual detection limit in the parts per billion range [203].

Although the major use of the SPR reflectivity technique is in the area of chemical and biochemical sensor devices, a number of other possible applications exist. For example, electro-optical light modulators based on SPR methods have been proposed [102–104]. In addition, multiwavelength SPR instruments or Fourier transform SPR instruments may soon compete with commercial ellipsometers to provide simultaneous determination of the thickness and index of refraction of films deposited on metal surfaces. In the future, the application of SPR methods to other spectroscopic measurements (e.g., SHG, fluorescence, Raman scattering) appears to be a very promising area of research.

ACKNOWLEDGMENTS

The authors gratefully acknowledge the support of the National Science Foundation in these studies and would like to thank Drs. R. Georgiadis, W. Knoll, D. Kolb, and K. Krischer for preprints and reprints of their SPR papers. The authors also acknowledge T. Frutos and A. Thiel for the oligonucleotide hybridization data in Fig. 35.

REFERENCES

1. A. E. Dowrey and C. Marcott, Appl. Spectrosc. 36:414 (1982).
2. M. J. Green, B. J. Barner, and R. M. Corn, Rev. Sci. Instrum. 62:1426 (1991).
3. B. J. Barner, M. J. Green, E. I. Saez, and R. M. Corn, Analytical Chemistry 63:55 (1991).
4. R. V. Duevel and R. M. Corn, Anal. Chem. 64:337 (1992).
5. B. L. Frey, D. G. Hanken, and R. M. Corn, Langmuir 9:1815 (1993).
6. C. E. Jordan, B. L. Frey, S. Kornguth, and R. M. Corn, Langmuir 10:3642 (1994).
7. B. Beden and C. Lamy, in *Spectroelectrochemistry; Theory and Practice* (R. J. Gale, ed.), Plenum Press, New York, 1988.
8. B. Beden, in *Spectroscopic and Diffraction Techniques in Interfacial Electrochemistry* (C. Gutierrez and C. Melendres, eds.), Kluwer Academic Publishers, Dordrecht, The Netherlands, 1990, p. 103.
9. S. M. Stole, D. D. Popenoe, and M. D. Porter, in *Electrochemical Interfaces: Modern Techniques for In-Situ Interface Characterization* (H. D. Abruna, ed.), VCH, New York, 1991, ch. 7.
10. R. M. Corn and D. A. Higgins, Chem. Rev. 94:107 (1994).
11. R. M. Corn and D. A. Higgins, *Characterization of Organic Thin Films*, Butterworth-Heineman, 1995.
12. Y. R. Shen, in *Spectroscopic and Diffraction Techniques in Interfacial Electrochemistry* (C. Gutierrez and C. Melendres, eds.), Kluwer Academic Publishers, Dordrecht, The Netherlands, 1990, p. 281–311.
13. C. D. Bain, J. Chem. Soc. Faraday Trans. 91:1281 (1995).
14. P. Guyot-Sionnest, R. Superfine, J. H. Hunt, and Y. R. Shen, Chem. Phys. Lett. 144:1 (1988).
15. T. H. Ong, R. N. Ward, P. B. Davies, and C. D. Bain, J. Am. Chem. Soc. 114:6243 (1992).
16. T. H. Ong, P. B. Davies, and C. D. Bain, Langmuir 9:1836 (1993).
17. D. M. Kolb, in *Spectroelectrochemistry: Theory and Practice* (R. J. Gale, ed.), Plenum Press, New York, 1988, ch. 4.
18. W. Plieth, in *Spectroscopic and Diffraction Techniques in Interfacial Electrochemistry* (C. Gutierrez and C. Melendres, eds.), Kluwer Academic Publishers, Dordrecht, The Netherlands, 1990, p. 223.
19. J. Pemberton, in *Electrochemical Interfaces: Modern Techniques for In-Situ Interface Characterization* (H. D. Abruna, ed.), VCH, New York, 1991, ch. 5.
20. R. K. Chang, in *Spectroscopic and Diffraction Techniques in Interfacial Electrochemistry* (C. Gutierrez and C. Melendres, eds.), Kluwer Academic Publishers, Dordrecht, The Netherlands, 1990, p. 155.
21. R. L. Birke and J. R. Lombardi, in *Spectroelectrochemistry: Theory and Practice* (R. J. Gale, ed.), Plenum Press, New York, 1988, ch. 6.

22. E. Burstein, W. P. Chen, Y. J. Chen, and A. Hartstein, J. Vac. Sci. Technol. *11*:1004 (1974).
23. H. Raether, in *Physics of Thin Films*, Vol. 9, Academic Press, New York, 1977, p. 145.
24. V. M. Agranovich and D. L. Mills, eds., *Surface Polaritons: Electromagnetic Waves at Surfaces and Interfaces*, North-Holland, Amsterdam, 1982.
25. H. Knobloch, C. Duschl, and W. Knoll, J. Chem. Phys. *91*:3810 (1989).
26. H. Knobloch, H. Brunner, A. Leitner, F. Aussenegg, and W. Knoll, J. Chem. Phys. *98*:10093 (1993).
27. H. Kano and S. Kawata, Optics Lett. *21*:1848 (1996).
28. S. Byahut and T. E. Furtak, Rev. Sci. Instrum. *61*:27–32 (1990).
29. B. Pettinger, A. Tadjeddine, and D. B. Kolb, Chem. Phys. Lett. *66*:544 (1979).
30. A. Girlando, M. R. Philpott, D. Heitmann, J. D. Swalen, and R. Santo, J. Chem. Phys. *72*:5187 (1980).
31. W. Knoll, M. R. Philpott, J. D. Swalen, and A. Girlando, J. Chem. Phys. *77*:2254–2260 (1982).
32. S. Ushioda and R. Loudon, in *Surface Polaritons: Electromagnetic Waves at Surfaces and Interfaces* (V. M. Agranovich and D. L. Mills, eds.), North-Holland, Amsterdam, 1982, p. 535.
33. S. Ushioda and Y. Sasaki, Phys. Rev. B *27*:1401 (1983).
34. R. M. Corn and M. R. Philpott, J. Chem. Phys. *80*:5245 (1984).
35. C. Duschl and W. Knoll, J. Chem. Phys. *88*:4062 (1988).
36. B. Rothenhäusler, C. Duschl, and W. Knoll, Thin Solid Films *159*:323 (1988).
37. J. Giergiel, C. E. Reed, J. C. Hemminger, and S. Ushioda, J. Phys. Chem. *92*:5357 (1988).
38. W. Wittke, A. Hatta, and A. Otto, Appl. Phys. A *48*:289 (1989).
39. M. G. Lee, J. H. Lee, and J. S. Chang, Surf. Sci. Lett. *271*:L362 (1992).
40. A. Nemetz, T. Fischer, A. Ulman, and W. Knoll, J. Chem. Phys. *98*:5912 (1993).
41. M. Futamata, Langmuir *11*:3894 (1995).
42. M. Futamata, J. Phys. Chem. *99*:11901 (1995).
43. M. Futamata, E. Keim, A. Bruckbauer, D. Schumacher, and A. Otto, Appl. Surf. Sci. *100/101*:60 (1996).
44. H. J. Simon, D. E. Mitchell, and J. G. Watson, Phys. Rev. Lett. 33:1531 (1974).
45. H. J. Simon, R. E. Benner, and J. G. Rako, Opt. Commun. *23*:245 (1977).
46. F. DeMartini, P. Ristori, E. Santamato, and A. C. A. Zammit, Phys. Rev. B *8*:3797 (1981).
47. Y. R. Shen and F. DeMartini, in *Surface Polaritons: Electromagnetic Waves at Surfaces and Interfaces* (V. M. Agranovich and D. L. Mills, eds.), North-Holland, Amsterdam, 1982, p. 629.

48. J. E. Sipe and G. I. Stegeman, in *Surface Polaritons: Electromagnetic Waves at Surfaces and Interfaces* (V. M. Agranovich and D. L. Mills, eds.), North-Holland, Amsterdam, 1982, p. 661.
49. R. T. Deck and D. Sarid, J. Opt. Soc. Am. *72*:1613 (1982).
50. R. M. Corn, M. Romagnoli, M. D. Levenson, and M. R. Philpott, Chem. Phys. Lett. *106*:30 (1984).
51. P. M. Adam, L. Salomon, F. de Fornel, and J. P. Goudonnet, Phys. Rev. B *48*:2680 (1993).
52. F. Chao, M. Costa, and A. Tadjeddine, J. Electroanal. Chem. *329*:313 (1992).
53. R. M. A. Azzam and N. M. Bashara, *Ellipsometry and Polarized Light*, North-Holland, Amsterdam, 1977.
54. W. Plieth, Kozlowski, and T. Twomey, in *Adsorption of Molecules at Metal Electrodes*, Vol. 239 (J. Lipkowski and P. N. Ross, eds.), VHC, New York, 1992.
55. A. Ulman, *An Introduction to Ultrathin Organic Films*, Academic Press, New York, 1991.
56. K. A. Peterlinz and R. Georgiadis, Langmuir *12*:4731 (1996).
57. S. Löfås, M. Malmqvist, I. Rönnberg, E. Stenberg, B. Liedberg, and I. Lundström, Sensors Actuators B *5*:79–84 (1991).
58. J. Spinke, M. Liley, H.-J. Guder, L. Angermaier, and W. Knoll, Langmuir *9*:1821 (1993).
59. J. Spinke, M. Liley, F.-J. Schmitt, H. J. Guder, L. Angermaier, and W. Knoll, J. Chem. Phys. *99*:7012 (1993).
60. M. Mrksich, G. B. Sigal, and G. M. Whitesides, Langmuir *11*:4383 (1995).
61. K. A. Peterlinz, R. Georgiadis, T. M. Herne, and M. J. Tarlov, J. Am. Chem. Soc. *119*:3401 (1997).
62. I. Pockrand, J. D. Swalen, J. G. I. Gordon, and M. R. Philpott, Surf. Sci. *74*:237 (1977).
63. D. G. Hanken and R. M. Corn, Anal. Chem. *67*:3767–3774 (1995).
64. D. G. Hanken, R. R. Naujok, J. M. Gray, and R. M. Corn, Anal. Chem. *69*:240 (1997).
65. R. Kotz, D. M. Kolb, and J. K. Sass, Surf. Sci. *69*:359 (1977).
66. J. G. Gordon and S. Ernst, Surf. Sci. *101*:499 (1980).
67. A. Tadjeddine, Electrochim. Acta *34*:29 (1989).
68. D. M. Kolb, in *Surface Polaritons: Electromagnetic Waves at Surfaces and Interfaces* (V. M. Agranovich and D. L. Mills, eds.), North Holland, Amsterdam, 1982, p. 299.
69. G. Flatgen, K. Krischer, B. Pettinger, K. Doblhofer, H. Junkes, and G. Ertl, Science *269*:668 (1995).
70. X. Chen, M. C. Davies, K. M. Shakesheff, S. J. B. Tendler, P. M. Williams, and J. Davies, J. Vac. Sci. Technol. B *14*:1582 (1996).
71. J. Davies, C. J. Roberts, A. C. Dawkes, J. Sefton, J. C. Edwards, T. O. Glasbey, A. G. Haymes, M. C. Davies, D. E. Jackson, M. Lomas, K. M. Shakesh-

eff, S. J. B. Tendler, M. J. Wilkins, and P. M. Williams, Langmuir *10*:2654 (1994).
72. X. Chen, K. M. Shakesheff, M. C. Davies, J. Heller, C. J. Roberts, S. J. B. Tendler, and P. M. Williams, J. Phys. Chem. *99*:11537 (1995).
73. K. M. Shakesheff, X. Chen, M. C. Davies, A. Domb, C. J. Roberts, S. J. B. Tendler, and P. M. Williams, Langmuir *11*:3921 (1995).
74. S. I. Bozhevolnyi, B. Vohnsen, I. I. Smolyaninov, and A. V. Zayats, Optics Communications *117*:417 (1995).
75. J. D. Swalen, J. G. Gordon, M. R. Philpott, A. Brillante, I. Pockrand, and R. Santo, Am. J. Phys. *48*:669 (1980).
76. A. Otto, Z. Phys. *216*:398 (1968).
77. E. Kretschmann and H. Raether, Z. Naturforsch. Teil A *23*:2135 (1968).
78. A. Brillante and I. Pockrand, J. Mol. Struct. *79*:169 (1982).
79. A. Hatta, S. Suzuki, and W. Suetaka, Appl. Surf. Sci. *40*:9 (1989).
80. I. Pockrand, J. D. Swalen, R. Santo, A. Brillante, and M. R. Philpott, J. Chem. Phys. *69*:4001 (1978).
81. G. H. Cross, N. A. Cade, I. R. Girling, I. R. Peterson, and D. C. Andrews, J. Chem. Phys. *86*:1061 (1987).
82. K. A. Peterlinz and R. Georgiadis, Optics Commun. *130*:260 (1996).
83. K. S. Johnston, S. R. Karlsen, C. C. Jung, and S. S. Yee, Mater. Chem. Phys. *42*:242 (1995).
84. R. C. Jorgenson, C. C. Jung, S. S. Yee, and L. W. Burgess, Sensors Actuators B *13–14*:721 (1993).
85. S. R. Karlsen, K. S. Johnston, R. C. Jorgenson, and S. S. Yee, Sensors Actuators B *24–25*:747 (1995).
86. C. R. Lawrence, A. S. Martin, and J. R. Sambles, Thin Solid Films *208*:269 (1992).
87. M. A. Kessler and E. A. H. Hall, J. Colloid Interface Sci. *169*:422 (1995).
88. J. G. Gordon and J. D. Swalen, Opt. Commun. *22*:374 (1977).
89. I. Pockrand, Surf. Sci. *72*:577–588 (1978).
90. I. Pockrand and J. D. Swalen, J. Opt. Soc. Am. *68*:1147 (1978).
91. J. C. Loulergue, M. Dumont, Y. Levy, P. Robin, J. P. Pocholle, and M. Papuchon, Thin Solid Films *160*:399 (1988).
92. W. M. K. P. Wijekoon, B. Asgharian, M. Casstevens, M. Samoc, G. B. Talapatra, P. N. Prasad, T. Geisler, and S. Rosenkilde, Langmuir *8*:135 (1992).
93. D. Morichere, V. Dentan, F. Kajzar, P. Robin, Y. Levy, and M. Dumont, Optics Comm. *74*:69 (1989).
94. V. Dentan, Y. Levy, M. Dumont, P. Robin, and E. Chastaing, Optics Comm. *69*:379 (1989).
95. H. Knobloch, H. Orendi, M. Buchel, M. Sawodny, A. Schmidt, and W. Knoll, Fresenius J. Anal. Chem. *349*:107 (1994).
96. Z. Sekkat, C.-S. Kang, E. F. Aust, G. Wegner, and W. Knoll, Chem. Mater. *7*:142 (1995).

97. B. L. Frey, C. E. Jordan, S. Kornguth, and R. M. Corn, Anal. Chem. *67*:4452 (1995).
98. G. H. Cross, I. R. Girling, I. R. Peterson, N. A. Cade, and J. D. Earls, Elect. Lett. *22*:1111 (1986).
99. G. H. Cross, I. R. Girling, I. R. Peterson, N. A. Cade, and J. D. Earls, J. Opt. Soc. Am. B *4*:962 (1987).
100. M. Dumont, Y. Levy, and D. Morichere, in *Organic Molecules for Nonlinear Optics and Photonics* (J. Messier, ed.), Kluwer Academic Publishers, Dordrecht, The Netherlands, 1991, p. 461.
101. E. F. Aust and W. Knoll, J. Appl. Phys. *73*:2705 (1993).
102. E. M. Yeatman and M. E. Caldwell, Appl. Phys. Lett. *55*:613 (1989).
103. T. Okamoto, T. Kamiyama, and I. Yamaguchi, Optics Lett. *18*:1570 (1993).
104. C. Jung, S. Yee, and K. Kuhn, Appl. Optics *34*:946 (1995).
105. D. G. Hanken and R. M. Corn, Anal. Chem. *69*:3665 (1997).
106. D. G. Hanken and R. M. Corn, Isr. J. of Chem. *37*:165 (1997).
107. M. Malmqvist, Nature *361*:186 (1993).
108. P. Nilsson, B. Persson, M. Uhlen, and P. A. Nygren, Anal. Biochem. *224*:400 (1995).
109. B. Cheskis and L. P. Freedman, Biochemistry *35*:3309 (1996).
110. P. Schuck and A. P. Minton, Anal. Biochem. *240*:262 (1996).
111. P. Schuck, Biophys. J. *70*:1230 (1996).
112. F.-J. Schmitt, L. Haussling, H. Ringsdorf, and W. Knoll, Thin Solid Films *210/211*:815 (1992).
113. L. Haussling, H. Ringsdorf, F.-J. Schmitt, and W. Knoll, Langmuir *7*:1837 (1991).
114. J. R. Rahn and R. B. Hallock, Langmuir *11*:650 (1995).
115. N. J. Geddes, A. S. Martin, F. Caruso, R. S. Urquhart, D. N. Furlong, J. R. Sambles, K. A. Than, and J. A. Edgar, J. Immunol. Methods *175*:149 (1994).
116. P. T. Leung, D. Pollardknight, G. P. Malan, and M. F. Finlan, Sensors Actuators B *22*:175 (1994).
117. D. J. v.-d. -. Heuvel, R. P. H. Kooyman, J. W. Drijfhout, and G. W. Welling, Anal. Biochem. *215*:223 (1993).
118. D. Piscevic, R. Lawall, M. Vieth, M. Liley, Y. Okahata, and W. Knoll, Appl. Surf. Sci. *90*:425 (1995).
119. A. A. Kruchinin and Y. G. Vlasov, Sensors Actuators B *30*:77 (1996).
120. G. B. Sigal, C. Bamdad, A. Barberis, J. Strominger, and G. M. Whitesides, Anal. Chem. *68*:490 (1996).
121. V. I. Silin, G. A. Balcytis, G. N. Zhizhin, and V. A. Yakovlev, Vibrational Spectr. *5*:133 (1993).
122. M. Stelzle, G. Weissmuller, and E. Sackmann, J. Phys. Chem. *97*:2974 (1993).
123. S. Terrettaz, T. Stora, C. Duschl, and H. Vogel, Langmuir *9*:1361 (1993).

124. H. Lang, C. Duschl, M. Gratzel, and H. Vogel, Thin Solid Films *210/211*:818 (1992).
125. E. L. Florin and H. E. Gaub, Biophys. J. *64*:375 (1993).
126. E. Stenberg, B. Persson, H. Roos, and C. Urbaniczky, J. Colloid Interface Sci. *143*:513 (1990).
127. A. Szabo, L. Stolz, and R. Granzow, Current Opin. Struct. Biol. *5*:699 (1995).
128. B. Rothenhäusler and W. Knoll, Nature *332*:615 (1988).
129. C. E. Jordan and R. M. Corn, Anal. Chem. *69*:1449 (1997).
130. W. Hickel and W. Knoll, J. Appl. Phys. *67*:3572 (1990).
131. F.-J. Schmitt and W. Knoll, Biophys. J. *60*:716 (1991).
132. B. Fischer, S. P. Heyn, M. Egger, and H. E. Gaub, Langmuir *9*:136 (1993).
133. J. S. Schildkraut, Appl. Optics *27*:3329 (1988).
134. H. E. de Bruijn, B. S. F. Altengurg, R. P. H. Kooyman, and J. Greve, Optics Commun. *82*:425 (1991).
135. H. E. de Bruijn, M. Minor, R. P. H. Kooyman, and J. Greve, Optics Commun. *95*:183 (1993).
136. E. D. Palik, ed., *Handbook of Optical Constants of Solids*, Academic Press, Orlando, 1985.
137. M. D. Porter, D. L. Allara, T. B. Bright, and C. E. D. Chidsey, J. Am. Chem. Soc. *109*:3559 (1987).
138. C. A. Goss, D. H. Charych, and M. Majda, Anal. Chem. *63*:85 (1991).
139. T. Hoshi, J. Anzai, and T. Osa, Anal. Chem. *67*:770 (1995).
140. N. M. Green, in *Advances in Protein Chemistry*, Academic Press, New York, 1975, p. 85.
141. L. Pugliese, A. Coda, M. Malcovati, and M. Bolognesi, J. Mol. Biol. *231*:698 (1993).
142. M. Wilchek and E. A. Bayer, Anal. Biochem. *171*:1 (1988).
143. R. C. Ebersole, J. A. Miller, J. R. Moran, and M. D. Ward, J. Am. Chem. Soc. *112*:3239 (1990).
144. S. Zhao and W. M. Reichert, Langmuir *8*:2785 (1992).
145. P. M. Nellen and W. Lukosz, Biosensors Bioelectronics *8*:129 (1993).
146. P. Pantano and W. G. Kuhr, Anal. Chem. *65*:623 (1993).
147. G. Decher, B. Lehr, K. Lowack, Y. Lvov, and J. Schmitt, Biosensors Bioelectronics *9*:677–648 (1994).
148. M. S. Ayyagari, R. Pande, S. Kamtekar, H. Gao, K. A. Marx, J. Kumar, S. K. Tripathy, J. A. Akkara, and D. L. Kaplan, Biotechnol. Bioeng. *45*:116 (1995).
149. H. Morgan, D. M. Taylor, and C. D'Silva, Thin Solid Films *209*:122 (1992).
150. K. Fujita, S. Kimura, Y. Imanishi, E. Rump, J. van Esch, and H. Ringsdorf, J. Am. Chem. Soc. *116*:5479 (1994).
151. R. M. Zimmerman and E. C. Cox, Nucleic Acids Res. *22*:492 (1994).

152. C. D. Bain, E. B. Troughton, Y.-T. Tao, J. Evall, G. M. Whitesides, and R. G. Nuzzo, J. Am. Chem. Soc. *111*:321 (1989).
153. C. E. D. Chidsey and D. N. Loiacono, Langmuir *6*:682 (1990).
154. E. L. Smith, C. A. Alves, J. W. Anderegg, M. D. Porter, and L. M. Siperko, Langmuir *8*:2707 (1992).
155. L. Sun, R. M. Crooks, and A. J. Ricco, Langmuir *9*:1775 (1993).
156. S. J. Stranick, A. N. Parikh, Y.-T. Tao, D. L. Allara, and P. S. Weiss, J. Phys. Chem. *98*:7636 (1994).
157. J. P. Folkers, P. E. Laibinis, and G. M. Whitesides, Langmuir *8*:1330 (1992).
158. O. Chailapakul and R. M. Crooks, Langmuir *9*:884 (1993).
159. H. Lee, H. G. Hong, T. E. Mallouk, and L. J. Kepley, J. Am. Chem. Soc. *110*:618 (1988).
160. H. Lee, T. E. Mallouk, L. J. Kepley, H. G. Hong, and S. Akhter, J. Phys. Chem. *92*:2597 (1988).
161. H. C. Yang, K. Aoki, H.-G. Hong, D. D. Sackett, M. F. Arendt, S.-L. Yau, C. M. Bell, and T. E. Mallouk, J. Am. Chem. Soc. *115*:11855 (1993).
162. T. M. Putvinski, M. L. Schilling, H. E. Katz, C. E. D. Chidsey, A. M. Mujsce, and A. B. Emerson, Langmuir *6*:1567 (1990).
163. A. C. Zeppenfeld, S. L. Fiddler, W. K. Ham, B. J. Klopfenstein, and C. J. Page, J. Am. Chem. Soc. *116*:9158 (1994).
164. M. B. Dines and P. M. DiGiacomo, Inorganic Chem. *20*:92 (1981).
165. W. Kohler, D. R. Robello, C. S. Willand, and D. J. Williams, Macromolecules *24*:4589 (1991).
166. J. Huang, D. A. Dahlgren, and J. C. Hemminger, Langmuir *10*:626 (1994).
167. M. J. Tarlov, D. R. F. Burgess, and G. Gillen, J. Am. Chem. Soc. *115*:5305 (1993).
168. D. Piscevic, W. Knoll, and M. J. Tarlov, Supramol. Sci. *2*:99 (1995).
169. C. E. Jordan, A. G. Frutos, A. J. Thiel, and R. M. Corn, Anal. Chem. *69*:4939 (1997).
170. A. Frutos, Q. Liu, A. Thiel, A. M. W. Sanner, A. E. Condon, L. M. Smith, and R. M. Corn, Nucleic Acids Res. *25*:4748 (1997).
171. A. J. Bard and L. R. Faulkner, *Electrochemical Methods*, J. Wiley, New York, 1980.
172. H. O. Finklea, in *Electroanalytical Chemistry: A Series of Advances*, Vol. 19 (A. J. Bard and I. Rubinstein, eds.), Marcel Dekker, Inc., New York, 1996, p. 109.
173. C. P. Smith and H. S. White, Anal. Chem. *64*:2398 (1992).
174. C. P. Smith and H. S. White, Langmuir *9*:1 (1993).
175. W. R. Fawcett, M. Fedurco, and Z. Kovacova, Langmuir *10*:2403 (1994).
176. W. R. Fawcett, J. Electroanal. Chem. *378*:117 (1994).
177. P. H. Schmidt and W. J. Plieth, J. Electroanal. Chem. *201*:163 (1986).
178. C. A. Widrig, C. Chung, and M. D. Porter, J. Electroanal. Chem. *201*:335 (1991).

179. J. M. Pope, Z. Tan, S. Kimbrell, and D. A. Buttry, J. Am. Chem. Soc. *114*:10085 (1992).
180. J. M. Pope and D. A. Buttry, in press.
181. D. J. Lockhart and S. G. Boxer, Chem. Phys. Lett. *144*:243 (1988).
182. D. J. Lockhart, C. Kirmaier, D. Holten, and S. G. Boxer, J. Phys. Chem. *94*:6987 (1990).
183. S. E. Creager and K. Weber, Langmuir *9*:844 (1993).
184. G. K. Rowe and S. E. Creager, J. Phys. Chem. *98*:5500 (1994).
185. X. Gao, H. S. White, S. Chen, and H. D. Abruna, Langmuir *11*:4554 (1995).
186. Y. R. Shen, *The Principles of Nonlinear Optics*, Wiley, New York, 1984.
187. K. D. Singer, S. L. Lalama, J. E. Sohn, and R. D. Small, in *Nonlinear Optical Properties of Organic Molecules and Crystals*, Vol. 1 (D. S. Chemla and J. Zyss, eds.), Academic Press, Orlando, 1987, p. 437.
188. K. D. Singer, M. G. Kuzyk, and J. E. Sohn, J. Opt. Soc. Am. B *4*:968 (1987).
189. D. R. Robello, P. T. Dao, J. Phelan, J. Revelli, J. S. Schildkraut, M. Scozzafava, A. Ulman, and C. S. Willand, Chem. Mater. *4*:425 (1992).
190. D. R. Robello, P. T. Dao, J. S. Schildkraut, M. Scozzafava, E. J. Urankar, and C. S. Willand, Chem. Mater. *7*:284 (1995).
191. H. E. Katz, G. Scheller, T. M. Putvinski, M. L. Schilling, W. L. Wilson, and C. E. D. Chidsey, Science *254*:1485 (1991).
192. H. E. Katz, W. L. Wilson, and G. Scheller, J. Am. Chem. Soc. *116*:6636 (1994).
193. H. E. Katz and M. L. Schilling, Chem. Mater. *5*:1162 (1993).
194. R. R. Naujok, D. A. Higgins, D. G. Hanken, and R. M. Corn, J. Chem. Soc., Faraday Trans.:1411 (1995).
195. D. M. Burland, R. D. Miller, and C. A. Walsh, Chem. Rev. *94*:31 (1994).
196. M. Ahlheim, M. Barzoukas, P. V. Bedworth, M. Blanchard-Desce, A. Fort, Z.-Y. Hu, S. R. Marder, J. W. Perry, C. Runser, M. Staehelin, and B. Zysset Science *271*:335 (1996).
197. T.-A. Chen, A. K.-Y. Jen, and Y. Cai, J. Am. Chem. Soc. *117*:7295 (1995).
198. T. Verbiest, D. M. Burland, M. C. Jurich, V. Y. Lee, R. D. Miller, and W. Volksen, Science *268*:1604 (1995).
199. L. J. Kepley, D. D. Sackett, C. M. Bell, and T. E. Mallouk, Thin Solid Films *208*:132 (1992).
200. S. D. Evans and A. Ulman, Chem. Phys. Lett. *170*:462 (1990).
201. A. M. Becka and C. J. Miller, J. Phys. Chem. *97*:6233 (1993).
202. D. A. Higgins, R. R. Naujok, and R. M. Corn, Chem. Phys. Lett. *213*:485 (1993).
203. C. C. Jung, S. B. Saban, S. S. Yee, and R. B. Darling, Sensors and Actuators B *32*:143 (1996).
204. B. L. Frey, Ph.D Thesis, University of Wisconsin, Madison, 1996.

ELECTROCHEMISTRY IN NEURONAL MICROENVIRONMENTS

Rose A. Clark, Susan E. Zerby, and Andrew G. Ewing

Pennsylvania State University, University Park, Pennsylvania

I. Introduction   228
II. Electrochemistry: Methods and Electrodes   234
    A. Potentiometric versus voltammetric measurements   234
    B. Carbon fiber versus platinum microelectrodes   238
    C. Ultrasmall carbon microelectrodes   239
III. Intracellular Voltammetry   252
    A. Introduction   252
    B. Monitoring intracellular dopamine   252
    C. Monitoring glucose in single cells   257
    D. Monitoring intracellular and extracellular oxygen   259
IV. Extracellular Voltammetry   259
    A. Introduction   259
    B. Brain studies   262
    C. Overview of single-cell systems used in electrochemical studies of exocytosis   266
    D. Bovine adrenal chromaffin cells   266
    E. Rat pheochromocytoma (PC12) cells   276
    F. Mast cells, pancreatic β cells, and rat melanotrophs   280
    G. Invertebrate and mammalian neurons   282
V. Concluding Remarks and Other Directions   286
    References   287

I. INTRODUCTION

The study of the brain and its function has been an area of research for many years and has involved the use of many different analytical techniques. These analytical techniques have provided unique information regarding the neurophysiology and neuropharmacology of the brain and so are complementary to each other. One technique, electroanalysis, has found extensive applications since many neurochemicals are easily oxidized or reduced. Electroanalytical techniques also possess unique characteristics that make them useful for the study of chemical processes occurring in complex biological matrices, like the brain. One property that is particularly advantageous is the ability to provide both qualitative and quantitative information while being specific for electroactive substances. Thus, interferences are minimized as only a relatively small number of molecules are easily oxidized or reduced in these systems [1]. Detection of other nonelectroactive compounds and discrimination between electrooxidizable species has also been addressed through the modification of electrode surfaces [2–6]. Electrochemical techniques continue to gain interest as new developments in the field of electrochemistry allow more sensitive, selective, and rapid detection of chemical events occurring in the brain and neuronal-like systems in culture.

The use of carbon fibers in the construction of electrodes has allowed the development of microelectrodes with dimensions in the micrometer range. The small size of these electrodes provides many advantages in terms of electrochemistry and the practical aspects of their use in biological environments. The microelectrodes can be placed in close proximity to the site of release or to a synapse without appreciable perturbation to the surrounding environment [7]. Intracellular measurements are also possible with microelectrodes [8]. There are several advantages of microelectrodes over conventional electrodes, most notably for neuroscience work is the small size. The obvious advantage here is that smaller electrodes allow access to smaller microenvironments. In addition, the reduced electroactive area of microelectrodes leads to a reduced double layer capacitance relative to electrodes of conventional size. This allows rapid voltammetric measurements to be carried out at these electrodes [9–11] so that millisecond to microsecond events (exocytosis) and kinetic processes (reuptake) can be monitored [7]. The improved faradaic-to-nonfaradaic current ratio at microelectrodes compared to macroelectrodes enhances the signal-to-noise ratio [12]. Another advantage of the small size of microelectrodes is the extremely efficient diffusional transport they exhibit due to the large diffusion layer relative to the electrode dimensions. This results in steady-state voltammetry when slow scan rate cyclic voltammetry is employed. The steady-state limiting current is directly proportional to the analyte concentration, which provides a direct means of quantitating electroactive substances. Another positive aspect of microelectrodes in brain studies is that the potential problem of catalytic regeneration reactions of the oxidized analyte are minimized since it is more difficult for the regenerated material to return to the surface of small electrodes [13–16]. Currents generated at microelectrodes typically lie in the pA$^+$ to nA range, thus providing immunity to the problem of "ohmic drop" experienced at larger electrodes [17,18]. The small currents also facilitate the use of a simpler two-electrode configuration in voltammetric studies [19]. In terms of the biological environment, smaller currents mean fewer products are generated at the electrode, which is beneficial since many of these products are detrimental to cells at high concentrations [20]. Finally, the small currents are less apt to interfere with the normal neuronal activity of the surrounding tissue or cells [21]. Electrochemical methods allow the detection of cellular function with minimal perturbation to the biological microenvironment. This is of utmost importance in trying to understand the function of neuronal cells in the complex matrix of the brain without any effects from the measurements.

The fundamental building block of the brain is the single nerve cell or neuron [22,23]. It has been estimated that the human brain contains 10^{10}–10^{12} neurons [22]. Nerve cells are unique in that they are specialized to communicate with each other as well as with other types of cells. A neuron is comprised of a central cell body with two different types of projections: the dendrites and the axon (Fig. 1). The dendrites are classically thought to be the input portion of the neuron where signals are received and the axon referred to as the output segment of the nerve cell. In general, signals are received at the dendrites, travel to the cell body where they are integrated, and then are sent down the axon to the terminal end where they are relayed to the next cell or cells. The site of information transfer between the axon terminal of the presynaptic neuron and the dendrite of the postsynaptic cell is called a synapse. The synapse is a very small gap between communicating cells ranging in size from 1 to 100 nm from one side to the other.

Electron microscopy has revealed that spherical storage granules or vesicles aggregate inside the presynaptic neuron near the cell membrane.

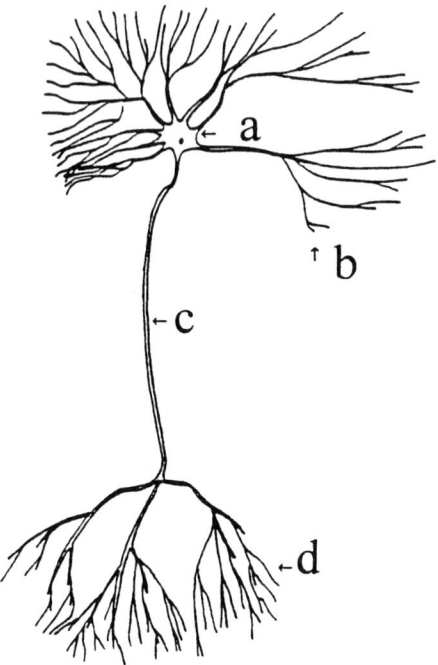

FIG. 1. Illustration of a "typical" neuron. (Adapted from Ref. 29.)

These storage compartments are where neurotransmitters are packaged, stored and sometimes synthesized. Concentrations of neurotransmitter as high as 1 M in vesicles have been reported within the presynaptic neuron [22]. An electrical signal called an action potential initiates vesicle mobilization to the cell membrane. Vesicle fusion with the cell membrane results in the expulsion of the vesicle contents into the synapse. This is the process termed exocytosis [24] (Fig. 2).

After exocytosis, neurotransmitters diffuse across the synapse and bind to receptors on the postsynaptic cell. The binding of neurotransmitters to receptors causes changes in the permeability of the membrane to Na^+ and other small ions leading to the generation of a small postsynaptic potential. The magnitude of the postsynaptic potential is dependent on the amount of neurotransmitter released from the presynaptic neuron. Postsynaptic potentials might be either excitatory or inhibitory, lasting between a

Electrochemistry in Neuronal Microenvironments

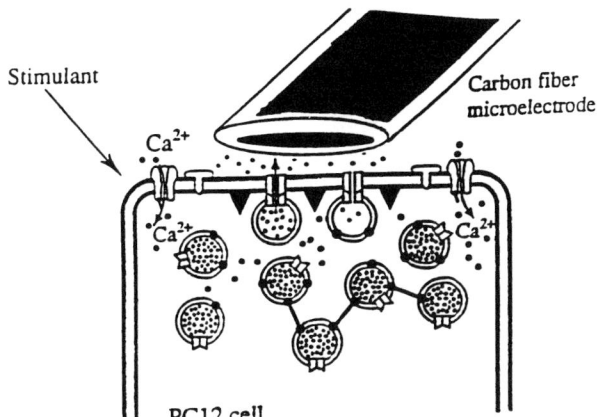

FIG. 2. Schematic of the exocytosis process at a cell membrane. During exocytosis, a vesicle containing a chemical messenger fuses with the cell membrane, resulting in expulsion of the messenger into the extracellular space.

few milliseconds to tens of seconds. Integration of hundreds of synapses at the postsynaptic cell determines whether the receiving neuron initiates a new action potential and so determines the fate of the net signal received by the cell. A schematic of the traditional view of neurotransmission is shown in Fig. 3. These models are being refined and updated as more evidence is gained through electrochemical and other methods.

As shown in Fig. 3, the excess released neurotransmitter needs to be deactivated to prevent continual stimulation of the postsynaptic cell. Deactivation usually occurs via three mechanisms: diffusion of the neurotransmitter out of the synapse, transport of the neurotransmitter into the presynaptic cell (this "reuptake" process is responsible for the majority of the deactivation), and metabolic breakdown of the neurotransmitter.

Neurons differ in the signaling agent utilized and in some cases use more than one type of transmitter. Behavioral connections for specific neurotransmitters make it desirable to define them in structural classes. The general classes of neurotransmitters include acetylcholine, catecholamines, indoleamines, amino acids, and peptides [22]. Acetylcholine is the primary transmitter in cholinergic systems, whereas the catecholamines dopamine and norepinephrine are specific for adrenergic systems (see Table 1). Dopamine is important in the regulation of motor

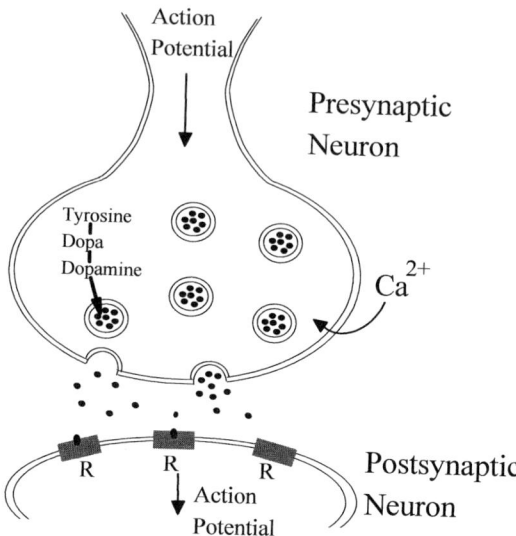

FIG. 3. An illustration showing the production and release of catecholamine neurotransmitters from a presynaptic nerve ending. Transmitters bind to the receptors (R) on the postsynaptic membrane causing the eventual development of an action potential in the postsynaptic cell.

function as well as the control of mood and emotion. It has also been shown to play a part in Parkinson's disease, schizophrenia, and depression. The indoleamine serotonin (5-hydroxytryptamine) is important in the regulation of sleep mechanisms [23]. Quantitation of these substances in the extracellular fluid of the brain has been conducted in an effort to relate concentration of neurotransmitter to an observed behavior. Both dopamine and serotonin, as well as the catecholamines norepinephrine and epinephrine, are easily oxidized and therefore easily detected electrochemically. In the 1970s, Adams and co-workers began to develop methods to monitor electroactive dopamine in a dopamine-rich portion of the mammalian brain [25] using voltammetry at small carbon electrodes. This work sparked interest in the use of electroanalysis in neuroscience-related research and has led to the continuing development of methods to more effectively monitor neurotransmitter action in neuronal microenvironments.

Electrochemistry in Neuronal Microenvironments

TABLE 1
Typical Neurotransmitter Systems

Cholinergic	
Acetylcholine	$(CH_3)_3\overset{+}{N}CH_2CH_2-O-\overset{O}{\underset{\|\|}{C}}-CH_3$
Adrenergic (Catecholaminergic)	
Dopamine	HO-C6H3(OH)-CH$_2$-CH$_2$-NH$_2$
Norepinephrine	HO-C6H3(OH)-CH(OH)-CH$_2$-NH$_2$
Epinephrine	HO-C6H3(OH)-CH(OH)-CH$_2$-NH-CH$_3$
Serotonergic (Tryptaminergic)	
5-Hydroxytryptamine or serotonin	HO-indole-CH$_2$CH$_2$NH$_2$
Amino Acidic	
γ-Aminobutyric acid (GABA)	$NH_2CH_2CH_2CH_2COOH$
Glutamate	$HOOC-CH_2CH_2CH(NH_2)-COOH$
Glycine	NH_2CH_2COOH
Peptidic	
Met-enkephalin	

In this review both the development of electrochemical techniques useful for investigation of neuronal communication and their application in vivo and in vitro are addressed. Various electrochemical methods and electrodes used in neuronal cell studies are discussed as well as electrode surface modifications employed in order to detect nonelectroactive substances relevant to neuronal function. Specific examples of electrochemical applications for monitoring dopamine, O_2, and glucose using intracellular voltammetry are presented. The final section examines voltammetric methods used in extracellular measurements in the rat brain and to study several single-cell systems, both in vivo and in vitro. The majority of the work presented deals with the monitoring of exocytosis from single nerve-like cells (model systems) in culture (bovine adrenal chromaffin cells, PC12 cells, mast cells, rat melanotrophs, and β-lymphocyte cells). One mammalian and two invertebrate systems are also highlighted. Discussion of the data both in a qualitative and quantitative sense is presented, as well as the usefulness of histograms and their meaning in terms of vesicle radii distribution. At the end of the chapter specific reviews dealing with electrochemistry in neuronal microenvironments are listed.

II. ELECTROCHEMISTRY: METHODS AND ELECTRODES

A. Potentiometric Versus Voltammetric Measurements

Potentiometry and voltammetry are two of the basic electroanalysis methods available for chemical analysis. These are broad categories that include many different techniques developed for use in different experimental situations.

Potentiometric ion-selective electrodes [26] are commonly used to measure concentrations of ions both outside and inside cells. The most common ion-selective electrodes are used to measure pH, K^+, Na^+, or Ca^{2+} [27] levels in both resting and stimulated cells [28,29]. These electrodes have been used to study diffusion characteristics as well as to determine the volume of the extracellular space in the brain [30,31]. All types of cells possess these ions, H^+, K^+, Na^+, and Ca^{2+}, therefore, potentiometry is a useful technique to study ion fluxes and concentrations. This technique is not capable of detecting neurotransmitters at neuronal cells, hence, voltammetry must be employed.

Several neuroactive species are easily oxidized (Fig. 4) and are, therefore, amenable to detection by voltammetry. The magnitude of the current is proportional to the analyte concentration and the integrated current, charge, is proportional to the mass of the analyte through Faraday's law:

$$\frac{Q}{nF} = N \qquad (1)$$

where Q is the charge in coulombs, n is the number of electrons transferred per mole of analyte oxidized or reduced, F is the Faraday constant (96,485

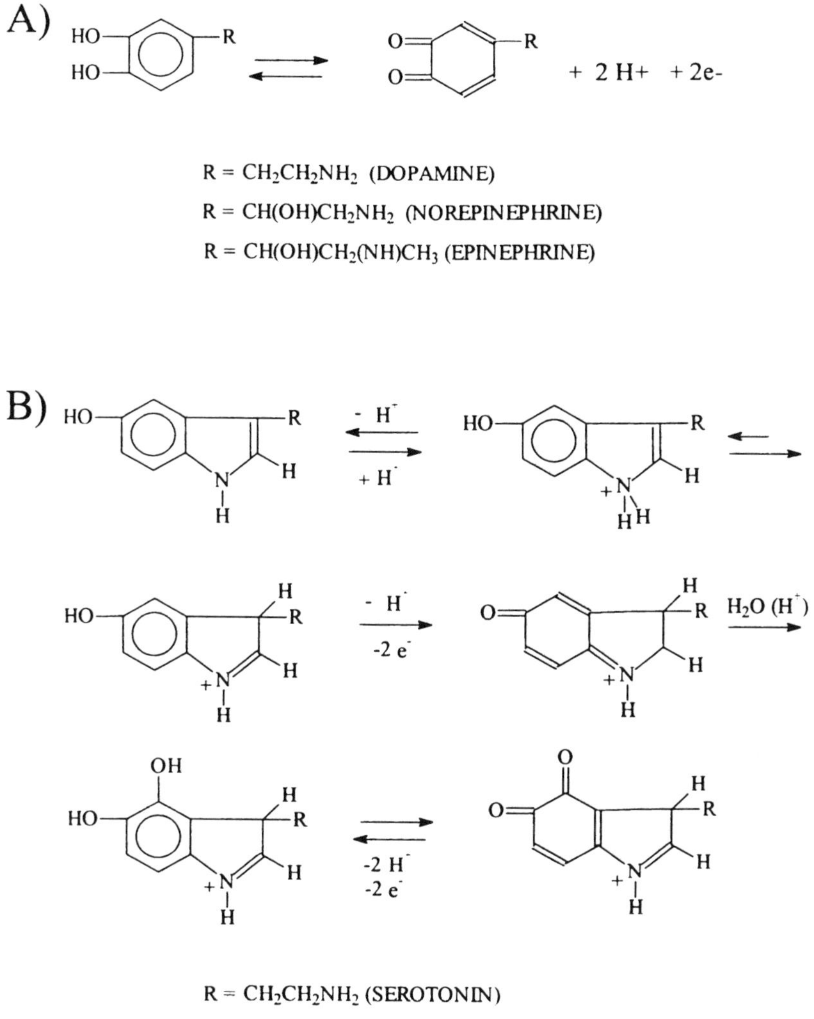

Electrochemistry in Neuronal Microenvironments

A)

HO–⟨ring⟩–R ⇌ O=⟨quinone⟩=O –R + 2 H+ + 2e−

R = CH$_2$CH$_2$NH$_2$ (DOPAMINE)
R = CH(OH)CH$_2$NH$_2$ (NOREPINEPHRINE)
R = CH(OH)CH$_2$(NH)CH$_3$ (EPINEPHRINE)

B)

R = CH$_2$CH$_2$NH$_2$ (SEROTONIN)

FIG. 4. Oxidation reactions for (A) catecholamines and (B) indoleamines.

C/equiv.), and N is the total number of moles of substance oxidized or reduced. The measurement of the current flow in response to an applied voltage is the basis of various voltammetric techniques. The dynamic nature of this technique in conjunction with the rapid response times of microelectrodes facilitates measurement of the time course of neurochemical processes. Broderick [32] points out that the utility of voltammetry in the neurosciences lies in this ability to follow the time course of events, which allows connections between neurophysiology and disease and/or behavior to be established.

Many different voltammetric techniques have been developed and successfully applied to a multitude of analytical problems. In this chapter, only those most pertinent to neuroscience applications are discussed. These techniques can be put into one of two broad categories: potential step or pulse and continuously applied potentials. The pulse techniques include chronoamperometry, normal pulse voltammetry, and differential pulse voltammetry, while linear sweep voltammetry, cyclic voltammetry, and amperometry can be categorized as continuous potential methods.

A comparison of various electrochemical methods for neurochemical experiments reveals that each method has specific advantages and disadvantages in particular protocols. Faster sampling rates are important for studies involving neurotransmitter secretion, which occurs on the millisecond time scale [33]. Thus chronoamperometry with short potential pulses has been used in many experiments. This technique also has the advantage of producing fewer electrolysis products than longer scanning methods because the oxidizing potential is applied for shorter bursts.

Charging current due to double layer capacitance decays more rapidly than the faradaic current and is usually negligible if the current is sampled at the end of a potential step experiment. At microelectrodes, frequently used in neurochemical studies, the small charging current contributes little to the net current, allowing faster measurements to be taken. Chronoamperometry is useful for obtaining rapid quantitative measurements of neurotransmitter concentrations but provides no qualitative information [4,34,35].

Normal pulse voltammetry can be used to obtain qualitative information about easily oxidized species. The height of the current plateau can be used for quantitation and the half-wave potential for identification. Ponchon et al. [36] applied normal pulse voltammetry at carbon fibers in vitro to resolve various electroactive substances found in neuronal environments ($E_{1/2}$ of dopamine = 0.2 V; homovanillic acid = 0.53 V; ascorbic acid = 0.45 V; serotonin = 0.34 V; 5-hydroxyindoleacetic acid = 0.5 V). This technique has also been used in vivo by Ewing et al. to discriminate between dopamine and ascorbate [21,37,38].

The most widely used quantitative pulse technique is differential pulse voltammetry. This technique virtually eliminates any charging current and provides a convenient peak shape for quantitation [33]. Differential pulse voltammetry is capable of resolving compounds having significantly different oxidation potentials, and considerable work has been done in an effort to increase this resolution through electrode surface modification [3,4,39,40]. This technique has also been used successfully to measure levels of 5-hydroxyindoles and 5-hydroxyindoleacetic acid in the rat brain [41–44].

More qualitative techniques that yield chemical information about the analyte are linear sweep voltammetry, staircase voltammetry, and cyclic voltammetry. When using these techniques at microelectrodes, slow scan rates (<100 V/sec, varies with system) produce sigmoidal curves in which the limiting current can be easily determined, while fast scan rates (>100 V/sec) produce more "fingerprint-like" voltammograms that can be used for qualitative purposes after background subtraction [45] and also provides greater selectivity [46]. One of the unique characteristics of fast scan rate cyclic voltammetry is its ability to provide information about the kinetics and mechanisms of chemical reactions because of its rapid sampling over a range of potentials. This has allowed discrimination between biogenic amines common to neuronal environments [47]. This technique was first used in vivo by Kissinger et al. [25] and has now been applied extensively in neuronal microenvironments [48–50].

Quantitation of analytes in single cells [16,51–54] has been carried out extensively with direct amperometry at oxidizing potentials. Current is sampled at high rates in order to capture fast events like cellular exocytosis that occur on the millisecond to microsecond time scale. The charge for each (integrated current) can be used in Faraday's law [Eq. (1)] to calculate number of molecules released per exocytotic event. This technique is useful for quantitation and monitoring the time course of exocytosis but has poor selectivity. Applications of amperometry for analysis of single cells are discussed in more detail in Sec. IV.

No single voltammetric technique is ideal for all applications due to the nature of different experiments, but with the wide variety of techniques and electrodes available, appropriate detection schemes can be found for many different types of investigations involving electroactive substances in neuronal systems. All of the methods described are typically used with microelectrodes for in vivo and in vitro cellular studies.

B. Carbon Fiber versus Platinum Microelectrodes

Microelectrodes have many advantages (see above) and also some disadvantages. The small currents generated require additional amplification and are more susceptible to electrical interference. Their response in biological environments can be altered by electrode fouling [37] and the effects of oxidation products [55] as well as drugs and drug metabolites [56,57]. The deleterious effects of electrode fouling can be minimized by postexperiment calibration of electrodes.

Platinum wire and carbon fibers are just two of the many materials that have been used in the construction of microelectrodes. Both platinum and carbon can be fabricated into durable electrodes with micrometer size or smaller tips and used in minute environments [58–60]. These two materials have also been combined to construct platinized carbon microelectrodes [61–65]. Selectivity, sensitivity, stability, and reproducibility of the signals are the major differences between these electrodes, especially when working in biological environments such as the brain or cell cultures. The electrode material of choice for a given experiment depends upon the nature of the substances being investigated.

Platinum possesses partially unsaturated surface d-orbitals that facilitate the adsorption stabilization of free radical products in slow electrooxidation reactions. This is especially beneficial in the detection of aliphatic alcohols and amines. Platinum electrodes are also very sensitive to the presence of hydrogen and oxygen as is clear from the observed peaks associated with the adsorption/desorption of these two gases [61]. This is advantageous if one wishes to detect H_2 or O_2, but is detrimental if these gases are background signals that obscure detection of the analyte. The strong adsorption characteristic of Pt makes it useful in a "clean" environment, but much less so in biological matrices where proteins and other biological molecules adsorb to the surface. Electrode fouling that results in poorer sensitivity, stability, and reproducibility of measurements has been addressed by depositing thin layers of platinum on carbon surfaces as well as by coating the platinum with various permselective membranes. These modifications have improved the performance of platinum electrodes, nevertheless there is still room for further improvements.

Electrochemistry in Neuronal Microenvironments

Carbon fibers have been widely used in the construction of microelectrodes [66,67] due to their resistance to drift [68] and inert nature in biological environments. Like platinum microelectrodes, some adsorption of large biomolecules occurs [55] causing deterioration of the electrode response over time. Usually there is a 30–50% decrease in electrode sensitivity following implantation into brain tissue, which levels off after approximately 2 hours, and reproducible, stable measurements can be collected for 6–12 hours [20,37,57,69]. In culture, sensitivity loss is much less and is evaluated through postexperiment calibration [68].

C. Ultrasmall Carbon Electrodes

The surfaces of graphitic materials, like carbon fibers, are quite complex [70]. Graphitic carbon consists of extensive sheets of fused aromatic rings, stacked in a planar fashion as shown in Fig. 5. The surface of an uninterrupted basal plane is hydrophobic, nonionic, has low polarity, and is rich in π-electron density with little if any chemical functionality [71]. Therefore, the basal plane of the fiber is almost electrochemically inert in aqueous solutions. The surface of a high-modulus carbon fiber that is parallel to the fiber axis has a high degree of basal plane character, whereas the cross-sectional area has a large amount of edge orientation [14]. It appears that most electron transfer occurs at the edges of the basal planes where there are a large number of carboxyl and hydroxyl functional groups (Fig. 5) [72]. The carbon fiber in its naked, untreated form has modest sensitivity and resolution; however, electrochemical treatment and various surface modi-

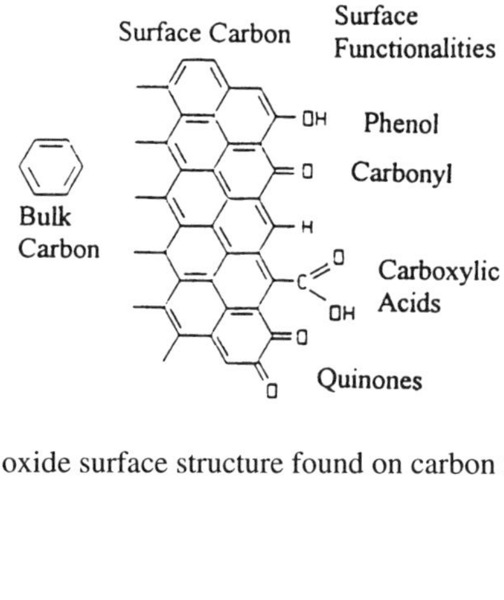

FIG. 5. Graphitic oxide surface structure found on carbon materials.

fications, both physical and chemical, have been used to enhance the electroactive edges of the basal plane.

1. Fabrication of Ultrasmall Carbon Electrodes

The fabrication of microelectrodes using carbon fibers typically involves several steps. The following serves as an example of how carbon fiber electrodes can be made. Carbon fibers of varying diameters (5–40 μm, with 5 μm or 10 μm being the most common) can be obtained from several manufacturers [14]. Fiber bundles are cut several centimeters longer than the glass capillaries and are separated into single fibers before aspirating under low pressure into a capillary (0.6 mm i.d., 1.2 mm o.d., A-M Systems, Everett, WA). The capillary is then pulled to a small tip using a micropipette puller (Ealing, Harvard Apparatus, Edenbridge, KY, or Narashige, Tokyo, Japan). When preparing disk electrodes, the fiber is sealed into the pulled capillary by epoxy (Epo-Tek 301, Epoxy Technology, Billerica, MA). When preparing cylinder electrodes, this epoxy must be prepared 1–2 hours in advance to increase viscosity. A drop of the epoxy is applied to the glass-fiber junction with minimal application to the fiber outside of the capillary. If the epoxy is not properly applied or cured, gaps between the carbon fiber and glass can occur, resulting in large residual currents due to leakage. An alternative method of applying epoxy is given by Kawagoe et al. [68].

The capillary is back-filled with a conductive material (colloidal graphite, mercury, gallium) to make contact with the carbon fiber. A nickel-chromium wire is inserted in the open end of the capillary to make the final electrical connection and secured with Duco (Devcon) cement. Cylindrical electrodes are ready to use once the connection is made. Electrodes can be stored for approximately 1 month without substantial loss of sensitivity. To complete disk electrodes, the fiber is trimmed at the glass-fiber junction to expose a fresh electroactive carbon surface. In single-cell experiments electrodes are often beveled [73] at 25–45° angles using a 0.25 μm diamond paste (Buehler, Lake Bluff, IL) slurry on a micropipette beveler (World Precision Instr., New Haven, CT). This provides a larger, more reproducible electroactive detection area with enhanced edge plane exposure for more sensitive detection. The best results are obtained when electrodes are beveled the day of use and calibrated in appropriate solutions to ensure adequate sensitivity and response time.

Carbon ring electrodes [74–76] are fabricated by depositing carbon on the inner surface of a pulled quartz capillary tube. Quartz tubes, approximately 1.3 mm in diameter, are first heated [methane-oxygen flame (Ealing, Harvard Instr., Edenbridge, KY)] and pulled with a vertical micropipetted puller (or manually) down to tip diameters of 1–4 μm. Carbon is then deposited on the inner surface of the capillary by pyrolyzing methane as it passes through the capillary using a Bunsen burner. The temperature of a Bunsen burner is often hot enough to soften or bend the quartz capillary tip. This can be avoided by placing the electrode tip into a 1/4 in. diameter quartz tube and heating the outer quartz tube with the Bunsen burner. Once a shiny black carbon ring is produced, the tip is filled with epoxy (Epon 828 with *m*-phenylenediamine chloride as hardener, Miller-Stephenson Chemical Co., Danbury, CT). The electrode tips are trimmed or beveled to expose a fresh electroactive carbon surface. Electrical contact with the carbon deposit is accomplished with mercury or colloidal graphite and a nickel-chromium wire. Tip diameters as small as 1 μm have been fabricated, making them very useful in microenvironments [74].

2. Microelectrode Geometries

Microelectrodes can be produced in various geometries accompanied by specific characteristics that make each type useful under different conditions. The three most common geometries for brain and single neuronal cell analysis are cylinder, disk, and ring.

Cylinder electrodes with fibers extending 100–200 μm beyond the glass are commonly used in in vivo studies [69]. They have also found much use in electrochemical detection at the end of high-performance liquid chromatography (HPLC) columns and capillary electrophoresis capillaries because they have the appropriate geometry and also have a large detection area. The large surface area of cylinder electrodes is advantageous when surface modifications such as selective polymer membranes or enzyme immobilization are carried out that coat part of the electrode surface. All points on a cylinder electrode are uniformly accessible, like a spherical electrode. Currents at these electrodes are time dependent even at long times, unlike that at disk and ring electrodes [12].

The disk geometry is also widely used in neuronal cell studies. These electrodes are usually beveled so that the electroactive surface is elliptical [73]. The tip size of these electrodes is dependent on the diameter of the carbon fiber with the most common sizes ranging from 7 to 13 μm across the minor axis. These dimensions make the electrodes well suited for the detection of catecholamines at single cells of similar dimensions

(5–200 μm diameter). Disk-shaped electrodes are fairly durable and reuseable upon rebeveling. Disk electrodes have the advantage of low capacitive currents and steady-state faradaic responses [32]. Electrolysis at the outer circumference of the disk diminishes the flux of electroactive material to the center of the electrode, but despite this, quasi-spherical diffusion is established in relatively short time periods, allowing a steady-state current to be established. The steady-state current for a shielded disk microelectrode is given by:

$$i_{ss} = 4rnFDC \qquad (2)$$

where r is the average of the minor and major radius [73].

Carbon ring electrodes (74) are very similar to the disk electrode except that there is no central electroactive area. Like disk electrodes, voltammograms obtained at slow scan rates are sigmoidal, indicating diffusion-limited steady-state current. The limiting current can be approximated (assuming quasi-spherical diffusion and a ring that is relatively unshielded) as:

$$i_l = 2\pi rnFDC \qquad (3)$$

3. Electrochemical Treatments and Preparation of Smaller Electrodes

To improve sensitivity and selectivity in the complex chemical matrix of biological systems, several electrochemical pretreatments have been developed. Blaedel and Mabbot [77] have reported that electrochemical treatment of pyrolytic carbon electrodes greatly enhances their selectivity and sensitivity. Gonon and co-workers [2,3] have gone on to develop a pretreatment for carbon fiber microelectrodes that applies a triangular wave potential (0–3 V at 70 Hz) for 20 seconds in phosphate-buffered saline, which also improves selectivity and sensitivity. This initial treatment is usually followed by secondary treatments that further improve the sensitivity and kinetics of the electrode response [3]. Secondary treatments often consist of application of either a continuous anodic or cathodic potential [57,78] or an alternating potential [42]. The amplitudes of these potentials are always less than that of the initial treatment.

The improved selectivity is due to the catalytic properties of the treated surface, resulting in a more reversible oxidation reaction. The oxidation peaks for ascorbic acid, dopamine, and dopamine metabolites shift to more negative values [33] and become well defined. This allows catechols that appear between +50 and +100 mV to be resolved from ascorbic

Electrochemistry in Neuronal Microenvironments 243

acid both in vivo and in vitro [2]. In addition, this treatment causes reversible adsorption of cations at the surface of the electrode, which leads to increased sensitivity for cations such as dopamine, norepinephrine, and serotonin [31,47,81]. Detection limits below 10 nM for dopamine and norepinephrine have been achieved. Low detection limits are essential since the basal or resting levels of biogenic amines in neuronal systems may be on the order of 5–50 nM or less [69].

There are some drawbacks to electrochemical treatment. At untreated electrodes, dopamine oxidation is more reversible than that of dihydroxyphenylacetic acid, and so this does not interfere with the detection of dopamine [79]. However, at treated electrodes resolution is compromised by dihydroxyphenylacetic acid oxidization at +50 mV and dopamine at +85 mV [57]. Treated electrodes also become more prone to fouling in vivo and can only provide reproducible results for approximately 5 hours. In addition, some drugs cause greater interference at treated electrodes, as seen in the case of caffeine both in vivo and in vitro [80]. Lastly, since the high sensitivity of the treated electrodes for catechols is caused by a relatively slow reversible adsorption process, the time constant of the electrode is increased (from 1 to 10–100 sec for dopamine) [3,69,81], thus, these electrodes respond sluggishly to changes in concentration [82]. Electrochemical treatment provides enhanced sensitivity and chemical resolution, but does not provide the needed temporal resolution. Temporal resolution is better attained at untreated electrodes using fast scan cyclic voltammetry [83]. Often it is necessary to choose between the speed of the measurement and selectivity. Therefore, the method of analysis and pretreatment procedures applied is dependent on the type information desired.

The effects of electrode treatment depend on the interaction of several factors: the type and size of the fiber, the range of voltages used, the duration and sequence of the applied waveforms, and also the treatment medium [84]. The mechanisms of the changes that occur upon treatment are unclear. A fivefold increase in capacitance observed by Kovach et al. [85] upon treatment correlates with the increased surface area of the treated carbon fiber electrodes observed with electron microscopy. The surface is roughened, and visible (0.1–1.0 μm) cracks are apparent. Such surface roughening has also been observed by Falat and Cheng [86] at treated graphite/epoxy electrodes.

The deterioration of the electrode surface by electrochemical treatment has also been used to decrease the size of the electrode tip through electrochemical etching [16]. In this case, cylindrical electrodes are made (see above) and then placed in a solution of 0.5 mM $K_2Cr_2O_7$ and 5 M H_2SO_4 [87] which is suspended in a small platinum loop. A 60 Hz sine wave, 6–15 V_{p-p}, is applied between the electrode and the loop. The taper of the etched fiber can be adjusted by changing its position in the drop. The electrodes are rinsed in distilled water and allowed to dry for 1 hour at 150°C. Electrodeposition of poly(oxyphenylene) is utilized to insulate the electrode tip and beveling or trimming to expose a carbon disk. Deposition is accomplished in a solution of 0.23 M 2-allylphenyl, 0.4 M allylamine, and 0.23 M 2-butoxyethanol in 1:1 (v:v) water-methanol using a platinum electrode at a potential of +4 V for 2 min. The electrodes are again rinsed with distilled water and cured for 2–4 hours at 150°C. Tip diameters accomplished by this technique are in the low micron range. When used in the amperometric mode at single bovine adrenal cells, these electrodes provide information identical to that from glass-encased carbon fiber microelectrodes. Greater adsorption of catecholamine is exhibited compared to conventional carbon fiber electrodes; however, this is not a problem when used for amperometry. The small size of these electrodes is advantageous for obtaining spatial information from surfaces where inhomogeneities in chemical composition exist. Electrodes with nanometer tip dimensions are the ultimate goal of this work so that monitoring of exocytosis can be carried out inside the synapse rather than outside as is presently done.

Electrodeposition of poly(oxyphenylene) has also been performed on electrodes etched in a oxygen-methane flame to obtain electrodes with overall physical dimensions of 400 nm [88]. Electrode tips approaching 100 nm, described by Wong and Xu, consist of pyrolyzed graphite in quartz capillaries pulled to a small tip by a horizontal micropipet puller [89]. These electrodes are also capable of detecting dopamine and are approaching the dimensions needed for analysis within a synapse.

4. Permselective Polymer-Modified Microelectrodes

Significant improvement in the selectivity and sensitivity of carbon fiber microelectrodes has been accomplished by coating the electroactive surface with permselective polymers. In this way, interfering substances can be blocked from the detecting surface while allowing the species of interest to selectively contact the surface. One commonly used polymer coating is Nafion [4,6,7,69,90,91], a perfluorosulfonated derivative of Teflon. It is an acid cation exchange polymer that excludes anions while allowing the permeation of cations. Thus in a neuronal environment, Nafion-coated electrodes block the electroactive anion, ascorbate, and the anionic metabolites of dopamine, norepinephrine, and serotonin while allowing the incorporation of dopamine, norepinephrine, and serotonin (cations at physiological pH). This greatly enhances the selectivity of these electrochemical probes since anions are often present in larger concentrations than the analytes (basal concentration in rat striatal extracellular fluid: 100 µM ascorbic acid, 5 µM dihydroxyphenylacetic acid, 50 nM dopamine, 1 µM 5-hydroxyindoleacetic acid, 60 nM serotonin [92]). Figure 6 compares the voltammograms of some of these species at both naked (Fig. 6A) and Nafion-coated electrodes (Fig. 6B). Nafion films also increase the sensitivity of carbon microelectrodes due to their preferential incorporation of large hydrophobic cations over smaller hydrophilic ones [93,94]. Endogenous cationic neurotransmitters like dopamine, norepinephrine, and serotonin readily diffuse into the Nafion film and become preconcentrated there even in the presence of relatively high concentrations of inorganic cations, common to biological systems. This results in several orders of magnitude lower detection limits for these neurotransmitters compared to uncoated electrodes [95]. Reliable detection has been accomplished down to approximately 50 nM with S/N = 3 [69]. Sensitivity, however, is directly related to the film hickness [90]. It has been shown at glassy carbon electrodes that thinner films provide greater sensitivity at the cost of a shorter linear response range due to film saturation.

The Nafion film thickness affects the response time of the carbon fiber microelectrode. Thicker coatings result in longer analyte diffusion times to the carbon surface, therefore increasing the response time of the electrode. The increased time constant of the electrode makes it less suitable for real-time monitoring of rapid concentration changes that occur in neurosecretion. The increase in electrode response time due to Nafion alone, however, is minimal (1–2 sec) in comparison to electrochemically treated Nafion-coated electrodes [69]. Therefore, the monitoring of dynamic events is best carried out with untreated Nafion coated electrodes. It has been established that no apparent distortion of the concentration pulse occurs with films of minimal thickness (<200 nm) since they allow sufficiently rapid dopamine permeation [6]. The film thickness can be controlled by the method used for the application of the Nafion. If dip-coating, thinner films can be made by increasing the time the disk electrode tip is in the Nafion solution. The coating on the tip appears to dissolve when placed in the Nafion solution [6]. Cylinder electrodes are more satisfactorily

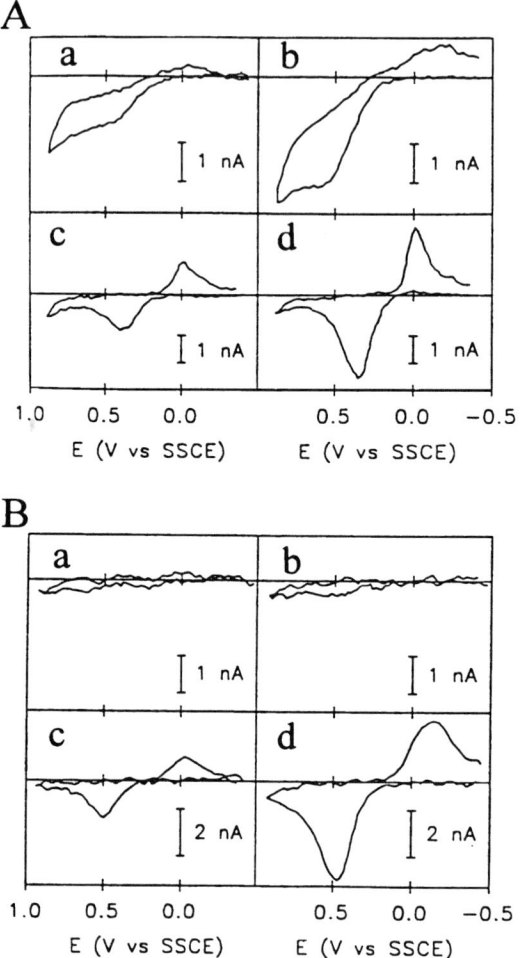

FIG. 6. (A) Background-subtracted voltammograms at naked carbon-fiber microelectrodes in pH 7.4 phosphate buffer recorded at 200 V/sec. a: 200 μM ascorbic acid; b: 200 μM dihydroxyphenylacetic acid; c: 10 μM norepinephrine; d: 10 μM dopamine. (B) Subtracted cyclic voltammograms at a Nafion-coated carbon-fiber microelectrode. All species as from (A). (From Ref. 68.)

Electrochemistry in Neuronal Microenvironments

coated by electroplating Nafion onto the surface [49,82,96]. In this procedure [49], a 4 µL drop of Nafion (5% DuPont 1100 EW Nafion, Solution Technology, Inc., Mendenhall, PA) is held in a 3-mm diameter loop of Ag/AgCl wire attached to the reference lead of a two-electrode potentiostat. The electrode tip is lowered into the Nafion drop for 30 seconds while +0.5 V is continuously applied. After removing the tip from the drop, it is allowed to air-dry and then baked for 10 minutes at 60°C. This method of applying Nafion is reported to be the most effective in terms of selectivity for dopamine relative to ascorbic acid (2000:1), whereas selectivity obtained by coating at higher potentials [96] has been reported as 1400:1. Dip-coating [6] selectivity has been reported as 500:1. The selectivity factor is the raio of current for equal concentrations of analyte versus that of the potential interferents [69]. It should be noted that pH dependence of dopamine voltammetric waves at Nafion-coated carbon fiber electrodes does differ significantly from that observed at uncoated electrodes [97]. The oxidation peak shows less pH dependence than the reduction peak at coated electrodes. This is important for the interpretation of electrochemical data collected in the context of neurotransmission, since metabolic activity as well as the neurotransmission process affect extracellular pH in a transient manner [98].

A carbon fiber microelectrode modified with Nafion and dibenzo-18 crown-6 (NA-CRO) has been reported to measure basal levels of dopamine in the striatum of anaesthetized rats with no interference from dihydroxyphenylacetic acid [92]. NA-CRO consists essentially of a ring of oxygens, which may serve as a chemical trap for positively charged substances such as dopamine, therefore increasing the amount of amine in contact with the sensor while enhancing the negative charge of the Nafion coating. NA-CRO–modified electrodes show a halved sensitivity for dihydroxyphenylacetic acid in comparison to Nafion-modified electrodes, while the sensitivity for dopamine increases to the range of 0.1–1 nM compared to 2–5 nM for Nafion electrodes.

Another type of polymer coating exhibiting rapid responses to dynamic concentration changes and antifouling properties has been applied to glassy carbon [99] and carbon ring [100] electrodes. This polymer, poly(ester sulfonic acid), is similar to Nafion in that it is permselective and excludes anions while binding counterionic species. The electrode surface is prepared by dipping the electrode tip into a diluted [1:20 (v:v) poly(ester sulfonic acid): acetone] polymer solution and allowing the film to dry in air. Like Nafion, the transport of selected ions is strongly affected by the film thickness. Faster response times are observed at poly(ester sulfonic acid)–coated electrodes, making them more suitable than Nafion for following transient events. In vivo use of these electrodes indicates a greater resistance to fouling than for uncoated electrodes, possibly due to restricted mass transfer of large molecules to the electrode [100]. Detection limits at these electrodes are a factor of two greater than naked electrodes when used to detect dopamine in the cytoplasm of the pond snail *Planorbis corneus*. The selectivity of poly(ester sulfonic acid)–coated electrodes for dopamine has been determined in the presence of the interferents ascorbic acid and dihydroxyphenylacetic acid, which did not significantly affect the sigmoidal voltammograms. The effectiveness of the cation exchange of the negatively charged poly(ester sulfonic acid) coating on carbon ring electrodes can be attributed to the preference of the hydrophobic polyester backbone for hydrophobic cations [101].

5. Enzyme-Modified Microelectrodes

In order to detect nonelectroactive substances relevant to neurochemical processes, microelectrodes have been modified by surface attachment of enzymes. The enzymatic products are electroactive and can be used to indirectly detect the substrate of interest, such as glucose, glutamate, choline, or acetylcholine. Enzymes can be attached to the electrode surface through entrapment in a polymer matrix like Nafion [102,103], covalently linking with carbodiimide [104] or avidin-biotin technology [105] and intermolecular cross-linking with a bifunctional reagent like glutaraldehyde [104]. These immobilization methods stabilize the enzymes by increasing their rigidity and thus making it more difficult to unfold and denature [106]. The addition of a protective polymer film has been incorporated in some instances to reduce electrode fouling, but this can also lead to increases in the response time [103] and sampling rate.

Glucose oxidase, an enzyme highly specific for glucose has often been used in enzyme-modified electrodes [64,65,102–104,107,108]. Glucose is oxidized in the presence of glucose oxidase by the following reaction:

$$\text{Glucose} + O_2 \longrightarrow \text{Gluconolactone} + H_2O_2$$

and can be detected indirectly by amperometric measurement of H_2O_2. Platinized carbon and platinum surfaces have been shown to give increased current responses due to H_2O_2 oxidation over that observed at naked carbon electrodes [104,108]. Partial platinization also allows the

Electrochemistry in Neuronal Microenvironments

electrodes to be operated at a more negative potential, resulting in less noise and lower background currents [109,110], while leaving some carbonaceous sites available for further electrode modification. Attachment of glucose oxidase by cross-linking with glutaraldehyde provides an enhanced response to glucose due to higher enzyme loading. These electrodes also exhibit a more reproducible response than those with covalently linked enzymes but owing to thicker sensing layers have a longer response time [64].

Abe and co-workers [65,107] have described the fabrication of enzymatic glucose sensors for measurements inside single nerve cells with platinized electrodes. The platinization process appears to result in a porous platinum coating as evidenced by an increase in the residual current following platinum deposition. The porous surface maximizes the amount of enzyme that can be attached due to the increased surface area. Platinization on carbon ring microelectrodes is accomplished by reducing hexachloroplatinate in the presence of lead acetate. Bovine serum albumin cross-linked with glutaraldehyde serves to immobilize the enzyme on the electrode surface. Electrode tip diameters from 2 to 10 µm in diameter have been examined and show a linear increase in response time with diameter suggesting that enzyme/albumin films placed on larger microelectrodes are thicker and thus restrict mass transport of glucose or H_2O_2 to the electroactive surface. These electrodes exhibit a linear response between 50 µM and 5 mM glucose. As is evident from the oxidation reaction of glucose, the amperometric response of this sensor is expected to be stoichiometrically dependent on the O_2 concentration. At low O_2 concentrations, O_2 is the limiting reagent for H_2O_2 production and shows that the sensor enzymatically measures both glucose and O_2 levels. These sensors are not affected by varying pH in the physiological region (6.0–8.0), but do experience interference from dopamine. Enzyme activity loss or electrode fouling most likely contribute to the modest loss of response (23–27%) observed over the course of 20 hours.

Glucose can also be detected through the used of a mono- and bi-enzyme sensor developed by Garguilo and co-workers [111]. In this method, glucose oxidase and horseradish peroxidase are co-immobilized on a single sensor and used to detect glucose in two different modes (Schemes I and II) due to their mutual ability to participate in electron transfer with the redox polymer poly(4-vinylpyridine)osmium 2,2′ bipyridine chloride (PVP-Os(bpy)$_2$Cl).

$$\text{Glucose} + \text{Med(ox)} \longrightarrow \text{Gluconolactone} + \text{Med(red)}$$
$$\text{Med(red)} \longrightarrow \text{Med(ox)} + e^-$$

Scheme I

$$\text{Glucose} + O_2 \longrightarrow \text{Gluconolactone} + H_2O_2$$
$$H_2O_2 + 2H^+ + 2\,\text{Med(red)} \longrightarrow 2\,H_2O + 2\,\text{Med(ox)}$$
$$\text{Med(ox)} + e^- \longrightarrow \text{Med(red)}$$

Scheme II

The Med(ox) and Med(red) represent the oxidized and reduced mediator in the redox polymer, respectively. At positive potentials, Scheme I is used, whereas Scheme II is employed at more negative potentials.

Choline biosensors have been produced in a similar fashion using choline oxidase and horseradish peroxidase [111,112]. Choline is important in neurotransmission studies because it is a metabolite of the neurotransmitter acetylcholine, which cannot be directly detected by electrochemical means. The choline sensor can only work in the bienzyme mode due to the high cosubstrate selectivity of choline oxidase for O_2. The coating solution contains choline oxidase, horseradish peroxidase redox polymer, and polyethylene glycol. Carbon fiber cylinder electrodes are dipped in this solution and dip-coated in Nafion. Use of the Nafion film eliminates possible interferences caused by ascorbic acid in in vivo studies. The scheme for the choline sensor is shown in Fig. 7. The response time reported for this sensor in an anaesthetized rat brain is 2 seconds for amperometric detection carried out at –0.1 V vs. SSCE. The detection limit has been determined to be 10 µM, which is a physiologically relevant choline concentration.

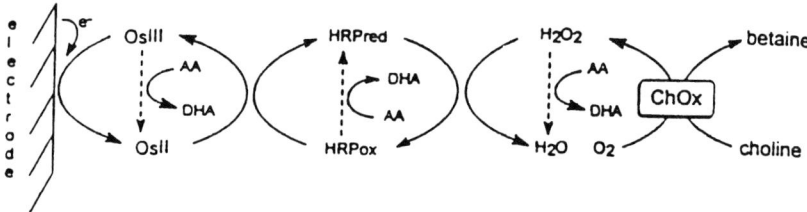

FIG. 7. The sequence of events by which an electrochemical signal is generated for choline, which itself is not electrochemcially active. (From Ref. 112.)

Electrochemistry in Neuronal Microenvironments

Huang and co-workers [113] have developed a choline sensor by immobilizing enzymes through cross-linking on platinum microelectrodes. The best response to choline is obtained by dip-coating the 25 μm platinum wire encased in glass in a 3% cellulose acetate solution in 1:1 acetone:cyclohexanone three times for 1-second intervals and then allowing it to air-dry for approximately 30 minutes. The choline oxidase combined with cross-linker (glutaraldehyde) is applied by dipping the electrode briefly in the mixture. Electrodes are inverted and air-dried. Linear responses from 5.0×10^{-7} to 1.0×10^{-4} M choline with response times of 15–20 seconds have been obtained when used in artificial brain extracellular fluid. The reaction for choline in the presence of choline oxidase is:

$$\text{Choline} + 2\,O_2 \longrightarrow \text{Betaine} + 2\,H_2O_2$$

Peroxide (H_2O_2) can be detected amperometrically at +650 mV vs. Ag/AgCl. Acetylcholine sensors have been produced in a similar manner with limited success by co-immobilization of acetylcholinesterase and choline oxidase on the electrode tip [113].

Dehydrogenase-modified carbon fiber microelectrodes have been reported for detection of other nonelectroactive substrates important in neuroscience [105]. Carbon fiber electrodes are modified by covalently attaching dehydrogenase enzymes through avidin-biotin coupling and utilized for the detection of ethanol, glucose-6-phosphate, and glutamate. Pantano and Kuhr have reported 300-msec response times and 0.5 mM detection limits for glutamate at glutamate dehydrogenase–modified electrodes.

Many electrode configurations with and without enzyme modification are possible. The size, shape, and surface structure can be manipulated to achieve the optimum sensitivity, selectivity, and stability for almost any application [7]. The optimized electrode can be combined with a variety of electrochemical techniques to further enhance detection of neurochemicals in microenvironments. Hopefully, the above brief descriptions of some of the existing electrodes and voltammetric methods will prove to be useful in designing detection methods for many experimental conditions. Applications of the techniques and electrodes described above to intracellular and extracellular neuroscience are discussed in the following sections.

III. INTRACELLULAR VOLTAMMETRY

A. Introduction

Detection of species inside single nerve cells is very important for gaining further insight into the metabolism, function, and regulation of individual neurons. Development of carbon and platinum microelectrodes with total overall dimensions in the micrometer range has allowed direct detection of several different electroactive substances in the cytoplasm of single nerve cells [60,100,114]. Electrode tips 10 µm or less in diameter are small enough to impale fairly large cells (100–200 µm diameter) without causing excessive damage to the cell. The small currents produced at these electrodes do not interfere with cell function, and only a small negative shift in potential is observed at the electrode upon implantation [62]. The rapid response times of voltammetric electrodes makes them ideally suited for monitoring the dynamic chemical changes that occur during neurotransmission and normal metabolism [67]. Electrode response times can vary depending on the electrode pretreatment and coating status (see above). In this section experiments will be described that use carbon and platinum electrodes for the measurement of ascorbic acid, dopamine, and oxygen intracellularly. Some of the results presented are obtained with electrodes coated to minimize electrode fouling, exclude common interferents, or facilitate the immobilization of enzymes.

Initial intracellular voltammetric studies have been conducted to measure the endogeneous ascorbic acid levels at single neurons. Meulmans and co-workers [60] have used the snail *Aplysia californica* as a model neuronal system to measure intracellular ascorbic acid voltammetrically. Studies have been carried out with 0.5- to 2-µm tip diameter carbon fiber or platinum electrodes using differential pulse voltammetry at single snail neurons. The endogenous ascorbic acid in these single neurons has been determined at approximately 100 µM. They have also been able to monitor the kinetics of uptake, clearance of two electroactive drugs [115], and serotonin levels in *Aplysia* neurons [60]. These initial studies of ascorbic acid demonstrate the potential of electrochemical methods for intracellular monitoring of electroactive neurochemicals.

B. Monitoring Intracellular Dopamine

Carbon and platinum microelectrodes and various electrochemical techniques have been used to monitor the easily oxidized neurotransmitter,

Electrochemistry in Neuronal Microenvironments

dopamine, inside the giant dopamine cell of the pond snail *Planorbis corneus*. Active transport of dopamine into the cell has been monitored with linear scan voltammetry using a poly(ester sulfonic acid)–coated carbon ring electrode implanted in the neuron [100]. The cell is bathed in 0.5 mM dopamine to provide dopamine for uptake. Oxidation current monitored at +0.78 V is plotted vs. time in Fig. 8. The rising portion of the peak represents the active transport of dopamine across the cell membrane, while the rapid decay represents dopamine clearance through metabolism and/or vesicularization of dopamine. In these experiments, the rate of clearance can be determined using the point of maximal rate decay following dopamine administration. The peak current for repeated bathing with dopamine gradually decreases over time. The average loss of response in the cell is 36% after 20 minutes of continuous voltammetry. This is most likely due to the loss of electrode sensitivity brought on by specific adsorption of large biomolecules rather than a physiological process. Loss of sensitivity caused by fouling of the electrode surface has been largely overcome by use of a different electrochemical technique, integrated pulse linear scan voltammetry, at platinized carbon ring electrodes [63]. This

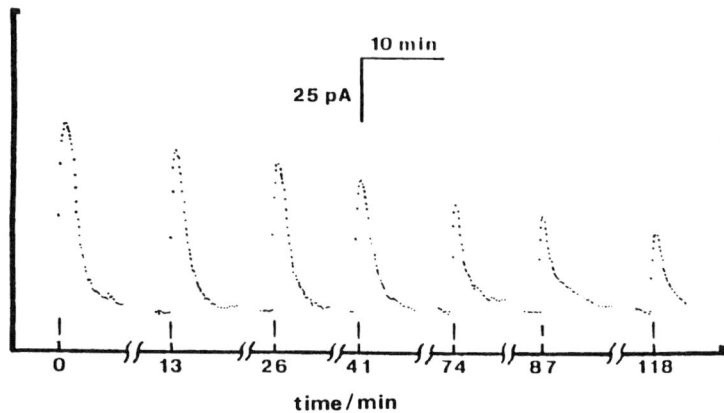

FIG. 8. Observed time course plot of repeated dopamine bathings (0.5 mM, 30 μL each arrow) with the electrode placed in the cell body of the identified dopamine neuron. The dopamine solution used for extracellular bathings was a modified snail saline solution. The current was monitored at +0.78 V vs. SSCE with 7 seconds between measurements. (From Ref. 100.)

method, also referred to as pulsed amperometric detection, minimizes electrode fouling while maintaining a reactive electrode surface. The applied waveform is based on the pulse techniques developed by Johnson's group for platinum and gold electrodes [116,117]. This consists of a rapid cyclic potential sweep followed by a large positive potential pulse to oxidatively clean the electrode surface and then a large negative potential pulse to restore the reactivity of the surface by stripping the metal oxide deposited during the previous cleaning step [63]. A schematic of the waveform is shown in Fig. 9. This technique allows continuous amperometric detection that is both sensitive and reproducible. The coulombic charge can be obtained by integrating the electrode current during the cyclic sweep. The average peak concentration of dopamine uptake in the *Planorbis* studies has been found to be 44 ± 2 µM, and the rate of dopamine clearance estimated from the linear portion of the declining limiting current is on average 0.29 µM/sec [63]. Microelectrodes coupled with cyclic voltammetry appear to be an ideal detection method for dynamic chemical processes inside single nerve cells.

Intracellular studies involving the measurement of basal dopamine levels have also been carried out to gain insight into the distribution of

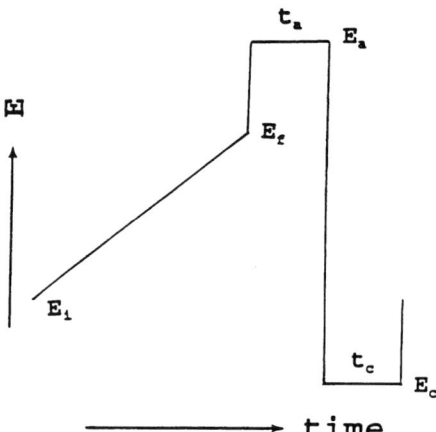

FIG. 9. Schematic illustrating the waveform employed for linear potential sweep pulsed voltammetry. The potential is scanned from E_i to E_f. E_a denotes an oxidative cleaning potential held for a time t_a, E_c an electrode reactivation potential held for t_c. (From Ref. 63.)

Electrochemistry in Neuronal Microenvironments 255

dopamine inside the cell. Carbon ring electrodes with electrode tips of approximately 2 μm have been placed inside the giant dopamine cell of *Planorbis* to monitor dopamine using staircase voltammetry [114]. Basal levels of free dopamine are below the detection limit of this method, so 0.5 mM dopamine is added extracellularly. The uptake of dopamine by the cell causes a sharp increase in the oxidation current detected in the cytoplasm, which agrees with results presented by Lau et al. [63,100]. The extracellular application of nominfensine (700 μL, 0.6 mM), an uptake inhibitor, diminishes the oxidation current, verifying that the rise in current observed is due to the uptake of extracellular dopamine, not free intracellular dopamine. Bathing the cell in 300 μL of 50% ethanol disrupts the cellular membranes and results in a large increase in intracellular dopamine (identified by comparison to a voltammogram of authentic dopamine). To verify the identification of the electroactive compound released upon exposure to ethanol, capillary electrophoresis has been carried out. The analysis involves sampling from the cell cytoplasm and injecting the pL volume (approx. 1% of total cell volume) onto a capillary. Capillary electrophoresis is useful for separating cytoplasmic samples as it is capable of carrying out high efficiency separations of extremely low volume samples. The separation is based on the differential electrophoretic mobilities of ionic and neutral species as they migrate in the potential field applied across the capillary [118]. A schematic of the capillary electrophoresis system used for acquiring, separating, and detecting cytoplasmic samples directly from intact neurons is shown in Fig. 10. Detection is accomplished via a carbon fiber electrode placed at the end of the separation capillary [119,20]. Species identification is determined by comparing the electrophoretic mobility of the analyte to standards under the same experimental conditions. In the ethanol experiment [114], no dopamine response is observed when the cell is resting (the intracellular dopamine concentration is below the detection limit of this method). Upon exposure to ethanol, the dopamine peak is quantified to be 14 fmol. An injection volume of 100–300 pL results in an estimated cellular dopamine concentration of 1.4×10^{-4}–4.7×10^{-5} M after lysing. This technique has also been coupled to off-column amperometric detection [121] with a detection limit of 10^{-8} M in 50 pL samples [122]. These lower detection limits are necessary for detection of basal levels of cytoplasmic free dopamine. Olefirowicz and Ewing [122] have determined basal dopamine concentration in *Planorbis* to be 2.2 ± 0.52 μM. Capillary electrophoresis with electrochemical detection has also allowed insight into the compartmentalization of dopamine in the giant

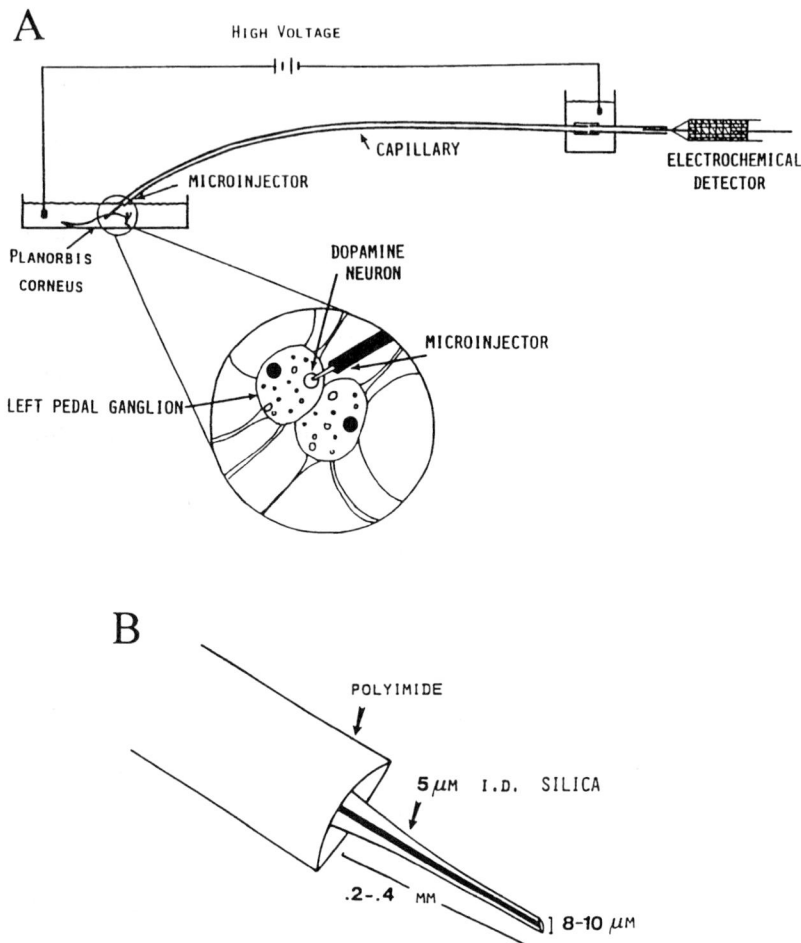

FIG. 10. (A) Schematic diagram of the capillary electrophoresis system used for the injection, separation, and detection of cytoplasmic samples. Also shown is an exploded view of the left and right pedal ganglia of *Planorbis corneus* with the microinjector inserted directly into the dopamine neuron. (B) A schematic of an etched microinjector. In order to construct the microinjector, approximately 1 cm of the polymide coating is removed from the high voltage end of the electrophoresis capillary. The exposed fused silica is then placed into a 40% solution of hydrofluoric acid for 35 minutes to etch the outer diameter to the appropriate size. (From Ref. 122.)

Electrochemistry in Neuronal Microenvironments

dopamine cell of *Planorbis* showing that the majority of dopamine (97%) is stored in vesicles. Electrochemical detection has been performed directly in single cells and combined with other techniques to provide additional insight into the structure and function of individual neuronal cells.

C. Monitoring Glucose in Single Cells

Development of glucose microsensors (see above) has provided the ability to qualitatively monitor glucose transients in cell cytoplasm. Transient changes that occur in single cell metabolic and respiratory processes during cell stimulation, neurotransmitter release, and neurotransmitter reuptake can now be investigated [119]. Glucose metabolism is of particular interest because it is a major energy source for the brain and energy is required for neurotransmitter secretion [62]. Abe and co-workers [65,107] have used 2-μm diameter platinized carbon ring electrodes with amperometric detection (0.6 V vs. SSCE) to monitor intentional manipulations of cytoplasmic glucose in the giant dopamine cell of *Planorbis*. The injection of 2 pL of glucose into the cell is detected as an immediate increase in oxidation current followed by a gradual decrease with a return to baseline after 2 minutes owing to glucose metabolism and/or storage in the cell. A second stimulation results in similar response. The intracellular response to glucose is shown in Fig. 11. Control experiments with the injection of only buffer result in no response, indicating that the signal observed is most likely due to glucose oxidation. To be sure that the observed current is not due to direct oxidation of glucose at the platinum surface, controls using uncoated platinized carbon ring electrodes have been carried out. No definitive response results, indicating that glucose is only being detected through the enzymatic breakdown product and also that compounds electro-oxidized at 0.6 V vs. SSCE such as dopamine do not increase inside the cell following glucose injection under these conditions. Response times as low as 270 ms have been observed with only moderate loss of response (36.5%) after the cell experiments. Based on calibration experiments after use in the cell, the peak change in cytoplasmic glucose concentration following injection is 0.8 mM. The low level of oxygen measured for this cell at 0.032 mM may introduce up to a one-third error in the glucose measurements [62]. Hence, accurate evaluation of glucose will require simultaneous measurement of oxygen levels. Although relative levels of cellular glucose can still be evaluated with these electrodes, in the next section the use of microelectrodes to monitor intracellular oxygen is discussed.

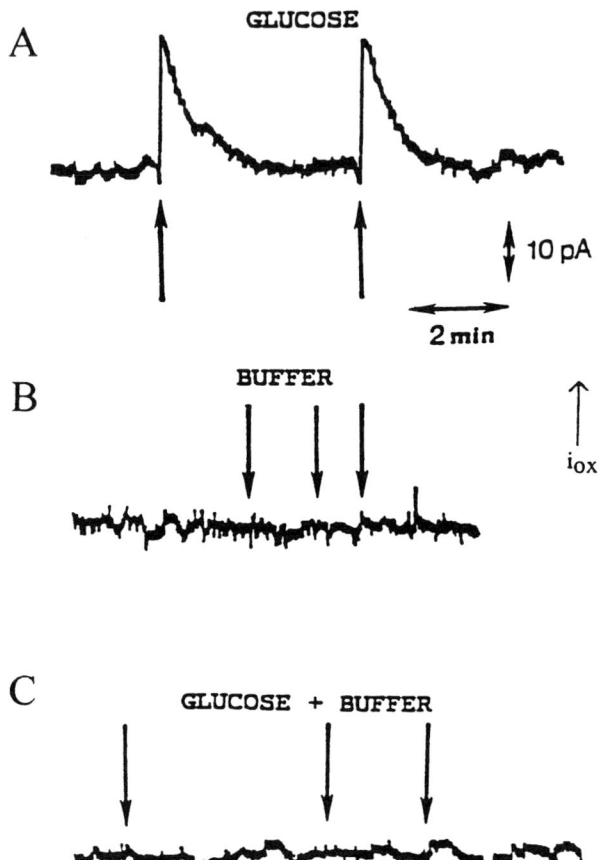

FIG. 11. Current observed at electrodes placed in the large dopamine neuron of *Planorbis corneus*. (A) Response to a 2 pL, 3 M intracellular glucose injection. (B) Response to an 8 pL intracellular injection of only pH 7.4 buffer. (C) Response of an electrode not coated with glucose oxidase to an 8 pL, 3 M intracellular glucose injection. The electrode potential was 0.6 V vs. SSCE in all cases. (From Ref. 65.)

D. Monitoring Intracellular and Extracellular Oxygen

Oxygen monitoring is important in neuronal microenvironments due to the close link to synthesis, metabolism, release, and uptake of neurotransmitters [62]. Increases in O_2 consumption following electrical stimulation of nerve tissue support its utility in the neurotransmission process [123–125]. Shibuki [123] has proposed that increases in O_2 consumption represent energy required to drive the exocytotic process. Oxygen levels in the giant dopamine cell of *Planorbis* have been measured using Nafion-coated platinized carbon ring electrodes with differential pulse voltammetry [62]. In contrast to carbon surfaces, platinum is a good catalyst for the reduction of O_2 to H_2O, allowing detection at more positive potentials (–0.4 V to +0.4 V vs. –1.2 V to +0.8 V at carbon electrodes) and minimizing the formation of cytotoxic H_2O_2. The Nafion coating minimizes fouling of the electrode in the cell and is permeable to O_2 [126]. Employing Nafion-coated platinized electrodes [62], it has been shown that the O_2 concentrations near the dopamine neuron (10 μm from the cell: 0.041 ± 0.001 mM) are significantly lower than further away (0.4 cm from the neuron: 0.16 ± 0.046 mM O_2). This suggests that O_2 is consumed by the cell and surrounding tissue. The intracellular O_2 concentration has been estimated to be 0.032 ± 0.004 mM from the limiting current of the cyclic voltammograms. Rapid transport of O_2 across the cell membrane has been observed when cells are bathed in O_2-containing solutions. A large increase in current that decreases within 2 minutes back to baseline suggests the rapid consumption of O_2 in cell respiration and/or metabolism. Lastly, air-saturated 3 M KCl (K^+ stimulates exocytosis) has been applied and the intracellular O_2 response compared to that seen with the application of air-saturated Ringer's solution (no exocytosis). The 3 M KCl/oxygen stimulated response is considerably less, possibly due to immediate consumption of the O_2 to drive exocytosis caused by the elevated extracellular K^+ levels, as proposed by Shibuki [123].

IV. EXTRACELLULAR VOLTAMMETRY

A. Introduction

Extracellular voltammetry has been and continues to be used extensively to gain a better understanding of the chemical nature of neuronal function. Since sampling is performed outside the cell, extracellular voltammetry would seem to be more easily implemented than intracellular studies. This is not necessarily correct. Probes sampling the volume surrounding a cell must still be small (μm) for placement at the chemical release site. The larger sampling environment inherent in extracellular monitoring requires very sensitive probes (μM to nM) to detect transient concentration pulses that undergo rapid dilution in the relatively unconfined volume outside the cell. Very rapid response times (msec) are required of electrodes due to rapid analyte diffusion and the almost equally rapid reuptake of analyte by the multitude of surrounding neurons or cells. Concentration pulses may last from 1 to 15 seconds [127,128] depending on the duration of the stimulation applied. Working outside the cell also introduces a whole new world of possible interferences, and so electrodes and techniques that are highly selective are necessary. A great number of extracellular studies have been carried out in the rat brain; however, more homogeneous and easily manipulated environments such as cultured cells have also been used to provide convenient neuronal models. Indeed, single cell measurements of exocytosis often involve events on the millisecond to microsecond time scale. The following discussion deals with in vivo voltammetry in the brain microenvironment, but many of the principles can be and are applied to voltammetry carried out on cultured cells in subsequent sections.

Several guidelines are important when working in an in vivo microenvironment. Placement of the microelectrode must be accomplished with minimal disturbance of the biological system while pinpointing the location of secretion. In the rat brain, specific regions are known to have higher concentrations of specific neurotransmitters. For example, much of the monitoring of dopamine is carried out in the nucleus accumbens or caudate nucleus, two areas with large populations of dopaminergic terminals. The stimulating and detection probes are placed in close proximity so that adequate stimulation can be obtained without disturbing the detected signal by injection [129]. Brain studies have shown that the amount of dopamine detected is dependent on the stimulus duration and frequency. Longer duration (up to 3 sec) and more frequent stimulation causes larger quantities of dopamine to be released [129]. Stimulation in these studies is commonly electrical or chemical. In the brain and also in culture to some extent, diffusion of the analyte to the electrode tip is affected by surrounding cells. It has been estimated that diffusion in the brain microenvironment is three times slower than in homogeneous aqueous solutions [130]. In certain protocols, this can lead to some delay in the analyte response at the electrode and is important when comparing in vivo voltammograms to in vitro calibrations. The sensitivity of electrodes is greatly affected by contact with biomolecules in the in vivo environment. A 40–50% decrease

Electrochemistry in Neuronal Microenvironments

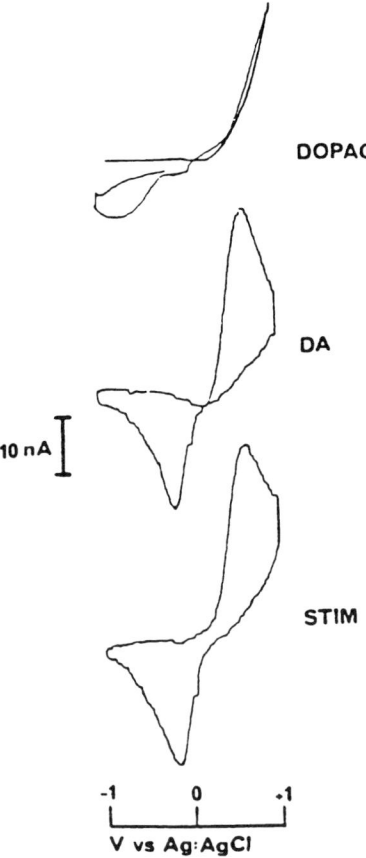

FIG. 12. Cyclic voltammograms obtained in the caudate nucleus from endogenous and exogenous material. The stimulation that invoked endogenous material is shown in the bottom trace, and iontophoretically applied dihydroxyphenylacetic acid and dopamine in the upper and middle traces, respectively. (From Ref. 128.).

in sensitivity is usually observed in the first 2–2.5 hours of implantation and then stabilizes for up to 12 hours [69]. The loss of electrode sensitivity in cell cultures is relatively minimal in comparison, but still needs to be considered for accurate quantitation. Another factor that must be kept in mind when working in the brain is that the signal observed is most likely from several cells. In culture, the ease of electrode placement, diffusion, and the sparser population of cells is conducive for single cell monitoring. Probably one of the most important factors in all these studies is the proper identification of the detected substance.

Qualitative analysis with in vivo measurements often requires both voltammetric and pharmacological considerations. One of the most useful and widely accepted electrochemical means of identifying substances is fast scan cyclic voltammetry [128]. The oxidation and reduction scan in fast scan voltammetry provides a more defined voltammogram for identification than those obtained using normal pulse voltammetry (Fig. 12), though these are useful in some cases [38]. The large charging current to faradaic current obtained with fast scan voltammetry necessitates background subtraction of prestimulation voltammograms from those obtained during neuronal stimulation [128]. The more reversible oxidation of protonated amines like dopamine and norepinephrine provides more defined voltammograms at fast scan rates than those for anions like ascorbic acid and dihydroxyphenylacetic acid, and so fast scan voltammetry provides added selectivity. For direct comparison of voltammograms of detected species in vivo with the voltammograms of the suspected substance, a standard injection is often performed [7]. The concentrations detected in vivo should also correlate with other methods of detection. In vivo studies should always include well-documented neuropharmacological interventions as part of the proof of the identity of the detected substance [46]. The most commonly used electrodes in extracellular voltammetry are Nafion-coated cylinders or beveled disk electrodes.

B. Brain Studies

In vivo experiments have constituted a great deal of work in the last 20 years. A few of these results are discussed here to sample the potential of this methodology from an electrochemical viewpoint. Detailed and exhaustive discussion of the neurochemistry examined is beyond the scope of this review. The first study involves the use of carbon fiber disk electrodes to investigate the effect of local application of KCl and amphetamine in the caudate nucleus of the rat brain [38]. The time course of the response has been followed using repeated chronoamperometric steps to a potential of +0.55 V. Characteristic voltammograms are used to identify the analyte after KCl stimulation using backstep-corrected normal pulse voltammetry. Release of neurotransmitter is expected since elevated extracellular K^+ levels depolarize cell membranes, therefore triggering exocy-

Electrochemistry in Neuronal Microenvironments

totic release. An interesting finding has been that the rise in current upon injection of the drug amphetamine (Fig. 13) is caused by increased levels of ascorbic acid as speculated by Gonon and co workers [131]. Using electrical stimulation and the same detection scheme, pharmacological manipulations have been used to indicate that there might be two compartments of dopamine storage in this area of the brain: one for immediate release and a second reserved for emergency release [79].

As previously discussed, fast scan rate voltammetry provides more defined and thus more characteristic voltammograms for the identification of in vivo species. Kuhr and Wightman [128] have used fast scan voltammetry at Nafion-coated beveled carbon disk electrodes with electrical stimulation of the rat medial forebrain to monitor the time course of stimulated dopamine release in the caudate nucleus. The maximum concentration of dopamine detected at a single electrode location during stimulation has been found to be directly proportional to the frequency of pulses in the stimulation train. Dopamine is observed in the extracellular fluid for approximately 1.5 seconds when short stimulation times are employed demonstrating the brief time that detectable dopamine levels are present

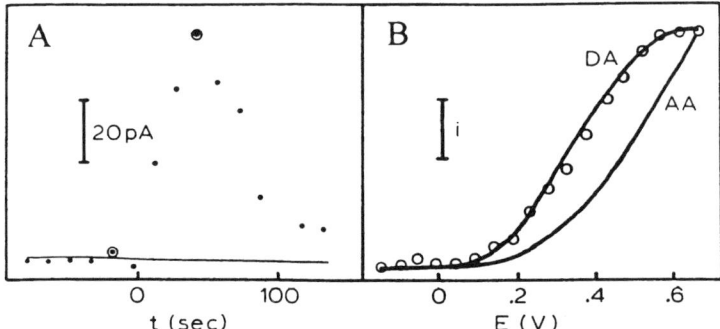

FIG. 13. Electrochemical response to injection at 0 time of 0.6 µmol of potassium chloride in 1.0 µL of physiological buffer into the caudate nucleus. Distance from electrode to syringe tip is 0.8 mm. (A) current vs. time response at 0.5 V. Solid line, injection of physiological buffer only; points, injection of potassium chloride. (B) difference voltammograms. Solid lines, dopamine (25 µM) and ascorbic acid (200 µM) in pH 7.4 buffer after in vivo use; circles, in vivo result from subtracting voltammograms obtained at the circles shown in (A). Current scales: i = 25 pA, dopamine; i = 28 pA, ascorbic acid; and i = 22 pA, in vivo. (From Ref. 38.)

following stimulation. The disappearance of dopamine appears to occur exponentially, which is characteristic of first-order uptake kinetics. Using an uptake inhibitor, nomifensine, a large increase in current is observed when cells are stimulated with a small electrical stimulus (15 pulses in 250 msec), which normally gives a barely detectable signal (Fig. 14). This confirms the large effect uptake has on the detected dopamine signal. Nomifensine allows measurement of dopamine overflow before uptake occurs and so allows detection of dopamine release as the result of a single stimulus pulse. Another method to investigate released dopamine without concern of uptake is to "knock out" the dopamine transporter. Experiments have been conducted by Wightman and co-workers [132] on "DAT (dopamine transporter) knockout" mice that do not have the dopamine transporter. Mice that are deficient by one allele of the dopamine transporter show slowed reuptake of dopamine, and mice that are missing both alleles demonstrate large amounts of dopamine release (Fig. 15). These results confirm that the dopamine uptake transporter functions to clear the released dopamine and provide a model to examine the role of the dopamine transporter in pharmacology. Additional pharmacological manipulations, not discussed here, have implicated the dopamine transporter as the target for the biochemical action of cocaine and amphetamine [132].

The simultaneous detection of two neurochemically relevant substances in vivo has also been accomplished using fast scan rate voltamme-

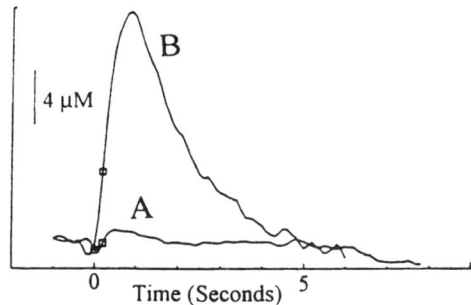

FIG. 14. Time course of dopamine release before (A) and 8 minutes after (B) administration of 20 mg/kg nomifensine. The release was induced by stimulation consisting of 15 300 µA biphasic pulses delivered at a frequency of 60 Hz. Although only one time course is shown, identical temporal profiles were obtained in three additional rats. (From Ref. 127.)

Electrochemistry in Neuronal Microenvironments

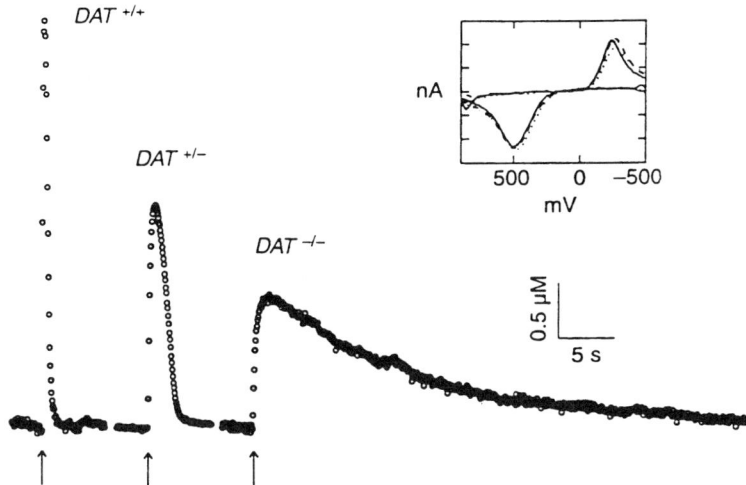

FIG. 15. Typical recordings of electrically stimulated dopamine efflux in striatal slices from genetically engineered mice. DAT$^{+/+}$ represents mice with both alleles of the dopamine transporter; DAT$^{+/+}$ is for mice missing one allele; DAT$^{-/-}$ is for mice missing both alleles. A single biphasic electrical pulse (4 msec, 350 µA) was applied at the time indicated by the arrows with a locally placed bipolar electrode. Data points (open circles) were collected from a recording electrode every 100 msec. Inset shows cyclic voltammograms recorded at the peak of dopamine efflux in each slice. Voltammograms recorded from DAT^{-} (solid line), DAT$^{???-/-}$ (broken line), and DAT$^{-/-}$ (dotted line) mice were scaled to approximately the same peak amplitude for comparison. These cyclic voltammograms are not significantly different from that obtained with authentic dopamine. The average time to clear released dopamine was 1, 3, and 100 sec in DAT$^{+/+}$, DAT^{--}, and DAT$^{-/-}$, respectively. (From Ref. 132.)

try. Zimmerman and Wightman [47] have demonstrated that both dopamine and O_2 can be independently but simultaneously monitored at a single Nafion-coated disk electrode by scanning from 0.0 to +0.8 V followed by a cathodic sweep to −1.4 V before returning to a resting potential at 0.0 V. This technique provides rapid analysis time with a scan rate of 400 V/sec.

A final example of in vivo voltammetry has been used to suggest that the brain is not "hard-wired," meaning that some neurons might and most likely do communicate by pathways outside of the synapse. Garris and co-workers [133] have used fast scan rate voltammetry with Nafion-coated beveled disk electrodes and electrical stimulation to examine dopamine efflux from the synaptic junction. It has been determined that binding of released dopamine to receptors and uptake sites does not appreciably alter its efflux from the synaptic cleft in the rat nucleus accumbens. This implies that the dopamine synapse is designed for efficient transmitter efflux from the synaptic gap possibly for the purpose of extrasynaptic transmission and the ability to communicate with a greater number of cells more quickly. The proposed scheme contradicts the established model of neurotransmission derived from experiments at neuromuscular junctions where communication occurs only inside the synapse.

As these few examples amply show, electrochemistry can be used for identifying released neurochemicals, monitoring the effects of drugs, studying the kinetics of release and uptake, and even investigating the basic process of transmission in the brain. Since measurements conducted in vivo are multiple cell (tissue) experiments, the next frontier has been to examine molecular events in singe-cell systems.

C. Overview of Single-Cell Systems Used in Electrochemical Studies of Exocytosis

Electrochemical techniques have been used to investigate many cells that release easily oxidized substances by exocytosis. Exocytosis can be summarized as the docking of a vesicles (storage compartment) to the cell membrane and subsequent release of the contents by fusion of the vesicle with the cell membrane. The small amount of neurotransmitter secreted from single vesicles can be detected electrochemically with microelectrodes. The nonsynaptic cell systems whose secretion of chemical messengers has been monitored using microelectrodes include bovine adrenal chromaffin cells, rat pheochromocytoma (PC12) cells, beige mouse mast cells, and human pancreatic β cells. Synaptic systems including invertebrate and mammalian neurons have also been investigated. An overview of some key studies using these cells will be presented in the sections that follow.

D. Bovine Adrenal Chromaffin Cells

The majority of work concerning single-cell exocytosis has been performed on bovine adrenal chromaffin cells. These cells are used as models of neurotransmitter biosynthesis, metabolism, and secretion due to their neuroectodermal origin and biochemical and functional similarities with sympathetic neurons [46,52]. They are obtained from the adrenal medulla of freshly slaughtered cattle and are easily maintained in culture for a couple of weeks [134]. The average cell radius is 8 μm [135] with secretory vesicles or granules (avg. diam. = 400 nm [136]) that store an average of 5–10 amol [135] of the catecholamines epinephrine and norepinephrine. Bovine adrenal cells have 20,000–30,000 vesicles and so contain approximately 150 fmol of catecholamine total [51–53,137,138]. Similar to sympathetic neurons, they have been shown to exhibit Ca^{2+} dependent exocytosis and are useful for investigations concerning this critically important process.

Though several different studies involving adrenal cells are discussed here, the experimental protocol is similar. Chemical stimulation is administered via a short pressure injection through a micropipette placed approximately 20 μm from the cell (nanoliter volumes ejected). It is estimated that the cell initially experiences an exposure to the stimulant that is approximately 90% of the injected concentration and this concentration of stimulant decreases at a rate proportional to the diffusion coefficient of the stimulant [139]. The electrode, which is usually beveled for these experiments, is manipulated via a micromanipulator onto the cell until the membrane is visibly deformed and then retracted to the desired position above the cell (Fig. 16A).

Direct chemical evidence for exocytosis has been reported by Wightman and co-workers at single adrenal cells with electrochemical detection at beveled carbon fiber microelectrodes [51,52]. The applied stimulant (100 μM nicotine) results in a series of sharp irregular concentration spikes superimposed on a secretion envelope (Fig. 16B). This response is consistent with the exocytotic theory of secretion [24] in which substances to be secreted are stored in intracellular vesicles and fusion of the vesicles with the cell membrane results in rapid expulsion of the contents into the extracellular space. Further support that the observed spikes are caused by exocytosis has been provided in a study by Chow et al. [53] in which the time-averaged signal of the carbon fiber microelectrode closely resembles the derivative of the membrane capacitance trace (Fig. 17). Membrane capacitance is proportional to the cell surface area and increases when vesicles fuse with the plasma membrane during exocytosis [140].

Background subtracted cyclic voltammograms obtained by the subtraction of data collected before stimulus from those collected immediately following maximal secretion show that the substance being detected fol-

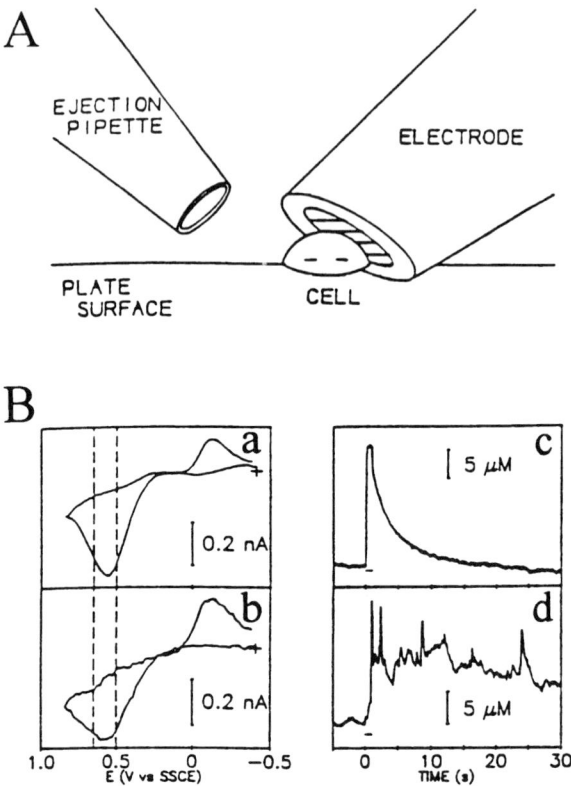

FIG. 16. (A) Experimental arrangement for measuring secretion from single adrenal medullary chromaffin cells. Drawing is approximately to scale. (B) Cyclic voltammetric response (200 V sec^{-1}, repeated at 100-msec intervals) of a carbon fiber electrode to norepinephrine ejection from an adjacent micropipette and to catecholamine secretion from a single chromaffin cell. Panels a and b are averaged background-subtracted voltammograms of the substances whose concentration changed during the measurement interval of panels c and d. Each time point in panels c and d is the integrated current recorded from 0.5–0.6 V from individual voltammograms (hatched lines in panels and b); bars to the right in panels c and d are the conversion of current to catecholamine concentration based on calibration curves constructed with standards. Panels a, c, electrode response to 1-sec 3 nL ejection of 20 μM norepinephrine applied at t = 0 with the ejection pipette 20 μm from the electrode. Panels b, d, electrochemical response obtained with the electrode tip adjacent to a single cell; at t = 0, a 1-sec ejection of nicotine (100 μM) was made 20 μm away from the cell. (From Ref. 51.)

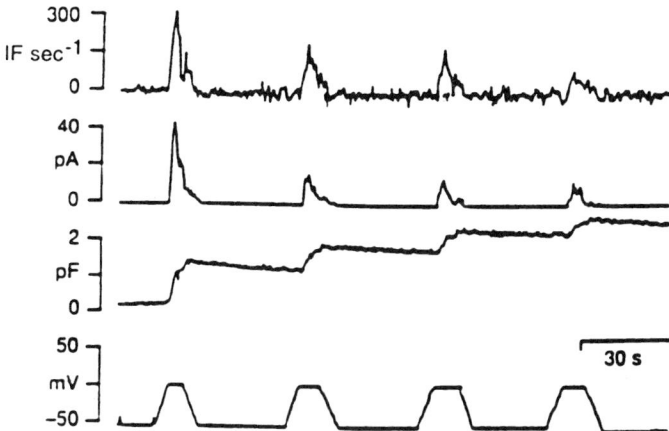

FIG. 17. Correlation between the amperometric signal and the rate of capacitance change. A cell was held at –60 mV and repetatively depolarized to 0 mV for about 8 seconds at a time (see lowest trace for voltage protocol). Capacitance (second trace from bottom) increased during depolarizations, and the amperometric signal (third trace from bottom) displayed peaks. The time derivative of the capacitance signal, which, like the amperometric signal, should be proportional to secretion rate, is displayed in the top trace (in femtofarads per second) for comparison with the latter. (From Ref. 53.)

lowing stimulation is a catecholamine. It is most likely epinephrine or norepinephrine since these are the catecholamines found in bovine adrenal cells, but due to the similarities in their voltammograms under the experimental conditions initially used (scan rate 200 V/sec) identification of the specific catecholamine is not possible. However, cyclic voltammetry carried at slower scan rates (10 V/sec) at Nafion coated carbon fiber microelectrodes provides the needed resolution to differentiate epinephrine from norepinephrine [141]. Though norepinephrine and epinephrine are structurally similar (see Table 1), the rate of cyclization of epinephrine is much faster and leads to the formation of an adrenochrome (not shown), which appears as a new voltammetric wave in slow scan cyclic voltammetry. Two criteria that can be used to distinguish the voltammograms from each other are the peak potentials and the peak separation. Figure 18 shows representative time traces obtained at a single cell following stimulation with 100 μM nicotine. In this case, the stimulated adrenal cell releases both epi-

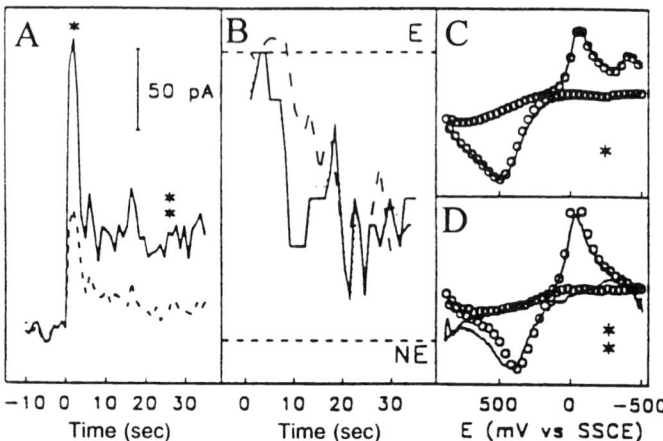

FIG. 18. Cosecretion of epinephrine and norepinephrine from an adrenal cell in culture. (A) Oxidative current (solid line) and adrenochrome current (dashed line) vs. time. (B) E_p (solid line), ΔE_p (dotted line), and i_{cyc}/i_{ox} (dashed line) vs. time. The scale for each of these parameters is adjusted so that the values will fall between the limits of pure epinephrine (upper dashed line) and pure norepinephrine (lower dashed line) as determined by the postcalibration. (C) Background-subtracted cyclic voltammograms (solid line) taken at the asterisk in (A). Circles represent the background-subtracted cyclic voltammogram of the epinephrine postcalibration. Voltammograms have been normalized to the oxidative peak currents. (D) Background-subtracted cyclic voltammogram (solid line) taken 25 sec after the stimulus. Circles represent the background-subtracted cyclic voltammogram of the norepinephrine postcalibration. (From Ref. 141.)

nephrine and norepinephrine as indicated by the voltammetry at 10 V/sec. Examining several cells, 75% of the cells release primarily either norepinephrine or epinephrine, while 25% release both norepinephrine and epinephrine. In further investigations using a cell-lysing agent and HPLC analysis [142], there appears to be a direct correlation between the proportions of epinephrine and norepinephrine contained in the cell and the proportions of these catecholamines released.

Measurements of nicotine-induced catecholamine secretion are attenuated as the electrode is moved further from the cell. The transport of catecholamines from the cell surface to the detecting electrode is thought to be a diffusion-controlled process [143], therefore, the observed amplitude and time course of the signal is determined by the distance between the electrode and the cell [144]. Kawagoe et al. [16] demonstrated that the amplitude of the induced spikes decreases with increasing distance between the cell and the sensing electrode for both 6- and 2-µm tip diameter electrodes. Though the spike shape changes with electrode-cell distance, the areas under these spikes do not change over a range from 1 to 10 µm. Recent studies with the electrode directly contacting the cell surface have suggested that diffusion is not the only process occurring after exocytosis. Chow et al. [53] reported a delay in vesicle fusion compared to neuronal systems as well as a small peak or "foot" at the onset of the current transient for vesicular release. Rapid fluctuations in transmitter release have also been found in a more detailed study of the foot region, attributed to opening and closing of the fusion pore prior to complete exocytosis [145]. In addition, Wightman et al. [146] studied the temporal characteristics of vesicular release in detail and describe a rate-limiting kinetic step that involves dissociation of the neurotransmitter from a protein matrix following vesicle fusion with the cell membrane. Three distinct stages of exocytosis have been proposed [147], which include formation of a fusion pore, pore expansion to release catecholamine, and finally dissociation of the intravesicular matrix. Adrenal cells appear to store catecholamines in a matrix of chromogranin A, which is solubilized at extracellular pHs but is polymeric at the more acidic pH inside the vesicle. When the vesicular matrix is expelled, the change in pH causes the chromogranin A to dissolve, thus releasing catecholamine.

To further prove that the changes in catecholamine concentration upon stimulation are due to exocytosis, the Ca^{2+} dependence of this change has been investigated. Ca^{2+} is generally considered to be required for exocytosis to occur. Wightman and co-workers [52] have carried out experiments in which 100 µM nicotine is administered to a cell in 2.0 mM Ca^{2+} and to a cell in Ca^{2+} free buffer. The response to nicotine exposure in the Ca^{2+}-free buffer is $1.4 \pm 0.7\%$ of the response for the same cells exposed to nicotine in the presence of 2.0 mM Ca^{2+}. It has also been established that blocking Ca^{2+} channels on the cell membrane with Cd^{2+} leads to a dramatically decreased response [134] Jankowski et al. [148] employed cells permeabilized with digitonin to more critically examine the role of Ca^{2+} inside the cell by bypassing events associated with transport across the cell membrane. At low concentrations (20 µM, 15–20 sec exposure), digitonin permeabilizes the cell membrane but not the granule membranes. In dishes of cells bathed with Ca^{2+} at concentrations greater than 0.6 µM, permeabilization with digitonin causes secretion by the influx of Ca^{2+} into the cell.

Thus, an increase in the concentration of extracellular Ca^{2+} results in an increased frequency of catecholamine secretion but does not alter the quantity of catecholamine released per event, implying that release is always from the same population of vesicles.

Different methods of stimulation have been investigated in order to see if they have an effort on the spike frequencies or the distribution of spike sizes [52]. Stimulation with nicotine (100 μM) that binds to a membrane receptor and triggers exocytosis through a receptor mediated mechanism results in the highest frequency response (1.2 ± 0.2 Hz), whereas at the lower concentration (10 μM) the response is only 0.5 ± 0.2 Hz. The use of a nicotinic agonist, carbamoylcholine (1 mM), results in a similar response at 0.7 ± 0.2 Hz. Stimulation with elevated levels of KCl (60 mM), which triggers exocytosis through direct depolarization of the cell membrane, also provides a similar duration response with a frequency of 0.6 ± 0.2 Hz. The frequency of the spike occurrence might be affected by stimulant concentration and also by the mechanism of release. However, the mean spike areas are not significantly different with varied stimulant administered (1.19 ± 0.9 pC with 60 mM K^+, 0.98 ± 0.13 pC with 1 mM carbamoylcholine, 1.04 ± 0.16 pC with 10 μM nicotine, and 1.19 ± 0.10 pC with 100 μM nicotine). These amounts, approximately 5 amol, correspond well with the vesicular content determined by other techniques (5–10 amol) [135].

Smaller flame-etched carbon fiber electrodes (1 μm) have allowed the mapping of the surface of adrenal cells to unveil spatially localized zones of exocytotic release [149]. Ba^{2+} has been used as the secretagogue owing to its ability to cause continuous release for several minutes. Two working electrodes have been employed to monitor release from distinct locations on the cell surface (Fig. 19). It has been estimated that the electrode samples events at the cell surface within a 2-μm radius of its tip. Release is detected at some locations and not observed at others, suggesting that secretion is spatially localized. These sites of release are stationary on a time scale of several minutes and are independent of the stimulant employed. The existence of these regions has been verified using immunocytochemistry and confocal microscopy in which fluorescent images of exocytotically deposited dopamine-β-hydroxylase in the cell membrane has been observed.

Several other studies employing amperometric detection of catecholamines at microelectrodes in conjunction with techniques to detect Ca^{2+} have been carried out in an effort to learn more about the role of Ca^{2+} in the

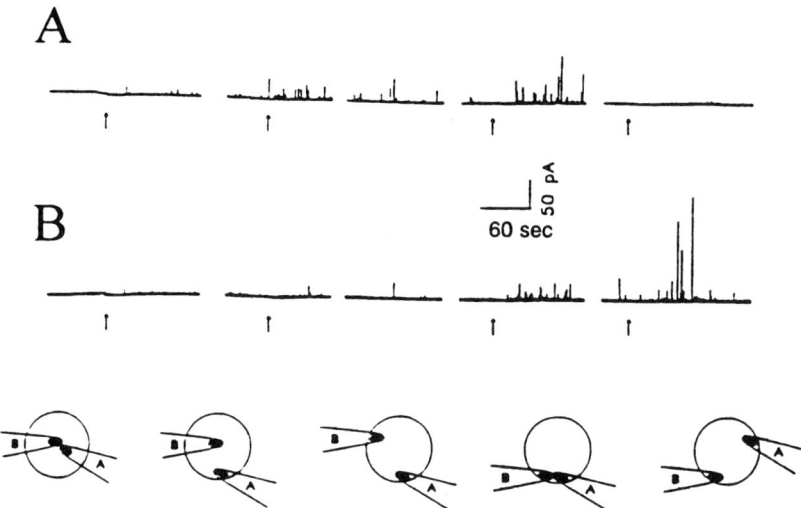

FIG. 19. Simultaneous amperometric measurements with flame-etched electrodes (≈1 μm radius), positioned 1 μm from the surface of individual bovine adrenal medullary cells. The breaks in the traces indicate the times when one of the electrodes was repositioned. The arrows below the traces indicate 5-second stimulations with 2 mm Ba^{2+}. Traces A and B correspond to the currents from the respective electrodes shown in the lower diagrams. The diagrams show a top view of the placement of the electrodes at the cell surface. Solid ellipses in the diagrams represent the active carbon surface of the microelectrodes. (From Ref. 149.)

exocytotic process. Fura-2 AM, a chelating fluorescent probe that allows detection of cytosolic free Ca^{2+} without perturbing the cell membrane [150], has been used in conjunction with electrochemical detection of catecholamine release to correlate responses to various chemical agents [151]. Secretion by exocytosis is strongly coupled to elevation in cytosolic Ca^{2+}. This elevation, whether induced by transmembrane entry or mobilization of intracellular Ca^{2+} stores, must cause the intracellular Ca^{2+} concentration to rise above a threshold value before release occurs. The threshold value is approximately 50% greater for mechanisms of release involving mobilization of Ca^{2+} from intracellular stores (128 ± 27 nM) than those involving transmembrane entry of Ca^{2+}. In another study, the distribution of intracellular Ca^{2+} levels has been measured using a pulsed laser imaging system and the fluorescent Ca^{2+} indicator dye, Rhod-2, which allows both spatial and temporal resolution [152]. Transient opening of voltage sensitive Ca^{2+} channels results in localized elevations of Ca^{2+} immediately beneath the plasma membrane that appear as either "hot spots" or partial rings. Catecholamine detection at flame-etched carbon fiber microelectrodes (~1 μm) has been used simultaneously to spatially map the release sites. From this study, it has been established that the sites of Ca^{2+} entry are colocalized with the active zones of catecholamine release in bovine adrenal chromaffin cells. In contrast to these results, Chow and co-workers [153] using amperometry coupled with capacitive measurements have suggested that calcium channels and vesicles are not closely opposed in the chromaffin cel, unlike synapses. These contrasting results are at present still controversial.

Amperometry at single bovine adrenal cells has also been used to study the pharmacology of drugs in regard to their regulation of exocytosis. Zhou et al. [154] used this technique to observe the role of autoreceptors in exocytosis. These results confirmed that adrenal cells do contain autoreceptors, which when relatively high levels of catecholamine are released inhibit subsequent release through a negative feedback mechanism. The action of amphetamine increases the extracellular concentration of catecholamine and seems to indicate that part of the pharmacology of amphetamine may include blockage of catecholamine autoreceptors.

Greater insight into the nature of the exocytotic process has been obtained by investigating the distribution of vesicle areas (charge), which is indicative of catecholamine content. A histogram illustrating the distribution of charge measured in the amperometric mode for individual spikes from bovine adrenal cells is shown in Fig. 20A. Several different mechanisms could lead to the skewed distributions obtained, with some spikes being much larger than average [52]. The most feasible mechanism based on the present information is exosytosis from vesicles with uniform catecholamine concentration but a large range of vesicle radii. Distribution of vesicle radii in adrenal chromaffin cells has been measured and is approximately Gaussian [136]. Using the assumption that vesicles are spherical and have a constant concentration of catecholamine, the observed spike areas in units of charge (Q) can be directly related to the vesicle radius, r, using Faraday's law, as shown below:

$Q = 4/3\ \pi nFCr^3$

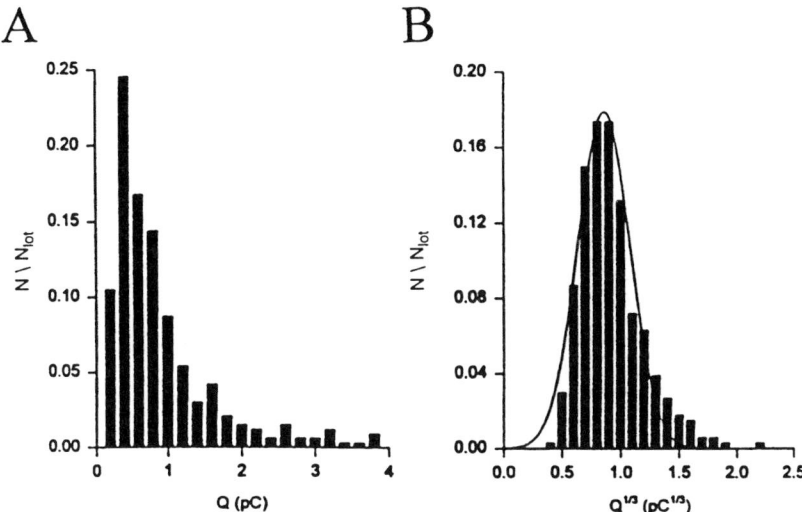

FIG. 20. Q and $Q^{1/3}$ histograms from a single bovine chromaffin cell. (A) Area histogram for spikes obtained during 15 repetitive 60 mM K$^+$ stimulations at a single chromaffin cell. The mean ± 1.03 pC. (B) Histogram of the cube root of the areas for the same data set. The Gaussian fit had a mean ± SD value of 0.86 ± 0.22 pC$^{1/3}$. Data are from 334 spikes. (From Ref. 157.)

where C is the vesicular concentration of catecholamine. The probability density function [155] for the distribution of spike areas based on these relationships using 156 nm as the mean vesicle radii and 42 nm as the variance of vesicle radii gives an estimated vesicular concentration of 0.19 M catecholamine with a correlation coefficient of 0.968, which is reasonably in accord with the estimate of 0.55 M by Hillarp [156]. If the cube root of the vesicular content ($Q^{1/3}$ or mole$^{1/3}$) is utilized for histogram generation, a normal (gaussian) distribution is obtained. The $Q^{1/3}$ histogram and a gaussian fit to the distribution is shown in Fig. 20B for a single bovine chromaffin cell [157]. Finnegan et al. also summarized the quantitation of vesicle size determined by amperometry at three additional cell types: mast cells, pancreatic β-cells, and pheochromocytoma cells [157].

E. Rat Pheochromocytoma (PC12) Cells

PC12 cells are an attractive model neuronal system due to their ability to differentiate into more neuronal-type cells upon treatment with nerve growth factor. The PC12 clonal cell line, derived from a cancerous rat adrenal gland [158] has been widely used as a model for adrenal chromaffin cells and sympathetic neurons. They are slightly smaller (5–12 μm diameter) than bovine adrenal cells [159]. Like sympathetic neurons, PC12 cells synthesize acetylcholine, as well as the catecholamines dopamine and norepinephrine [160,161]. They also store, release, and reuptake these neurotransmitters and exhibit Ca^{2+}-dependent exocytosis [54,62,163], making them very useful in studies of exocytosis.

Using amperometric detection at carbon fiber electrodes, catecholamine release has been observed at the zmol level from single PC12 cells in culture [54]. Detection at this low level suggests the possibility of applying this technique to monitor quantal release from synaptic vesicles that contain from 1.8 to 96.4 zmol of catecholamine [164]. Time-resolved individual exocytotic events have been detected and analyzed at single PC12 cells with 1.2 msec time resolution. The amperometric response at a single PC12 cell following three successive stimulations with both 1 mM nicotine and 105 mM KCl is compared to that of a control in Fig. 21. The shape of each transient is similar to those observed from bovine adrenal cells [52] except that the average charge for each transient is significantly smaller. Quantitation of vesicle content has been performed and histograms have been generated (Fig. 22). Individual current spikes have an average half-width of 9.3 ± 0.1 msec, and the average catecholamine content calculated using the area under each current transient and Faraday's law is 190 ± 3.5 zmol [54]. The average content of each vesicle can be modulated by chemical and pharmacological means. Vesicular dopamine levels can be modulated with a 50% decrease resulting from treatment with amphetamine [165] and increases in dopamine levels following incubation with L-3,4-dihydroxyphenylalanine [166].

PC12 cells are unique in that they respond to nerve growth factor (NGF) [167] by becoming more neuron-like both morphologically and physiologically. This makes them ideal for studying neuronal differentiation and the changes that occur during development and maturation. Zerby and Ewing [159] have used PC12 cells treated with NGF to probe the site of exocytosis during the differentiation process. Following treatment with NGF, PC12 cells extend processes within 18 hours of initial ex-

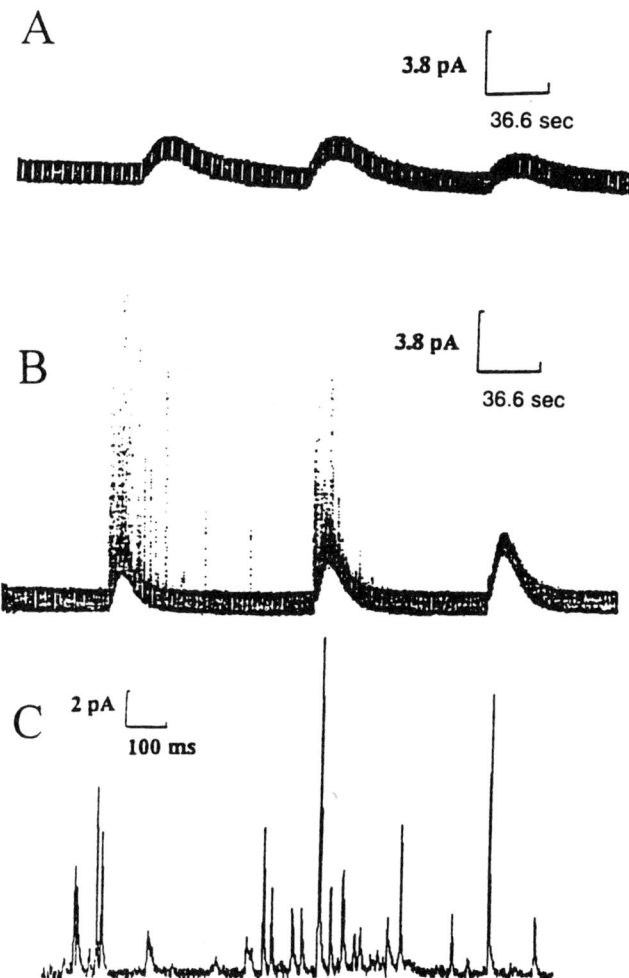

FIG. 21. (A) Amperograms of individual vesicluar exocytosis from PC12 cells. Amperogram of a control experiment with the tip of the working electrode 200 μm from the cell. (B) Amperogram of vesicular exocytosis induced by bathing a cell with 1 mM nicotine in 105 mM K^+ balanced salt solution while the electrode was placed on top of the cell. (C) Enlargement of a 1-second period of the amperogram from the first stimulation in B. Data displayed correspond to the time period near the middle of the first baseline disturbance of B. Traces A and B were obtained from computer screen dumps; each printed line represents 600 data points. Arrows represent the time at which stimulations were applied. (From Ref. 54.)

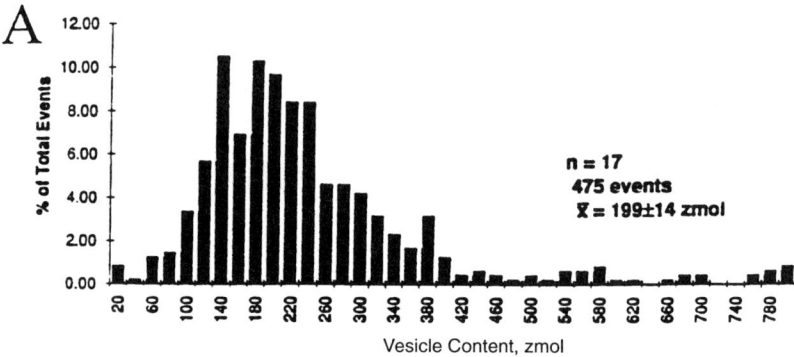

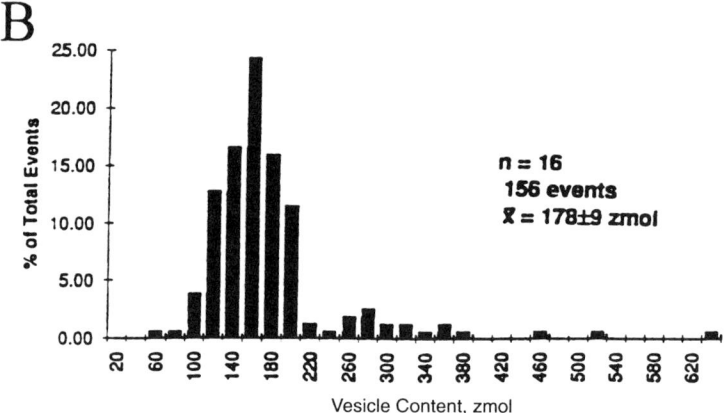

FIG. 22. (A) Distribution of the amount of catecholamine released following potassium stimulation of undifferentiated PC12 cells. The area under each current transient observed in the first 40 sec of potassium-stimulated release was converted into moles of dopamine detected per vesicle, using Faraday's law ($Q = nNF$). The total number of moles of dopamine detected for each exocytosis event was collected into bins having increments of 20 zmol and plotted as the percent of the total number of vesicles undergoing exocytosis in the first 40 sec following initiation of release. (B) Distribution of vesicle content for potassium-stimulated release at varicosities. Percent of total events observed for 16 cells (156 total events) is plotted vs. the vesicle content of dopamine for all current transients detected in the first 40 sec following initiation of release (whether immediate or delayed) in 20 zmol increments. (From Ref. 159.)

posure. Along these processes or neurites, bulbous regions (1–2 μm diameter), referred to as varicosities, are formed. These regions have been previously shown to contain aggregates of small vesicles (20–70 nm diameter) [158]. Beveled carbon fiber microelectrodes have been placed at several different locations on the differentiated cell to determine the site of exocytosis. Experiments carried out between days 10 and 14 of culture show no release from the cell body (n = 3), only occasional responses from the smooth regions of the neurites (n = 5) and frequent release when the electrode is located at a varicosity (n = 16). This response is most frequently observed at varicosities located at the intersections of several neurites. These results indicate that exocytosis from differentiated PC12 cells occurs preferentially at the varicosities with very little, if any, from the smooth processes or cell body. The average vesicular catecholamine content observed at varicosities is 178 ± 9 zmol, not significantly different from that observed at undifferentiated cells. This study lends some insight into functional changes that occur during differentiation, specifically the relocation of the site of exocytosis without alteration in the mean vesicle catecholamine content, though the distribution of vesicle content is slightly narrower (Fig. 22B).

PC12 cells have also been used in studies investigating the effect of different mechanisms of stimulated release on the time course or latency of release [168]. PC12 cells possess both nicotinic and muscarinic receptors that trigger exocytosis through two different mechanisms upon activation. Activation of nicotinic receptors by the application of 1 mM nicotine causes a change in the conformation of the receptor that opens channels permeable to Na^+. The influx of Na^+ causes sufficient depolarization of the cell membrane to open voltage-sensitive Ca^{2+} channels, allowing rapid influx of Ca^{2+} to cause exocytosis [169]. The application of 1 mM muscarine activates exocytosis through muscarinic receptors that act through intracellular second messengers to release Ca^{2+} from intracellular stores [170]. A KCl solution (105 mM) has been applied to cause exocytosis through rapid membrane depolarization. One study [168] has shown that the average vesicle catecholamine content is unaltered by the mechanism of release, but the latencies (time between application of the stimulant and secretion events) vary significantly. The mean latencies for each type of stimulation are reported as: 6 ± 1 sec (105 mM K^+); 37 ± 5 sec (1 mM nicotine); 103 ± 11 sec (1 mM muscarine). Figure 23 is representative of the amperograms obtained from the three mechanisms of stimulated release. The time resolution of carbon fiber microelectrodes not only allows the detection of sin-

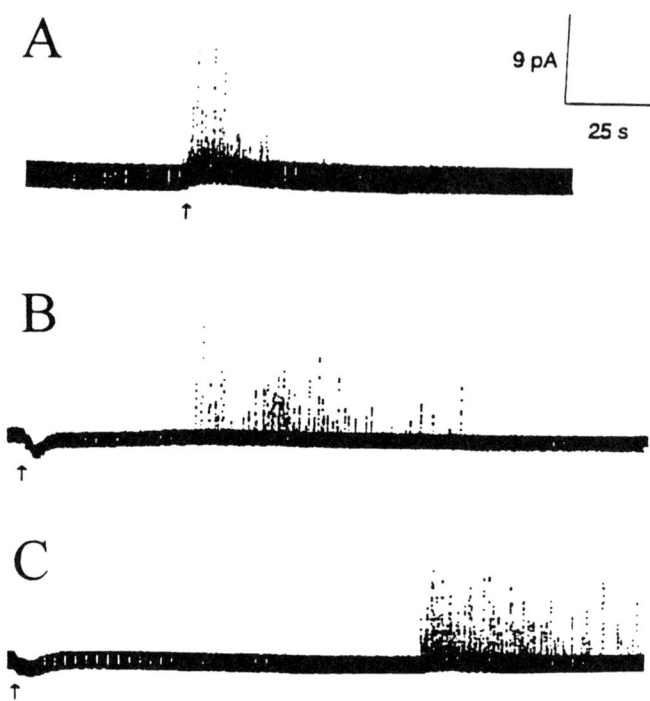

FIG. 23. Current-time traces for exocytosis at single PC12 cells. A 6-second ejection of stimulant (A) 105 mM K^+; (B) 1 mM nicotine; (C) 1 mM muscarine from a microinjector was administered at each arrow. The resulting current transients correspond to the oxidation of dopamine at the electrode tip as it is released from the cell. Detection was performed in the amperometric mode at 650 mV vs. SSCE. (From Ref. 168.)

gle events, but, as shown here, also provides an excellent means by which to monitor the time course of the stimulation-secretion process.

F. Mast Cells, Pancreatic β-Cells, and Rat Melanotrophs

Beige mouse mast cells have also been utilized for amperometric investigations of the secretory agents histamine and serotonin. Histamine and serotonin are easily oxidizable molecules that are important in the nervous and immune systems of mammals. One advantage of using this cell system is the large size of the vesicles present. The larger vesicles allow release to be correlated with simultaneous changes in membrane capacitance [171–173]. Thus, mast cells are especially useful for studies involving the fusion process of exocytosis. Quantitation of both histamine and serotonin has been described for whole mast cells using capillary chromatography with direct amperometry yielding 150 ± 18 fmol and 3.8 ± 1.3 fmol per cell, respectively [174]. Histamine and serotonin have also been shown to co-release from a single vesicle. The average concentration of serotonin in mast-cell vesicles has been reported at 3.8 mM by Finnegan et al. [157]. The mast cell appears to be a promising model system for which amperometric detection of release events can be used to further understand exocytosis.

Amperometric techniques have also been extended to secretory cells not linked to neurotransmission, human pancreatic β-cells, and rat melanotrophs. Pancreatic β-cells secrete insulin based on the levels of glucose in the body. Membrane potential experiments with intracellular microelectrodes [175] and amperometric monitoring [176] of insulin secretion from individual cells have been carried out in efforts to gain insight into the function of these important cells in the regulation of blood sugar levels in the body. Kennedy and co-workers have continued to investigate secretion from β-cells and have confirmed that the secreted products are indeed insulin using chromatographic methods [177]. The common link between exocytosis from β-cells and neurotransmission cell lines is that insulin is stored in vesicles. Exocytosis from these cells has also shown a calcium ion dependence similar to the other secretory cells described [177]. Kennedy et al. [178] have used amperometry to detect released insulin while varying the extracellular pH to determine if extracellular composition plays a role in the matrix dissociation. Their results show that changing the extracellular pH does not affect vesicle docking or release, but does affect the amount of soluble insulin available for detection. The average insulin quantity of a single vesicle has been determined to be 1.7 amol for a human β-cell, which is similar to the values reported for catecholamines in adrenal cells [157].

Kennedy and co-workers have also demonstrated the detection of peptide exocytosis at single rat melanotrophs. Peptide hormones are necessary for many biological functions, so understanding their secretion and storage characteristics is extremely important. The secretion of peptide hormones has been detected electrochemically by oxidation of tryptophan and tyrosine residues found in the proopiocortin cleavage products. Calcium mediated release of δ-melanocyte stimulating hormone has been measured at a carbon fiber electrode placed on a single rat melanotroph [179].

G. Invertebrate and Mammalian Neurons

Events at single neurons of the pond snail *Planorbis corneus* [180,181] and the leech *Hirudo medicinalis* [182] have been studied using extracellular voltammetry. The primitive neuronal nature of these invertebrates makes observations at these cells easier to interpret than those in the mammalian brain but, in the case of *Planorbis*, more difficult than those carried out in cultures since the cell monitored is connected to other nerve cells.

Bruns and Jahn [182] have described electrochemical detection of serotonin from synaptic vesicles in the leech *Hirudo medicinalis*. The leech neuron has been co-cultured with a postsynaptic neuron and a carbon-fiber electrode placed at the axonal stump. Two types of exocytotic responses have been observed: one interpreted as resulting from small clear and the second from large dense-core vesicles. Release of the small clear vesicles appears to occur more rapidly (faster time constant) than the large dense-core vesicles, which are more randomly distributed throughout the cell. Serotonin is released from the small clear and large dense vesicles in average amounts of 8 zmol (4800 molecules) and 13 amol (7.8×10^6 molecules), respectively [182]. For the large dense-core vesicles a "foot" is observed, suggesting the formation of a fusion pore (see above). It has been suggested that the faster rates observed for the small clear vesicles are due to the discharge of their contents on a submillisecond time scale through the undilated fusion pore.

Chen et al. [180] have found with amperometric measurements that stimulation of the identified giant dopamine neuron of *Planorbis* with elevated KCl results in massive exocytotic release from the cell body (Fig. 24). The current transients observed have rise times between 2 and 5 msec and an average base width of 14 ± 0.8 msec (n = 13 cells, 12,324 transients), which is consistent with the time scale expected for exocytosis. These events are not observed in the absence of Ca^{2+} from the cell medium. The dopamine cell of *Planorbis* has been estimated to contain approximately 5.4 ± 0.6 pmol of dopamine [183]. The average vesicle dopamine content observed in this study is 1.36 amol, inferring that approximately 4×10^6 vesicles are contained in this cell. Unlike the release events observed at cells in culture,

Electrochemistry in Neuronal Microenvironments

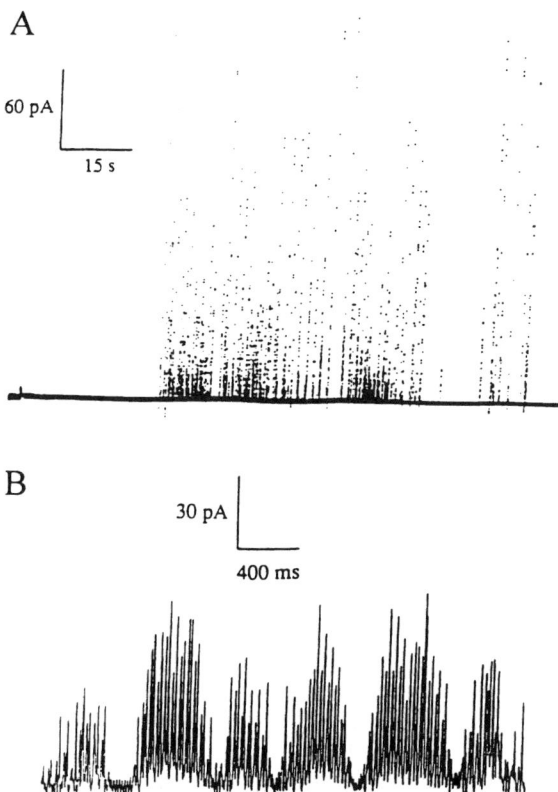

FIG. 24. (A) An example of current transients recorded with the amperometric constant-voltage method. A large *Planorbis* dopamine cell (diameter about 100 μm) was stimulated with a 4-sec potassium chloride (1 M) pulse (87 nL) delivered from a glass pipette which was placed about 15 μm from the cell body. The stimulation is shown by the horizontal bar below the trace. (B) Bursting release events were observed in 24 out of 29 cells that showed release transients. The overall success rate for observation of current transients from cells sampled was approximately 50%. (From Ref. 180.)

exocytosis at the giant dopamine cell in *Planorbis* exhibits a bursting pattern (Fig. 24B). This behavior is likely the result of interactions between the intact neuron and other neurons in this primitive neuronal system. The *Planorbis* system appears to be highly interesting pharmacologically as amphetamine has been shown to differentially alter catecholamine content in the two classes of vesicles [181] and physiologically where it appears that stimulus-release coupling does not result in a random burst of exocytosis, but rather in a controlled mechanism with regular time intervals between individual events [184]. Identification of the released transmitter has been accomplished by comparing cyclic voltammograms for stimulated release with those for the reverse transport of dopamine. These voltammograms have similar half-wave potentials and voltammetric shapes. Since norepinephrine and epinephrine have not been detected in this cell [183], it appears that the observed current transients are due to the release of dopamine from the cell body. Capillary electrophoresis with electrochemical detection has been used to verify that it is, indeed, dopamine being released. A fused silica capillary has been used to collect solution immediately outside the cell body for 30 seconds following a 6 second stimulation of the cell with elevated K^+. The only significant material detected following release has an electrophoretic mobility identical to that of dopamine.

In addition to the large number of vesicles released from the cell body, the amount of catecholamine released from these vesicles has an interesting distribution [180,181]. A bimodal distribution in which each phase appears to drop off exponentially with fewer events having a large catecholamine content is observed in the histogram (Fig. 25A). In order to correlate the content of release events with the radii of vesicles, the histogram has been plotted vs. the cube root of vesicle content. Again, as for adrenal cell measurements, an assumption that the catecholamine concentration within each vesicle remains constant has been made [52]. Figure 25B clearly shows a bimodal distribution as two distinct Gaussians. It appears that there are at least two distinct classes of vesicles or release events occurring at this cell. The smaller Gaussian makes up only 9% of the total release events and has a considerably smaller average vesicle catecholamine content (4%) than the average of the dominant Gaussian. This is in contrast to earlier studies [185] where it was assumed that the two apparent compartments of vesicles represented releasable vs. reserve stores of dopamine.

Electrochemistry in Neuronal Microenvironments

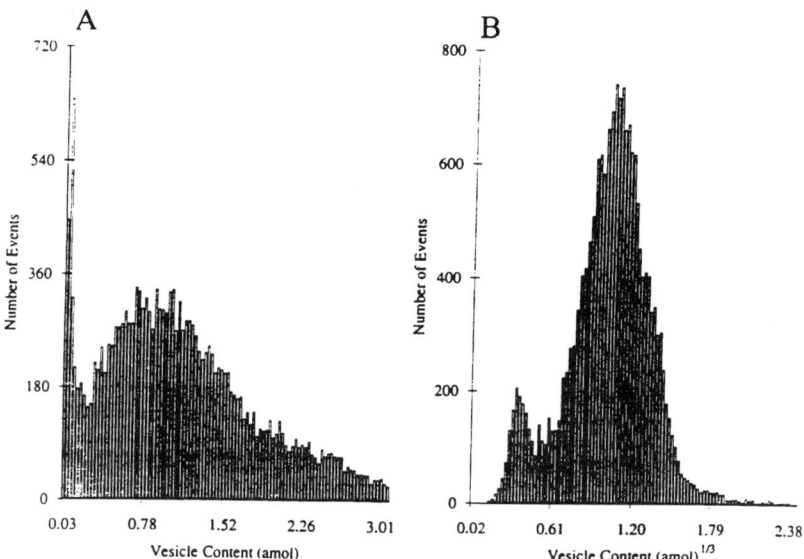

FIG. 25. Histograms of the frequency of release events vs. the amount of neurotransmitter released. (A) Frequency vs. attomoles of dopamine released (16 cells, 18,456 events). The average vesicle content was 1.36 ± 0.53 amol (mean ± s.d.), equivalent to 818,000 ± 319,000 molecules of dopamine. (B) Frequency vs. the curbed root of attomoles of dopamine released (16 cells and 18,456 events). (From Ref. 180.)

Developing superior cervical ganglion neurons from neonatal rats have also been investigated with amperometric methods [186]. Real-time electrochemical detection has been accomplished at varicosities on the axons of extending neurons with carbon fiber electrodes. Local application of potassium or black widow spider venom has been shown to initiate quantal release of catecholamines. The electrochemical system and electrodes used in this work have been optimized to detect the minute amounts released. Zhou and Misler [186] report a median spike charge of 11.3 femtocoulombs, which corresponds to 35,000 catecholamine molecules per release event (58 zmol). These values are on the order of those obtained for small clear vesicles in the in the leech *Hirudo medicinalis* [182]. Electrochemical systems are continually being optimized to decrease the amount of catecholamine that can routinely be detected.

IV. CONCLUDING REMARKS AND OTHER DIRECTIONS

Understanding communication between mammalian neurons is the ultimate goal in most neuronal studies; however, major advances are still necessary to approach the detection limits needed for measuring exocytosis at nerve terminals. The conventional size of a vesicle in a nerve terminal is approximately 50 nm and the estimated neurotransmitter is 1000–2000 molecules [187]. An estimate of 1000 molecules has also been used for modeling the efflux of dopamine measured electrochemically in the rat striatum [133]. Mammalian neurons have an added complexity of the cellular environment in vivo and the inconvenience that mature neurons stop proliferation. Added to this is the need for nanometer-size electrochemical probes to access the small synaptic gap and it is evident that technology is not yet available to achieve this goal. Improvements in instrumentation and in the ability to fabricate nanoelectrodes are issues that are continually being addressed.

Until these issues are resolved, model systems with larger vesicles and higher concentrations of neurotransmitter are under investigation to measure chemical signaling via individual exocytotic events. The capability of obtaining both qualitative and quantitative, temporally, and spatially resolved information regarding neurotransmission at single cells using carbon fiber microelectrodes has certainly made an impact on the current understanding of the regulation and function of exocytosis and is sure to continue to do so as this method of detection is applied in new biological systems.

Another area of high interest for electrochemical measurements involves monitoring nitric oxide (NO). The role of NO as a neurotransmitter or neuromodulator is unclear. However, NO appears to modulate a number of biological functions [188–191]. Real-time monitoring of NO is of critical importance to our further understanding of its role as a molecule involved in chemical communication. Malinski's group has reported a carbon fiber electrode coated with a p-type semiconducting porphyrin as being responsive to NO [192–194]. These electrodes have been used to monitor a putative NO signal from isolated endothelial cells. Although these experiments are not in a neuronal microenvironment, successful characterization of the porphyrin-based electrode could make this method amenable to further understanding NO as a transmitter in the brain. There has been some controversy about the NO electrode selectivity; however, a more recent characterization of Nafion-coated sensors demonstrates selective detection down to 1 nM NO [195].

Combined with enzyme-based electrodes and other chemically modified electrodes, a new group of electrochemical sensors promises to move the field into the area of monitoring molecules that are normally nor easily oxidized in the microenvironement of the brain. It seems that key molecules in the future will be the major excitatory neurotransmitter, glutamate [196], as well as acetylcholine, various peptides, and NO.

As a wrap up to this review, we include a few key reviews on specific areas of electrochemistry in the neuronal microenvironment. An excellent review of measuring dopamine dynamics in the brain has been provided by Wightman et al. [197]. Reviews on electrochemistry at single cells have been published in several formats [118, 198–202]. The later reviews [198–202] deal primarily with monitoring single exocytosis at individual cells.

ACKNOWLEDGMENTS

The contributions by our co-workers that are referenced herein are gratefully acknowledged. This work was supported, in part, by grants from the National Science Foundation, the National Institutes of Health, and the Office of Naval Research. R.A.C. is a National Science Foundation Postdoctoral Fellow.

REFERENCES

1. R. N. Adams, Anal. Chem. *48*:1128A (1976).
2. F. Gonon, M. Bua, R. Cespuglio, M. Jouvet, and J. F. Pujol, Nature *286*:902 (1980).
3. F. G. Gonon, C. M. Fombarlet, M. J. Bua, and J. F. Pujol, Anal. Chem. *53*:1386 (1981).
4. G. A. Gerhardt, A. I. Oke, G. Nagy, B. Moghaddam, and R. N. Adams, Brain Res. *290*:390 (1984).
5. L. D. Whiteley and C. R. Martin, Anal. Chem. *59*:1746 (1987).
6. E. W. Kristensen, W. G. Kuhr, and R. M. Wightman, Anal. Chem. *59*:1752 (1987).
7. C. A. Marsden, M. H. Joseph, Z. L. Kruk, N. T. Maidment, R. D. O'Neill, J. O. Schenk, and J. A. Stamford, Neuroscience *25*:389 (1988).

8. J. F. Cassidy and M. B. Foley, Chem. Br. *29*:764 (1993).
9. C. A. Amatore, A. Jutand, and F. Pflüger, J. Electroanal. Chem. *218*:361 (1987).
10. D. O. Wipf, E. W. Kristensen, M. R. Deakin, and R. M. Wightman, Anal. Chem. *60*:306 (1988).
11. M. I. Montenegro and D. Pletcher, J. Electroanal. Chem. *200*:371 (1986).
12. R. M. Wightman and D. O. Wipf, in *Electroanalytical Chemistry*, Vol. 15 (A. J. Bard, ed.), Marcel Dekker, New York, 1989, p. 267.
13. J. A. Stamford, Anal. Chem. *58*:1033 (1986).
14. M. A. Dayton, J. C. Brown, K. J. Stutts, and R. M. Wightman, Anal. Chem. *52*:946 (1980).
15. M. A. Dayton, A. G. Ewing, and R. M. Wightman, Anal. Chem. *52*:2392 (1980).
16. K. T. Kawagoe, J. A. Jankowski, and R. M. Wightman, Anal. Chem. *63*:1589 (1991).
17. R. J. Forster, Chem. Soc. Rev. *23*:289 (1994).
18. R. M. Wightman, Science *240*:415 (1988).
19. A. Fitch, and D. H. Evans, J. Electroanal. Chem. *202*:83 (1986).
20. J. A. Stamford, J. Neurosci. Meth. *17*:1 (1986).
21. A. G. Ewing, K. D. Alloway, S. D. Curtis, M. A. Dayton, R. M. Wightman, and G. V. Rebec, Brain Res. *261*:101 (1983).
22. G. M. Shepherd, *Neurobiology*, 2nd ed., Oxford University Press, New York 1988, p. 145.
23. J. R. Cooper, F. E. Bloom, and R. H. Roth, *The Biochemical Basis of Neuropharmacology*, Oxford University Press, Oxford, England, 1986.
24. F. Valtorta, R. Fesce, F. Grohovaz, C. Haimann, W. P. Hurlburt, N. Iezzi, F. Torri Tarelli, A. Villa, and B. Ceccarelli, Neuroscience *35*:477 (1990).
25. P. T. Kissinger, J. B. Hart, and R. N. Adams, Brain Res. *55*:209 (1973).
26. C. Nicholson, J. Neurosci. Meth. *48*:199 (1993).
27. S. Baudet, L. Hove-Madsen, and D. M. Bers, Meth. Cell Biol. *40*:93 (1994).
28. J. O. Schenk, E. Miller, and R. N. Adams, J. Chem. Ed. *60*:311 (1983).
29. R. M. Wightman, in *Electrochemistry in Research and Development* (R. Kalvoda and R. Parsons, eds.), Plenum Press, 1985.
30. C. Nicholson and J. M. Phillips, J. Physiol. *321*:225 (1981).
31. M. E. Rice and C. Nicholson, J. Neurophysiol. *65*:264 (1991).
32. P. A. Broderick, Electroanalysis *2*:241 (1990).
33. J. B. Justice, Jr., in *Voltammetry in the Neurosciences* (J. B. Justice, Jr., ed.), Humana Press, Clifton, NJ, 1987.
34. J. O. Schenk and R. N. Adams, in *Measurement of Neurotransmitter Release In Vivo* (C. A. Marsden, ed.), Wiley, New York, 1984.
35. G. A. Gerhardt, G. M. Rose, and B. J. Hoffer, Brain Res. *413*:327 (1987).
36. J. L. Ponchon, R. Cespuglio, F. Gonon, M. Jouvet, and J. F. Pujol, Anal. Chem. *51*:1483 (1979).

37. A. G. Ewing, M. A. Dayton, and R. M. Wightman, Anal. Chem. *53*:1842 (1981).
38. A. G. Ewing, R. M. Wightman, and M. A. Dayton, Brain Res. *249*:361 (1982).
39. F. Crespi, K. F. Martin, and C. A. Marsden, Neuroscience *27*:885 (1988).
40. C. D. Blaha and R. F. Lane, Brain Res. Bull. *10*:861 (1983).
41. R. Cespuglio, H. Faradji, J. L. Ponchon, R. Riou, M. Buda, F. Gonon, J. F. Pujol, and M. Jouvet, J. Physiol. (Paris) *77*:327 (1981).
42. R. Cespuglio, H. Faradji, J. L. Ponchon, M. Buda, F. Riou, F. Gonon, J. F. Pujol, and M. Jouvet, Brain Res. *223*:287 (1981).
43. R. Cespuglio, H. Faradji, J. L. Ponchon, M. Buda, F. Riou, F. Gonon, J. F. Pujol, and M. Jouvet, Brain Res. *223*:299 (1981).
44. T. Sharp, N. T. Maidment, M. P. Brazell, T. Zetterstrom, U. Ungerstadt, G. W. Bennet, and C. A. Marsden, Neuroscience *12*:1213 (1984).
45. J. O. Howell, W. G. Kuhr, R. E. Ensman, and R. M. Wightman, J. Electroanal. Chem. Interfacial Electrochem. *209*:77 (1986).
46. R. N. Adams, Prog. Neurobio. *35*:297 (1990).
47. J. E. Baur, E. W. Kristenson, L. J. May, D. J. Wiedemann, and R. M. Wightman, Anal. Chem. *60*:1268 (1988).
48. J. A. Stamford, Z. L. Kruk, and J. Millar, Brain Res. *454*:282 (1988).
49. M. E. Rice and C. Nicholson, Anal. Chem. *61*:1805 (1989).
50. J. B. Zimmerman and R. M. Wightman, Anal. Chem. *63*:24 (1991).
51. D. J. Leszczyszyn, J. A. Jankowski, O. H. Viveros, E. J. Dilberto, Jr., J. A. Near, and R. M. Wightman, J. Biol. Chem. *265*:14736 (1990).
52. R. M. Wightman, J. A. Jankowski, R. T. Kennedy, K. T. Kawagoe, T. J. Schroeder, D. J. Leszczyszyn, J. A. Near, E. J. Diliberto, Jr., and O. H. Viveros, Proc. Natl. Acad. Sci. USA *88*:10754 (1991).
53. R. H. Chow, L. von Rüden, and E. Neher, Nature *356*:60 (1992).
54. T. K. Chen, G. Luo, and A. G. Ewing, Anal. Chem. *66*:3031 (1994).
55. P. M. Plotsky, Brain Res. *235*:179 (1982).
56. R. C. Koile and D. C. Johnson, Anal. Chem. *51*:741 (1979).
57. F. Gonon, F. Navane, and M. Buda, Anal. Chem. *56*:573 (1984).
58. B. D. Pendley and H. D. Abruña, Anal. Chem. *62*:782 (1990).
59. R. M. Penner, M. J. Heben, T. L. Longin, and N. S. Lewis, Science *250*:1118 (1990).
60. A. Meulemans, B. Poulain, G. Baux, and L. Tauc, Brain Res. *414*:158 (1987).
61. N. Georgolios, D. Jannakoudaks, and P. J. Karabinas, J. Electroanal. Chem. *264*:235 (1989).
62. Y. Y. Lau, T. Abe, and A. G. Ewing, Anal. Chem. *64*:1702 (1992).
63. Y. Y. Lau, P. K. Y. Wong, and A. G. Ewing, Microchem. J. *47*:308 (1993).
64. E. R. Reynolds and A. M. Yacynych, Electroanalysis *5*:405 (1993).
65. T. Abe, Y. Y. Lau, and A. G. Ewing, Anal. Chem. *64*:2160 (1992).
66. R. M. Wightman, Anal. Chem. *53*:1125A (1981).

67. R. M. Wightman, L. J. May, and A. C. Michael, Anal. Chem. *60*:769A (1988).
68. K. T. Kawagoe, J. B. Zimmerman, and R. M. Wightman, J. Neurosci. Meth. *48*:225 (1993).
69. P. Capella, B. Ghasemzadeh, K. Mitchell, and R. N. Adams, Electroanalysis *2*:175 (1990).
70. R. L. McCreery, in *Electroanalytical Chemistry*, Vol. 17 (A. J. Bard, ed.), Marcel Dekker, New York, 1991, p. 221.
71. R. W. Murray, in *Electroanalytical Chemistry*, Vol. 13 (A. J. Bard, ed.), Marcel Dekker, New York, 1984, p. 191.
72. R. M. Wightman, E. C. Paik, S. Borman, and M. A. Dayton, Anal. Chem. *50*:1410 (1978).
73. R. S. Kelly and R. M. Wightman, Anal. Chim. Acta *187*:79 (1986).
74. Y. T. Kim, D. M. Scarnulis, and A. G. Ewing, Anal. Chem. *58*:1782 (1986).
75. G. Gerhardt and R. N. Adams, Anal. Chem. *54*:2618 (1982).
76. R. A. Saraceno and A. G. Ewing, Anal. Chem. *60*:2016 (1988).
77. H. J. Blaedel and G. A. Mabbott, Anal. Chem. *50*:933 (1978).
78. C. Mermet and F. Gonon, Neurosci. *19*:829 (1986).
79. A. G. Ewing, J. C. Bigelow, and R. M. Wightman, Science *221*:169 (1983).
80. F. Gonon, M. Buda, and J. F. Pujol, in *Measurement of Neurotransmitter Release In Vivo* (C. A. Marsden, ed.), Wiley, New York 1984.
81. M. F. Suaud-Chagny, R. Cespuglio, J. P. Rivot, M. Buda, and F. Gonon, J. Neurosci. Meth. *48*:241 (1993).
82. J. X. Feng, M. Brazell, K. Renner, R. Kasser, and R. N. Adams, Anal. Chem. *59*:1863 (1987).
83. J. A. Stamford, Trends Neurosci. *12*:407 (1989).
84. Z. Hahn, R. Cespuglio, H. Faradji, and M. Jouvet, J. Biochem. Biophys. Meth. *11*:265 (1985).
85. P. M. Kovach, M. R. Deakin, and R. M. Wightman, J. Phys. Chem. *90*:4612 (1986).
86. L. Falat and H. Y. Cheng, Anal. Chem. *54*:2108 (1982).
87. M. Armstrong-James, K. Fox, and J. Millar, J. Neurosci. Meth. *2*:431 (1980).
88. T. G. Strein and A. G. Ewing, Anal. Chem. *64*:1368 (1992).
89. D. K. Y. Wong and L. Y. F. Xu, Anal. Chem. *67*:4086 (1995).
90. L. D. Whiteley and C. R. Martin, Anal. Chem. *59*:1746 (1987).
91. M. E. Rice and C. Nicholson, Anal. Chem. *61*:1805 (1989).
92. F. Crespi and C. Möbius, J. Neurosci. Meth. *42*:149 (1992).
93. M. N. Szentirmay and C. R. Martin, Anal. Chem. *56*:1898 (1984).
94. R. B. Moore III, J. E. Wilkerson, and C. R. Martin, Anal. Chem. *56*:2572 (1984).
95. G. Nagy, G. A. Gerhardt, A. F. Oke, M. E. Rice, R. N. Adopaminems, R. B. Moore III, M. N. Szentirmay, and C. R. Martin, J. Electroanal. Chem. Interfacial Electrochem. *188*:85 (1985).

96. M. P. Brazell, R. J. Kasser, K. J. Renner, J. Feng, B. Moghaddopaminem, and R. N. Adopaminems, J. Neurosci. Meth. 22:167 (1987).
97. K. T. Kawagoe, P. A. Garris, and R. M. Wightman, J. Electroanal. Chem. 359:193 (1993).
98. M. Chesler, Prog. Neurobiol. 34:401 (1990).
99. J. Wang and T. Golden, Anal. Chem. 61:1397 (1989).
100. Y. Y. Lau, J. B. Chien, D. K. Y. Wong, and A. G. Ewing, Electroanalysis 3:87 (1991).
101. J. Wang and M. S. Lin, Electroanalysis 2:253 (1990).
102. J. Rishpon, S. Gottesfeld, C. Campbell, J. Dopaminevey, and T. A. Zawodzinski, Jr., Electroanalysis 6:17 (1994).
103. D. J. Harrison, R. F. B. Turner, and H. P. Balter, Anal. Chem. 60:2002 (1988).
104. S. V. Sasso, R. J. Pierce, R. Walla, A. M. Yacynych, Anal. Chem. 62:1111 (1990).
105. P. Pantano, W. G. Kuhr Anal. Chem. 65:623 (1993).
106. J. J. O'Malley and R. W. Ulmer, Biotech. Bioeng. 15:917 (1973).
107. T. Abe, Y. Y. Lau, and A. G. Ewing, J. Am. Chem. Soc. 113:7421 (1991).
108. Y. Ikariyama, S. Yamauchi, T. Yukiashi, and H. Ushioda, Anal. Lett. 20:1407 (1987).
109. T. W. Sohn, P. W. Stoecker, W. Carp, and A. M. Yacynych, Electroanalysis 3:763 (1991).
110. G. Heider, S. V. Sasso, K. Huang, H. J. Wieck, and A. M. Yacynych, Anal. Chem. 62:1106 (1990).
111. M. G. Garguilo, N. Huynh, A. Proctor, and A. C. Michael, Anal. Chem. 65:523 (1993).
112. M. G. Garguilo and A. C. Michael, J. Am. Chem. Soc. 115:12218 (1993).
113. Z. Huang, R. Villarta-Snow, G. I. Lubrano, and G. G. Guilbault, Anal. Biochem. 215:31 (1993).
114. J. B. Chien, R. A. Wallingford, and A. G. Ewing, J. Neurochem. 54:633 (1990).
115. A. Meulemans, B. Poulain, G. Baux, and L. Tauc, Anal. Chem. 58:2088 (1986).
116. W. R. LaCourse, D. A. Mead, and D. C. Johnson, Anal. Chem. 62:220 (1990).
117. G. G. Neuberger and D. C. Johnson, Anal. Chem. 60:2288 (1988).
118. A. G. Ewing, T. G. Strein, and Y. Y. Lau, Acc. Chem. Res. 25:440 (1992).
119. R. A. Wallingford and A. G. Ewing, Anal. Chem. 60:258 (1988).
120. R. A. Wallingford and A. G. Ewing, Anal. Chem. 60:1972 (1988).
121. R. A. Wallingford and A. G. Ewing, Anal. Chem. 59:1762 (1987).
122. T. M. Olefirowicz and A. G. Ewing, J. Neurosci. Meth. 34:11 (1990).
123. K. Shibuki, Brain Res. 487:96 (1989).
124. M. Erecinski, D. Nelson, and B. Chance, Proc. Natl. Acad. Sci. USA 88:7600 (1991).

125. A. Fein and M. Tsacopoulos, Nature *331*:437 (1988).
126. D. Lawson, L. D. Whiteley, C. R. Martin, M. N. Szentirmay, and J. I. Song, J. Electrochem. Soc. *135*:2247 (1988).
127. W. G. Kuhr and R. M. Wightman, Brain Res. *381*:168 (1986).
128. J. Millar, J. A. Stamford, Z. L. Kruk, and R. M. Wightman, Eur. J. Pharmac. *109*:341 (1985).
129. R. M. Wightman, C. Amatore, R. C. Engstrom, P. D. Hale, E. W. Kristensen, W. G. Kuhr, and L. J. May, Neuroscience *25*:513 (1988).
130. C. A. Nicholson and M. E. Rice, Ann. N. Y. Acad. Sci. *481*:55 (1986).
131. F. Gonon, M. Buda, R. Cespugio, M. Jouvet, and J. F. Pujol, Brain Res. *223*:69 (1981).
132. B. Giros, M. Jaber, S. R. Jones, R. M. Wightman, and M. G. Caron, Nature *379*:606 (1996).
133. P. A. Garris, E. D. Ciolkowski, P. Pastore, and R. M. Wightman, J. Neurosci. *14*:6084 (1994).
134. D. J. Leszczyszyn, J. A. Jankowski, O. H. Viveros, E. J. Diliberto, Jr., J. A. Near, and R. M. Wightman, J. Neurochem. *56*:1855 (1991).
135. M. Winkler and E. Westhead, Neuroscience *5*:1803 (1980).
136. R. E. Coupland, Nature *217*:384 (1968).
137. J. J. Nordmann, J. Neurochem. *42*:434 (1984).
138. J. H. Phillips, Neuroscience *7*:1595 (1982).
139. C. Nicholson, Brain Res. *333*:325 (1985).
140. E. Neher and A. Marty, Proc. Natl. Acad. Sci. USA *79*:6712 (1982).
141. E. L. Ciolkowski, B. R. Cooper, J. A. Jankowski, J. W. Jorgenson, and R. M. Wightman, J. Am. Chem. Soc. *114*:2815 (1992).
142. J. E. Baur and R. M. Wightman, J. Chrom. *482*:65 (1989).
143. T. J. Schroeder, J. A. Jankowski, K. T. Kawagoe, and R. M. Wightman, Anal. Chem. *64*:3077 (1992).
144. R. C. Engstrom, R. M. Wightman, and E. W. Kristensen, Anal. Chem. *60*:652 (1988).
145. Z. Zhou, S. Misler, and R. H. Chow, Biophys. J. *70*:1543 (1996).
146. R. M. Wightman, T. J. Schroeder, J. M. Finnegan, E. L. Ciolkowski, and K. Pihel, Biophys. J. *68*:383 (1995).
147. T. J. Schroeder, R. Borges, J. M. Finnegan, K. Pihel, C. Amatore, and R. M. Wightman, Biophys. J. *70*:1061 (1996).
148. J. A. Jankowski, T. J. Schroeder, R. W. Holz, and R. M. Wightman, J. Biol. Chem. *267*:18329 (1992).
149. T. J. Schroeder, J. A. Jankowski, J. Senyshyn, R. W. Holz, and R. M. Wightman, J. Biol. Chem. *269*:17215 (1994).
150. G. Gryznkiewicz, M. Poenie, and R. Y. Tsien, J. Biol. Chem. *260*:3440 (1985).
151. J. M. Finnegan and R. M. Wightman, J. Biol. Chem. *270*:1 (1995).

152. I. M. Robinson, J. M. Finnegan, J. R. Monck, R. M. Wightman, and J. M. Fernandez, Proc. Natl. Acad. Sci. USA 92:2474 (1995).
153. R. H. Chow, J. Klingauf, and E. Neher, Proc. Natl. Acad. Sci. (USA) 91, 12765 (1994).
154. R. Zhou, G. Luo and A. G. Ewing, J. Neurosci. 14:2402 (1994).
155. A. Papoulis, *Probability, Random Variables and Stochastic Processes*, McGraw-Hill, New York, 1965, p. 116.
156. N.-A. Hillarp, Acta Physiol. Scand. 47:271 (1959).
157. J. M. Finnegan, K. Pihel, P. S. Cahill, L. Huang, S. E. Zerby, A. G. Ewing, R. T. Kennedy and R. M. Wightman, J. Neurochem. 66:1914 (1996).
158. L. A. Greene and A. S. Tischler, Proc. Natl. Acad. Sci. USA, 73:2424 (1976).
159. S. E. Zerby and A. G. Ewing, Brain Res. 712:1 (1996).
160. D. Schubert and F. G. Klier, Proc. Natl. Acad. Sci. USA 74: 5184 (1977).
161. J. A. Wagner, J. Neurochem. 45:1244 (1985).
162. L. A. Greene and G. Rein, Brain Res. 129:247 (1977).
163. L. Baizer and N. Weiner, J. Neurochem. 44:495 (1985).
164. A. Dahlström, J. Höggendal, and T. Hökfelt, Acta Physiol. Scand. 67:289 (1966).
165. D. Sulzer, T. K. Chen, Y. Y. Lau, H. Kristensen, S. Rayport, and A. G. Ewing, J. Neurosci. 15:4102 (1995).
166. E. Pothos, M. Desmond, and D. Sulzer, J. Neurochem. 66:629 (1996).
167. R. Levi-Montalcini and P. U. Angeletti, Physiol. Rev. 48:534 (1968).
168. S. E. Zerby and A. G. Ewing, J. Neurochem. 66:651 (1996).
169. W. B. Stallcup, J. Physiol. (London) 286:525 (1979).
170. M. J. Berridge and R. F. Irvine, Nature 312:315 (1984).
171. G. A. deToledo, R. Fernández-Chacón, and J. M. Fernández, Nature 363:554 (1993).
172. J. M. Fernández, E. Neher, and B. D. Gomperts, Nature 312:453 (1984).
173. L. J. Breckenridge and W. Almers, Nature 328:814 (1987).
174. K. Pihel, S. Hsieh, J. W. Jorgenson, and R. M. Wightman, Anal. Chem. 67:4514 (1995).
175. H. P. Meissner, Meth. Enzymol. 192:235 (1990).
176. R. T. Kennedy, L. Huang, M. A. Atkinson, and P. Dush, Anal. Chem. 65:1882 (1993).
177. L. Huang, H. Shen, M. A. Atkinson, and R. A. Kennedy, Proc. Natl. Acad. Sci. USA 92:9608 (1995).
178. R. T. Kennedy, L. Huang, and C. A. Aspinwal, J. Am. Chem. Soc. 118:1795 (1996).
179. C. D. Paras and R. T. Kennedy, Anal. Chem. 67:3633 (1995).
180. G. Chen, P. F. Gavin, G. Luo, and A. G. Ewing, J. Neurosci. 15:7747 (1995).
181. G. Chen and A. G. Ewing, Brain Res. 701:167 (1995).
182. D. Bruns and R. Jahn, Nature 377:62 (1995).

183. B. Powell and G. A. Cottrell, J. Neurochem. 22:605 (1974).
184. G. Chen, D. Gutman, and A. G. Ewing, Brain Res. 733:119 (1996).
185. H. K. Kristensen, Y. Y. Lau, and A. G. Ewing, J. Neurosci. Meth. 51:183 (1994).
186. Z. Zhou and S. Misler, Proc. Natl. Acad. Sci. USA, 92:6938 (1995).
187. G. J. Siegel, B. W. Agranoff, R. W. Albers, and P. B. Molinoff, *Basic Neurochemistry*, 5th ed., Raven Press, New York, 1994, p. 190.
188. T. J. O'Dell, R. D. Hawkins, E. R. Kandel, and O. Arancio, Proc. Natl. Acad. Sci. USA 88:11285 (1991).
189. R. M. J. Palmer, A. G. Ferrige, and S. Moncada, Nature 327:526 (1987).
190. M. W. Radomski, R. M. J. Palmer, and S. Moncada, Proc. Natl. Acad. Sci. USA 87:5193 (1990).
191. J. B. Hibbs, R. R. Tainton, Z. Vavrin, and E. M. Rachlin, J. Immunol. 138:550 (1987).
192. T. Malinski, A. Ciszewski, J. Bennett, J. R. Fish, and L. Czuchajowski, J. Electrochem. Soc. 138:2008 (1991).
193. T. Malinski and Z. Taha, Nature 358:676 (1992).
194. T. Malinski, Z. Taha, S. Grunfield, A. Burewicz, P. Tomboulain, and F. Kiechle, Anal. Chim. Acta 279:135 (1993).
195. F. Lantoine, S. Trevin, F. Bedioui, and J. Devynck, J. Electroanal. Chem. 392:85 (1995).
196. Y. Hu, K. M. Mitchell, F. N. Albahadily, E. K. Michaelis, and G. S. Wilson, Brain Res. 659:117 (1994).
197. R. M. Wightman, L. C. May, and A. C. Michael, Anal. Chem. 60:769A (1988).
198. A. G. Ewing, T. K. Chen, and G. Chen, in *Neuromethods: In Vivo Monotoring I* (R. Adams, ed.), Humana Press, Clifton, NJ, 1995, p. 269.
199. R. M. Wightman, J. M. Finnegan, and K. Pihel, Tr. Anal. Chem. 14:154 (1995).
200. L. Huang and R. T. Kennedy, Tr. Anal. Chem. 14:158 (1995).
201. G. Chen and A. G. Ewing, Crit. Rev. Neurobiol. 11:59 (1997).
202. R. A. Clark and A. G. Ewing, Mol. Neurobiol. 15:1 (1997).

AUTHOR INDEX

Note: *Italicized* page numbers locate names in the References sections.

Abe, T., 238, 248, 249, 252, 257, 258, 259, 276, *289, 291*
Abruña, H.D., 196, *225*, 238, *289*
Ackerlof, G., 92, 93, *137*
Adam, P.M., 142, 143, *220*
Adams, R.N., 228, 232, 234, 236, 237, 239, 240, 241, 243, 244, 245, 247, 261, 267, *287, 288, 289, 290*
Adopaminems, R.N., 245, 247, *290, 291*
Agranoff, B.W., 286, *294*
Agranovich, V.M., 142, 143, 144, *219*
Ahlheim, M., 203, *225*
Akhter, S., 169, 171, 198, *224*
Akkara, J.A., 163, *223*
Albahadily, F.N., 287, *294*
Albers, R.W., 286, *294*
Allara, D.L., 159, 168, *223, 224*
Alloway, K.D., 229, 237, *288*
Almers, W., 281, *293*
Altengurg, B.S.F., 149, 156, *223*
Alves, C.A., *224*
Amatore, C.A., 228, 260, 271, *288, 292*
Anderegg, J.W., *224*
Andrews, D.C., 146, 156, *221*
Angeletti, P.U., *293*
Angermaier, L., 142, 147, 163, 166, 167, *220*
Anzai, J., 163, *223*

Aoki, K., 171, 174, *224*
Arancio, O., 286, *294*
Arendt, M.F., 171, 174, *224*
Armstrong-James, M., 244, *290*
Arndt-Jovin, D.J., 128, *139*
Asgharian, B., 147, 199, *221*
Aspinwal, C.A., 281, *293*
Atkinson, M.A., 281, *293*
Atruplová Bartácková, M., 8, *83*
Atwood, E.S., 115, *138*
Aubrt, J., 12, 17, *84*
Aussenegg, F., 142, 143, *219*
Aust, E.F., 147, 196, 198, 199, 203, *221, 222*
Ayyagari, M.S., 163, *223*
Azzam, R.M.A., 142, *220*

Bagchi, R.N., 19, *84*
Bain, C.D., 142, 166, *218, 224*
Baizer, L., 276, *293*
Balcytis, G.A., 147, *222*
Balter, H.P., 248, *291*
Bamdad, C., 147, *222*
Barberis, A., 147, *222*
Bard, A.J., 196, 211, 213, *224*
Barner, B.J., 142, *218*
Bartoll, J., 52, *85*
Barzoukas, M., 203, *225*
Bashara, N.M., 142, *220*
Bates, R.G., 90, 93, 95, *137*
Baudet, S., 234, *288*
Bauer, D., 6, *82*

Author Index

Baur, J.E., 237, 243, 265, *289, 292*
Bautsch, H.-J., 70, *85*
Baux, G., 238, 252, *289, 291*
Bayer, E.A., 163, *223*
Becka, A.M., 213, *225*
Beden, B., 142, *218*
Bedioui, F., 287, *294*
Bedworth, P.V., 203, *225*
Bedzyk, M.J., 128, *138*
Behnert, J., 45, *85*
Bell, C.M., 171, 174, 211, *224, 225*
Belyi, V.I., 5, *82*
Bender, P., 92, 93, *137*
Benner, R.E., 142, *219*
Bennet, G.W., 237, *289*
Bennett, J., 286, *294*
Berrettoni, M., 9, *83*
Berridge, M.J., *293*
Bers, D.M., 234, *288*
Bertelli, L., 63, *85*
Betzig, E., 128, *138, 139*
Bigelow, J.C., 243, 263, *290*
Birke, R.L., 142, *218*
Blaedel, H.J., 242, *290*
Blaha, C.D., 237, *289*
Blanchard-Desce, M., 203, *225*
Bloom, F.E., 229, 232, *288*
Blum, D., 11, *84*
Bobrowski, A., 9, 11, *83*
Bockris, J.O'M., 77, *86*
Bohacek, J., 12, 17, *84*
Bolognesi, M., 163, 166, *223*
Bommarito, G.M., 128, *138*
Bond, A.M., 9, 10, 11, 12, 16, 26, 27, 28, 29, 30, 31, 32, 33, 41, 52, 58, 64, 65, 66, 68, 69, 70, 71, 74, 77, *83, 84, 85*

Borges, R., 271, *292*
Borman, S., 239, *290*
Bousfield, W.R., 105, *138*
Boxer, S.G., 196, *225*
Bozhevolnyi, S.I., 142, *221*
Brainina, K.Z., 2, 4, 6, *82*
Braunstein, H., 106, *138*
Braunstein, J., 106, *138*
Brazell, M.P., 237, 243, 247, *289, 290, 291*
Breckenridge, .J., 281, *293*
Bright, T.B., 159, *223*
Brillante, A., 143, 144, 146, *221*
Broderick, P.A., 236, 242, *288*
Brown, J.C., 229, 239, 240, *288*
Bruce, P.G., 2, *82*
Bruckbauer, A., 142, *219*
Brunner, H., 142, 143, *219*
Bruns, D., 282, 285, *293*
Bua, M.J., 228, 237, 242, 243, *287*
Buchel, M., 147, *221*
Buda, M., 237, 238, 239, 242, 243, 263, *289, 290, 292*
Burewicz, A., 286, *294*
Burgess, D.R.F., 182, *224*
Burgess, L.W., 146, 156, *221*
Burland, D.M., 203, *225*
Burstein, E., 142, 143, 144, 146, *219*
Butler, J.N., 106, *138*
Buttry, D.A., 196, 208, *225*
Byahut, S., 142, *219*

Cade, N.A., 146, 147, 156, 199, *221, 222*
Caffrey, M., 128, *138*
Cahill, P.S., 275, 281, *293*
Cai, Y., 203, *225*

Author Index

Caldwell, M.E., 147, 199, 217, 222
Cammarata, V., 128, *139*
Campbell, C., 248, *291*
Capella, P., 239, 241, 243, 244, 245, 247, 261, *290*
Caron, M.G., 264, 265, *292*
Carp, W., 249, *291*
Caruso, F., 147, *222*
Cassidy, J.F., 228, *288*
Casstevens, M., 147, 199, *221*
Cauquil, O., 6, *82*
Ceccarelli, B., 230, 267, *288*
Cespuglio, R., 228, 236, 237, 242, 243, 263, *287, 288, 289, 290, 292*
Chailapakul, O., 168, *224*
Chance, B., 259, *291*
Chang, J.S., 142, *219*
Chang, R.K., 142, *218*
Chanturiya, V.A., 4, 8, *82*
Chao, F., 142, *220*
Charych, D.H., 161, *223*
Chastaing, E., 147, 166, 199, *221*
Chazalviel, J.-N., 128, 129, *138*
Chen, G., 282, 283, 284, 285, 287, *293, 294*
Chen, J., 9, *83*
Chen, S., 196, *225*
Chen, T.-A., 203, *225*
Chen, T.K., 237, 276, 277, 287, *289, 293, 294*
Chen, W.P., 142, 143, 144, 146, *219*
Chen, X., 142, 147, *220, 221*
Chen, Y.J., 142, 143, 144, 146, *219*
Cheng, H.Y., 243, *290*
Cheskis, B., 147, *222*

Chesler, M., 247, *291*
Chidsey, C.E.D., 159, 166, 171, 198, 199, *223, 224, 225*
Chien, J.B., 247, 248, 252, 253, 255, *291*
Chouaib, F., 6, *82*
Chow, R.H., 174, 237, 267, 269, 271, *289, 292, 293*
Chung, C., 196, 213, *224*
Ciolkowski, E.D., 266, 286, *292*
Ciolkowski, E.L., 269, 271, *292*
Ciszewski, A., 286, *294*
Coda, A., 163, 166, *223*
Colton, R., 10, 12, 27, 28, 29, 74, *83, 84*
Condon, A.E., 192, 193, *224*
Cooper, B.R., 269, *292*
Cooper, J.B., 27, *84*
Cooper, J.R., 229, 232, *288*
Corn, R.M., 142, 147, 148, 149, 159, 161, 162, 163, 164, 165, 167, 169, 170, 171, 172, 173, 175, 178, 179, 180, 182, 183, 184, 185, 187, 188, 189, 190, 191, 192, 193, 196, 197, 199, 201, 202, 203, 204, 205, 207, 208, 209, 210, 211, 212, 214, 215, 216, *218, 219, 220, 222, 223, 224, 225*
Costa, M., 142, *220*
Cottrell, G.A., 282, 284, *294*
Coupland, R.E., 174, 267, *292*
Coury, L.A., 12, *84*
Cox, E.C., 163, 167, *223*
Creager, S.E., 196, *225*
Crespi, F., 237, 245, 247, *289, 290*

Crooks, R.M., 166, 168, *224*
Cross, G.H., 146, 147, 156, 199, *221*, *222*
Curtis, S.D., 229, 237, *288*
Czuchajowski, L., 286, *294*

Dahlgren, D.A., 182, *224*
Dahlström, A., 276, *293*
Damaschun, F., 3, 9, 10, 11, 46, 70, *82*, *84*, *85*
Daniels, F., 12, 28, 29, 74, *84*
Dao, P.T., 198, 203, *225*
Darling, R.B., 217, *225*
Darlington, J., 8, *83*
Dausheva, M.R., 7, *83*
Davies, J., 142, 147, 163, *220–21*
Davies, M.C., 142, 147, 163, *220–21*
Davies, P.B., 142, *218*
Dawkes, A.C., 142, 147, 163, *220–21*
Dayton, M.A., 229, 237, 238, 239, 240, 262, 263, *288*, *289*, *290*
Deakin, M.R., 228, 243, *288*, *290*
Dean, J.A., 103, *138*
de Bruijn, H.E., 149, 156, *223*
Decher, G., 163, 166, *223*
Deck, R.T., 142, *220*
de Fornel, F., 142, 143, *220*
de Levie, R., 79, *86*
DeMartini, F., 142, *219*
de Mul, F.F.M., 128, *139*
Dentan, V., 147, 166, 199, *221*
de Oliveira, C.A.N., 63, *85*
Desmond, M., 276, *293*
Desnoyers, J.E., 106, *138*
deToledo, G.A., 281, *293*
Deutscher, R.L., 30, *85*

Devynck, J., 287, *294*
Dickens, P.G., 26, *84*
DiGiacomo, P.M., 174, *224*
Dilberto, E.J., Jr., 174, 237, 267, 271, 272, 276, 284, *289*, *292*
Dines, M.B., 174, *224*
Doblhofer, K., 8, 83, 142, 146, *220*
Domb, A., 142, 147, *221*
Dopaminevey, J., 248, *291*
Dostal, A., 11, 26, 41, 55, 56, 57, 58, 59, 60, 61, 62, 63, 68, *84*, *85*
Downard, A.J., 64, 65, 66, *85*
Dowrey, A.E., 142, *218*
Drijfhout, J.W., 147, *222*
Driscoll, J., 115, *138*
D'Silva, C., 163, 166, *223*
Dueber, R.E., 26, *84*
Duevel, R.V., 142, *218*
Dumont, M., 147, 166, 199, *221*, *222*
Duschl, C., 142, 143, 147, *219*, *222*, *223*
Dush, P., 281, *293*
Duty, R.C., 8, *83*

Earls, J.D., 147, 199, *222*
Ebersole, R.C., 163, 166, *223*
Edgar, J.A., 147, *222*
Edwards, J.C., 142, 147, 163, *220–21*
Egger, M., 148, 179, *223*
Eguren, M., 7, *82*
Eisenman, G., 91, 93, *137*
Emerson, A.B., 171, 198, 199, *224*
Engstrom, R.C., 260, 271, *292*

Author Index

Ensman, R.E., 237, *289*
Erecinsky, M., 259, *291*
Ernst, S., 142, 146, *220*
Ertl, G., 142, 146, *220*
Evall, J., 166, *224*
Evans, D.H., 229, *288*
Evans, S.D., 213, *225*
Ewing, A.G., 174, 229, 237, 238, 239, 240, 241, 242, 243, 244, 247, 248, 249, 252, 253, 254, 255, 256, 257, 258, 259, 262, 263, 275, 276, 277, 278, 279, 280, 281, 282, 283, 284, 285, 287, *288*, *289*, *290*, *291*, *293*, *294*

Falat, L., 243, *290*
Faradji, H., 237, 242, 243, *289*, *290*
Faulkner, L.R., 9, 83, 196, 211, 213, *224*
Fawcett, W.R., 196, 213, *224*
Fedurco, M., 196, 213, *224*
Feess, H., 77, *86*
Fein, A., 259, *292*
Feng, J.X., 243, 247, *290*, *291*
Fernández, J.M., 174, 281, *293*
Fernández-Chacón, R., 281, *293*
Fernando, D.R., 12, 28, 29, 74, *84*
Ferrige, A.G., 286, *294*
Fesce, R., 230, 267, *288*
Fiddler, S.L., 174, *224*
Finklea, H.O., 196, 213, *224*
Finlan, M.F., 147, *222*
Finnegan, J.M., 174, 271, 275, 281, 287, *292*, *293*, *294*
Fischer, B., 148, 179, *223*

Fischer, T., 142, *219*
Fish, J.R., 286, *294*
Fitch, A., 229, *288*
Flatgen, G., 142, 146, *220*
Fleischfresser, C., 15, 35, 37, *85*
Fletcher, S., 10, 11, 30, 31, 32, 33, 74, 77, *83*, *85*
Florin, E.L., 147, *223*
Foley, M.B., 228, *288*
Folkers, J.P., 168, *224*
Fombarlet, C.M., 228, 237, 242, 243, *287*
Forouzan, F., 89, 108, 109, 111, 112, 118, 121, 128, 135, *137*, *138*
Forster, R.J., 229, *288*
Fort, A., 203, *225*
Fortune, J.M., 12, *84*
Fox, K., 244, *290*
Franklin, T.C., 8, *83*
Freedman, L.P., 147, *222*
French, W.G., 5, *82*
Frey, B.L., 142, 145, 146, 147, 149, 150, 152, 153, 154, 155, 156, 157, 159, 160, 161, 162, 163, 164, 165, 167, 178, 182, 185, 190, *218*, *222*, *225*
Frutos, T.G., 192, 193, *224*
Fujita, K., 163, *223*
Furlong, D.N., 147, *222*
Furtak, T.E., 142, *219*
Futamata, M., 142, *219*

Gaillochet, M.P., 6, *82*
Galus, Z., 8, *83*
Gandolfi, G., 44, *85*
Gao, H., 163, *223*
Gao, X., 196, *225*

Garguilo, M.G., 249, 250, *291*
Garris, P.A., 247, 266, 286, *291, 292*
Gaub, H.E., 147, 148, 179, *223*
Gavin, P.F., 282, 283, 284, 285, *293*
Geddes, N.J., 147, *222*
Geisler, T., 147, 199, *221*
Georgiadis, R., 142, 146, 147, 149, 156, *220, 221*
Georgolios, N., 238, *289*
Gerhardt, G.A., 228, 236, 237, 240, 244, 245, *287, 288, 290*
Ghasemzadeh, B., 239, 241, 243, 244, 245, 247, 261, *290*
Giergiel, J., 142, *219*
Giggenbach, W., 128, *138*
Gillen, G., 182, *224*
Girlando, A., 142, *219*
Girling, I.R., 146, 147, 156, 199, *221, 222*
Giros, B., 264, 265, *292*
Glasbey, T.O., 142, 147, 163, *220–21*
Golden, T., 247, *291*
Goldfrank, L.R., 63, *85*
Gomperts, B.D., 281, *293*
Gonon, F., 228, 236, 237, 238, 239, 242, 243, 263, *287, 288, 289, 290, 292*
Gordon, G.I., 142, 146, *220*
Gordon, J.G., 142, 143, 144, 146, *220, 221*
Goss, C.A., 161, *223*
Gottesfeld, S., 248, *291*
Goudonnet, J.P., 142, 143, *220*
Granzow, R., 148, *223*
Gratzel, M., 147, *223*

Gray, J.M., 142, 147, 196, 199, 201, 205, *220*
Green, M.J., 142, *218*
Green, N.M., 163, 166, *223*
Greene, L.A., 276, 279, *293*
Greve, J., 128, 139, 149, 156, *223*
Grohovaz, F., 230, 267, *288*
Gruner, W., 6, *82*
Grunfield, S., 286, *294*
Grygar, T., 12, 17, 18, 19, 20, *84*
Gryznkiewicz, G., 273, *292*
Guder, H.-J., 142, 147, 163, 166, 167, *220*
Guibault, G.G., 251, *291*
Gupta, Y.M., 128, *138*
Gutman, D., 284, *294*
Guyot-Sionnest, P., 142, *218*

Hagan, C.R.S., 12, *84*
Hahn, Z., 243, *290*
Haimann, C., 230, 267, *288*
Hale, P.D., 260, *292*
Hall, E.A.H., 146, *221*
Hallock, R.B., 147, *222*
Ham, W.K., 174, *224*
Hammer, W.J., 95, *137*
Hanken, D.G., 142, 147, 169, 170, 171, 172, 173, 175, 178, 179, 196, 197, 199, 201, 202, 203, 204, 205, 207, 208, 209, 210, 211, 212, 215, 216, *218, 220, 222, 225*
Hanton, L.R., 64, 65, 66, *85*
Harootunian, A., 128, *138, 139*
Harrison, D.J., 248, *291*
Hart, J.B., 232, 237, *288*
Hartstein, A., 142, 143, 144, 146, *219*

Author Index

Harush, S., 128, *138*
Hass, D., 19, *84*
Hatta, A., 142, 146, 149, *219*, *221*
Haussling, L., 147, 163, *222*
Hawkins, R.D., 286, *294*
Haymes, A.G., 142, 147, 163, *220–21*
Heath, G.A., 64, 65, 66, *85*
Heben, M.J., 238, *289*
Heider, G., 249, *291*
Heinze, J., 53, *85*
Heitmann, D., 142, *219*
Heller, J., 142, 147, *221*
Hemminger, J.C., 142, 182, *219*, *224*
Hennig, H., 6, *82*
Henrion, G., 3, 9, 10, 11, 12, 14, 16, 19, 34, 35, 36, 45, 46, *82*, *84*, *85*
Hermes, M., 44, 58, 59, 60, 61, 63, *85*
Herne, T.M., 142, 146, 147, 156, *220*
Herrmann, R., 19, *84*
Heuvel, D.J.v.-d., 147, *222*
Heyn, S.P., 148, 179, *223*
Heyrovsky, J., 8, *83*
Heyrovsky, M., 8, *83*
Hibbs, J.B., 286, *294*
Hickel, W., 148, *223*
Higgins, D.A., 142, 203, 214, *218*, *225*
Hillarp, N.-A., 275, *293*
Hoffer, B.J., 236, *288*
Hoffman, R.S., 63, *85*
Höggendal, J., 276, *293*
Hökfelt, T., 276, *293*
Holten, D., 196, *225*
Holz, R.W., 271, 272, 273, *292*

Holze, R., 11, *84*
Hong, H.G., 169
Hong, H.-G., 171, 174, 198, *224*
Hoshi, T., 163, *223*
Hove-Madsen, L., 234, *288*
Howell, J.O., 237, *289*
Hsieh, S., *293*
Hu, Y., 287, *294*
Hu, Z.-Y., 203, *225*
Huang, J., 182, *224*
Huang, K., 249, *291*
Huang, L., 275, 281, 287, *293*, *294*
Huang, Z., 251, *291*
Hubbard, D., 100, *138*
Hunt, J.H., 142, *218*
Hurlburt, W.P., 230, 267, *288*
Huynh, N., 249, 250, *291*

Iezzi, N., 230, 267, *288*
Ikariyama, Y., 248, *291*
Imanishi, Y., 163, *223*
Inagaki, T., 128, *138*
Inzelt, G., 8, *83*
Irvine, R.F., *293*
Isaacson, M., 128, *138*, *139*

Jaber, M., 264, 265, *292*
Jackson, D.E., 142, 147, 163, *220–21*
Jahn, R., 282, 285, *293*
Jankowski, J.A., 174, 229, 237, 244, 267, 269, 271, 272, 273, 276, 284, *288*, *289*, *292*
Jannakoudaks, D., 238, *289*
Jaworski, A., 6, 9, 11, 34, 42, 77, *82*, *83*, *86*
Jayarama Reddy, S.J., 44, *85*

Jen, A.K.-Y., 203, *225*
Jirkovsky, J., 8, *83*
Johnson, D.C., 238, 254, *289, 291*
Johnston, K.S., 146, 156, *221*
Jones, J.R., 91, *137*
Jones, S.R., 264, 265, *292*
Jordan, C.E., 142, 147, 148, 149, 159, 161, 162, 163, 164, 165, 167, 178, 180, 182, 183, 184, 185, 187, 188, 189, 190, 191, 192, *218, 222, 223, 224*
Jorgenson, J.W., 269, *292, 293*
Jorgenson, R.C., 146, 156, *221*
Joseph, M.H., 228, 244, 251, 262, *287*
Jouvet, M., 228, 236, 237, 242, 243, 263, *287, 288, 289, 290, 292*
Jovin, T.M., 128, *139*
Jung, C.C., 146, 147, 156, 199, 217, *221, 222, 225*
Junkes, H., 142, 146, *220*
Jurich, M.C., 203, *225*
Justice, J.B., Jr., 236, 237, 242, *288*
Jutand, A., 228, *288*

Kaesche, H., 2, 4, *82*
Kahlert, H., 60, 63, *85*
Kajzar, F., 147, 199, *221*
Kalcher, K., 6, *82*
Kalnishevskaja, L.N., 6, *82*
Kalvoda, R., 7, *83*
Kamiyama, T., 147, 199, 217, *222*
Kamtekar, S., 163, *223*
Kandel, E.R., 286, *294*
Kang, C.-S., 147, 196, 198, 199, 203, *221*
Kano, H., 142, *219*
Kaplan, D.L., 163, *223*
Karabinas, P.J., 238, *289*
Karlsen, S.R., 146, 156, *221*
Kasser, R.J., 243, 247, *290, 291*
Katz, H.E., 171, 198, 199, 200, 211, *224, 225*
Kauffman, J.-M., 6, *82*
Kauschka, G., 26, 63, *85*
Kawagoe, K.T., 174, 229, 237, 239, 240, 244, 246, 247, 267, 271, 272, 276, 284, *288, 289, 290, 291, 292*
Kawata, S., 142, *219*
Keim, E., 142, *219*
Kelly, R.S., 240, 241, 242, *290*
Kemnitz, E., 12, 19, *84*
Kennedy, R.A., 281, *293*
Kennedy, R.T., 174, 237, 267, 271, 272, 275, 276, 281, 282, 284, 287, *289, 293, 294*
Kepley, L.J., 169, 171, 198, 211, *224, 225*
Kessler, M.A., 146, *221*
Kiechle, F., 286, *294*
Kim, Y.T., 240, 241, 242, *290*
Kimbrell, S., 196, 208, *225*
Kimura, S., 163, *223*
Kirgintsev, A.N., 103, *138*
Kirmaier, C., 196, *225*
Kissinger, P.T., 232, 237, *288*
Klemperer, W.G., 9, *83*
Klier, F.G., 276, *293*
Klingauf, J., 174, *293*
Klopfenstein, B.J., 174, *224*
Knobloch, H., 142, 143, 147, *219, 221*

Author Index

Knoll, W., 142, 143, 147, 148, 163, 166, 167, 179, 182, 196, 198, 199, 203, *219, 220, 221, 222, 223, 224*
Koeser, H.J.K., 133, *139*
Kohler, W., 174, *224*
Koile, R.C., 238, *289*
Kolb, D.B., 142, 143, *219*
Kolb, D.M., 142, 146, 206, *218, 220*
Kolthoff, I.M., 7, *83*
Komorsky-Lovric, A., 63, 75, *85, 86*
Komorsky-Lovric, S., 16, 34, 52, *84, 85*
Kooyman, R.P.H., 147, 149, 156, *222, 223*
Kopelman, R., 128, *138*
Kornguth, S., 142, 147, 149, 159, 161, 162, 163, 164, 165, 167, 178, 182, 185, 190, *218, 222*
Kotz, R., 142, 146, *220*
Kovach, P.M., 243, *290*
Kovacova, Z., 196, 213, *224*
Kozlowski, 142, *220*
Kratschmer, E., 128, *138*
Kratz, L., *138*
Kretschmann, E., 144, *221*
Krischer, K., 142, 146, *220*
Kristensen, E.W., 228, 237, 243, 244, 245, 247, 260, 265, 271, *287, 288, 289, 292*
Kristensen, H.K., 276, 284, *293, 294*
Kruchinin, A.A., 147, *222*
Kruk, Z.L., 228, 237, 244, 251, 260, 261, 262, 263, *287, 289, 292*

Kuhn, K., 147, 199, 217, *222*
Kuhr, W.G., 163, 223, 228, 237, 244, 245, 247, 248, 251, 260, 264, *287, 289, 291, 292*
Kulesza, P.J., 8, 9, *83*
Kumar, J., 163, *223*
Kunath, J., 6, *82*
Kuwana, T., 5, *82*
Kuzyk, M.G., 198, 203, *225*

LaCourse, W.R., 254, *291*
Laibinis, P.E., 168, *224*
Lalama, S.L., *225*
Lamache, M., 6, *82*
Lamy, C., 142, *218*
Lane, R.F., 237, *289*
Lang, H., 147, *223*
Lange, B., 2, 4, 9, 11, 34, 38, 39, 42, 45, 70, 79, 80, *82, 83, 84, 85*
Lantoine, F., 287, *294*
Lau, Y.Y., 238, 247, 248, 249, 252, 253, 254, 255, 257, 258, 259, 276, 284, 287, *289, 291, 293, 294*
Lawall, R., 147, 148, *222*
Lawrence, C.R., 146, 147, 156, *221*
Lawson, D., *292*
Lee, H., 169, 171, 198, *224*
Lee, J.H., 142, *219*
Lee, M.G., 142, *219*
Lee, V.Y., 203, *225*
Lehr, B., 163, 166, *223*
Leitner, A., 142, 143, *219*
Leszczyszyn, D.J., 174, 237, 267, 271, 272, 276, 284, *289, 292*

Leung, P.T., 147, *222*
Levenson, M.D., 142, *220*
Levi-Montalcini, R., *293*
Levy, Y., 147, 166, 199, *221*, *222*
Lewis, A., 128, *138*, *139*
Lewis, N.S., 238, *289*
Leyffer, W., 11, *84*
Licht, S., 89, 91, 92, 93, 94, 95, 96, 97, 98, 99, 100, 101, 102, 103, 108, 109, 111, 112, 114, 115, 117, 118, 121, 128, 135, *137*, *138*, *139*
Lieberman, K., 128, *138*
Liedberg, B., 142, 147, *220*
Light, T.S., 114, 115, *138*
Liley, M., 142, 147, 148, 163, 166, 167, *220*, *222*
Lin, M.S., 248, *291*
Lipkowski, J., 2, *82*
Lipsztein, J.L., 63, *85*
Liu, Q., 192, 193, *224*
Livanova, L., 15, 35, 37, *85*
Lockhart, D.J., 196, *225*
Löfås, S., 142, 147, *220*
Loiacono, D.N., 166, 199, *224*
Lomas, M., 142, 147, 163, *220–21*
Lombardi, J.R., 142, *218*
Longin, T.L., 238, *289*
Longmire, M.L., 8, *83*
Longo, K., 89, 111, 118, 121, *137*, *138*
Loudon, R., 142, *219*
Loulergue, J.C., 147, 199, *221*
Lovell, M.W., 91, *137*
Lovric, M., 11, 16, 60, 74, 76, 77, 78, 79, 80, *84*, *85*, *86*
Lowack, K., 163, 166, *223*

Lubrano, G.E., 251, *291*
Lukosz, W., 163, 166, *223*
Luk'yanov, A.V., 103, *138*
Lundström, I., 142, 147, *220*
Luo, G., 174, 237, 276, 277, 282, 283, 284, 285, *289*, *293*
Lvov, Y., 163, 166, *223*

Mabbott, G.A., 242, *290*
McCreery, R.L., 239, *290*
McDonald, R.S., 128, *138*
Mackay, R.A., 8, *83*
Madigan, N.A., 12, *84*
Mahon, P.J., 27, *84*
Maidment, N.T., 228, 237, 244, 251, 262, *287*, *289*
Majda, M., 161, *223*
Malan, G.P., 147, *222*
Malcovati, M., 163, 166, *223*
Malik, M.A., 9, *83*
Malinski, T., 286, *294*
Mallouk, T.E., 169, 171, 174, 198, 211, *224*, *225*
Malmqvist, M., 142, 147, *220*, *222*
Manassen, J., 102, *138*
Marassi, R., 9, *83*
Marcott, C., 142, *218*
Marder, S.R., 203, *225*
Marken, F., 10, 11, 12, 27, 28, 29, 30, 31, 32, 33, 58, 74, 77, *83*, *84*, *85*
Marsden, C.A., 228, 237, 244, 251, 262, *287*, *289*
Martin, A.S., 146, 147, 156, *221*, *222*
Martin, C.R., 228, 244, 245, *287*, *290*, *292*
Martin, K.F., 237, *289*

Marty, A., 267, *292*
Marx, K.A., 163, *223*
Mattlock, G., 95, *138*
May, L.C., 287, *294*
May, L.J., 237, 239, 243, 252, 260, 265, *289, 290, 292*
Mead, D.A., 254, *291*
Meggs, W.J., 63, *85*
Meissner, H.P., 281, *293*
Melo, D.R., 63, *85*
Mermet, C., 242, *290*
Meulemans, A., 238, 252, *289, 291*
Meyer, B., 2, 4, 6, 11, 21, 22, 23, 26, 39, 40, 41, 45, 47, 51, 52, 58, 70, 77, *82, 84, 85*
Michael, A.C., 239, 249, 250, 252, 287, *290, 291, 294*
Michaelis, E.K., 287, *294*
Micka, K., 7, *83*
Millar, J., 237, 244, 260, 261, 262, 263, *289, 290, 292*
Miller, C.J., 213, *225*
Miller, E., 234, *288*
Miller, E.C., 130, *139*
Miller, J.A., 163, 166, *223*
Miller, R.D., 203, *225*
Mills, D.L., 142, 143, 144, *219*
Minor, M., 149, 156, *223*
Minton, A.P., 147, *222*
Misler, S., 271, 285, *292, 294*
Mitchell, D.E., 142, *219*
Mitchell, K., 239, 241, 243, 244, 245, 247, 261, 287, *290, 294*
Möbius, C., 245, 247, *290*
Moghaddam, B., 228, 236, 237, 244, *287*
Moghaddopaminem, B., 247, *291*

Moh, G., 11, 40, 51, 52, *84*
Molinoff, P.B., 286, *294*
Moncada, S., 286, *294*
Monck, J.R., 174, *293*
Monk, P.M.S., 26, *84*
Montenegro, M.I., 228, *288*
Moore, R.B., III, 245, *290*
Moran, J.R., 163, 166, *223*
Morgan, H., 163, 166, *223*
Morichere, D., 147, 199, *221, 222*
Mortimer, R.J., 26, *84*
Mrksich, M., 142, 147, 173, *220*
Mujsce, A.M., 171, 198, 199, *224*
Müller, B.R., 8, *83*
Müller, W.-D., 14, 15, 16, 34, 35, 37, *84, 85*
Murphy, T.J., 12, *84*
Murray, R.W., 8, *83*, 239, *290*

Nagaosa, Y., 12, 28, 29, 74, *84*
Nagy, G., 228, 236, 237, 244, 245, *287, 290*
Naturforsch, Z., 144, *221*
Naujok, R.R., 142, 147, 196, 199, 201, 203, 205, 214, *220, 225*
Navane, F., 238, 239, 242, 243, *289*
Near, J.A., 174, 237, 267, 271, 272, 276, 284, *289, 292*
Neher, E., 174, 237, 267, 269, 271, 281, *289, 292, 293*
Nellen, P.M., 163, 166, *223*
Nelson, D., 259, *291*
Nemetz, A., 142, *219*
Neuberger, G.G., 254, *291*
Neuhold, C., 6, *82*
Neyman, E., 2, 4, *82*

Nicholson, C., 234, 237, 243, 244, 247, 260, 267, *288, 289, 290, 292*
Nilsson, P., 147, *222*
Nitschke, L., 3, 9, 10, 11, 12, 14, 15, 16, 19, 34, 35, 36, 37, 45, 46, *82, 84, 85*
Nnodimele, R., 8, *83*
Nordmann, J.J., 267, *292*
Nuzzo, R.G., 166, *224*
Nygren, P.A., 147, *222*

O'Dell, T.J., 286, *294*
Okahata, Y., 147, 148, *222*
Okamoto, T., 147, 199, 217, *222*
Oke, A.F., 245, *290*
Oke, A.I., 228, 236, 237, 244, *287*
Olefirowicz, T.M., 255, 256, *291*
Olesch, T., 12, 19, *84*
O'Malley, J.J., 248, *291*
O'Neill, R.D., 228, 244, 251, 262, *287*
Ong, T.H., 142, *218*
Orendi, H., 147, *221*
Osa, T., 163, *223*
Otto, A., 142, 144, *219, 221*
Otto, C., 128, *139*
Oznam, F., 128, 129, *138*

Page, C.J., 174, *224*
Paik, E.C., 239, *290*
Palik, E.D., 151, *223*
Palmer, R.M.J., 286, *294*
Pande, R., 163, *223*
Pantano, P., 163, *223*, 248, 251, *291*
Papoulis, A., 275, *293*
Papuchon, M., 147, 199, *221*
Paras, C.D., 282, *293*

Parikh, A.N., 168, *224*
Pastore, P., 266, 286, *292*
Pelzer, F., 9, 11, 34, 42, *83*
Pemberton, J., 142, *218*
Pendley, B.D., 238, *289*
Penner, R.M., 238, *289*
Penner, T.L., 128, *138*
Peramunage, D., 89, 109, 112, 115, 117, 118, 121, 128, 135, *137, 138*
Perry, J.W., 203, *225*
Persson, B., 147, 148, *222, 223*
Peterlinz, K.A., 142, 146, 147, 149, 156, *220, 221*
Peterson, I.R., 146, 147, 156, 199, *221, 222*
Pettinger, B., 142, 143, 146, *219, 220*
Pflüger, F., 228, *288*
Phelan, J., 198, 203, *225*
Phillips, J.H., 267, *292*
Phillips, J.M., 234, *288*
Philpott, M.R., 142, 143, 144, 146, *219, 220, 221*
Picknett, R.G., 95, 97, *138*
Pierce, R.J., 248, *291*
Pihel, K., 271, 275, 281, 287, *292, 293, 294*
Piscevic, D., 147, 148, 182, *222, 224*
Platford, R.F., 106, *138*
Pletcher, D., 228, *288*
Plieth, W., 142, 196, *218, 220, 224*
Plotsky, P.M., 238, 239, *289*
Pocholle, J.P., 147, 199, *221*
Pockrand, I., 142, 143, 144, 146, *220, 221*
Poenie, M., 273, *292*

Author Index

Pollardknight, D., 147, *222*
Ponchon, J.L., 236, 237, 242, *288, 289*
Pope, J.M., 196, 208, *225*
Popenoe, D.D., 142, *218*
Porter, M.D., 142, 159, 196, 213, *218, 223, 224*
Posokin, J.V., 6, *82*
PospiEil, L., 79, *86*
Pothos, E., 276, *293*
Poulain, B., 238, 252, *289, 291*
Powell, B., 282, 284, *294*
Prasad, P.N., 147, 199, *221*
Proctor, A., 249, 250, *291*
Pruss, N., 19, *84*
Pugliese, L., 163, 166, *223*
Pujol, J.F., 228, 236, 237, 242, 243, 263, *287, 288, 289, 290, 292*
Puppels, F.G., 128, *139*
Putvinski, T.M., 171, 198, 199, *224, 225*

Rabi, F., 14, 15, 16, 34, 35, 37, *84, 85*
Rachlin, E.M., 286, *294*
Radomski, M.W., 286, *294*
Raether, H., 142, 143, 144, *219, 221*
Raezke, K.-P., 45, *85*
Rahn, J.R., 147, *222*
Rako, J.G., 142, *219*
Raleigh, D.O., 2, *82*
Rayport, S., 276, *293*
Rebec, G.V., 229, 237, *288*
Reddy, A.K.N., 77, *86*
Reddy, S.J., 26, 41, 62, 63, *85*
Reed, C.E., 142, *219*
Reichert, W.M., 163, 169, *223*

Rein, G., *293*
Renner, K.J., 243, 247, *290, 291*
Revelli, J., 198, 203, *225*
Reynolds, E.R., 238, 248, 249, *289*
Ricco, A.J., 166, *224*
Rice, M.E., 234, 237, 243, 244, 245, 247, 260, *288, 289, 290, 292*
Rickert, H., 2, *82*
Ringsdorf, H., 147, 163, *222, 223*
Riou, F., 237, 242, *289*
Riou, R., 237, *289*
Rishpon, J., 248, *291*
Ristori, P., 142, *219*
Rivot, J.P., 243, *290*
Robello, D.R., 174, 198, 203, *224, 225*
Robert-Nicoud, M., 128, *139*
Roberts, C.J., 142, 147, 163, *220–21*
Robin, P., 147, 166, 199, *221*
Robinson, I.M., 174, *293*
Robinson, R.A., 106, *138*
Romagnoli, M., 142, *220*
Rönnberg, I., 142, 147, *220*
Roos, H., 148, *223*
Rose, G.M., 236, *288*
Rosenkilde, S., 147, 199, *221*
Roslonek, G., 9, *83*
Ross, P.N., 2, *82*
Rosseinsky, D.R., 26, *84*
Rothenhäusler, B., 142, 148, 179, *219, 223*
Rowe, G.K., 196, *225*
Rump, E., 163, *223*
Runser, C., 203, *225*
Rusling, J.F., 77, *86*
Rynders, G.F., 100, *138*

Saban, S.B., 217, *225*
Sackett, D.D., 171, 174, 211, *224, 225*
Sackmann, E., 147, *222*
Saez, E.I., 142, *218*
Salomon, L., 142, 143, *220*
Sambles, J.R., 146, 147, 156, *221, 222*
Samoc, M., 147, 199, *221*
Sánchez Batanero, P., 7, *82, 83*
Sanner, A.M.W., 192, 193, *224*
Santamato, E., 142, *219*
Santo, R., 142, 143, 144, 146, *219, 221*
Saraceno, R.A., 240, *290*
Sarid, D., 142, *220*
Sasaki, Y., 142, *219*
Sass, J.K., 142, 146, *220*
Sasso, S.V., 248, 249, *291*
Sawodny, M., 147, *221*
Scarnulis, D.M., 240, 241, 242, *290*
Scheller, G., 198, *225*
Schenk, J.O., 228, 234, 236, 244, 251, 262, *287, 288*
Scheurell, S., 12, 19, *84*
Schildkraut, J.S., 149, 156, 159, 198, 203, *223, 225*
Schilling, M.L., 171, 198, 199, 200, 211, *224, 225*
Schmidt, A., 147, *221*
Schmidt, P.H., 196, *224*
Schmitt, F.-J., 142, 147, 148, 163, 166, 167, 179, *220, 222, 223*
Schmitt, J., 163, 166, *223*
Scholz, F., 2, 3, 4, 6, 9, 10, 11, 12, 14, 15, 16, 19, 21, 22, 23, 26, 34, 35, 36, 37, 38, 39,

[Scholz, F.]
40, 41, 42, 44, 45, 46, 51, 52, 55, 56, 57, 58, 59, 60, 61, 62, 63, 68, 69, 70, 71, 73, 74, 76, 77, 78, 79, 80, *82, 83, 84, 85, 86*
Schöollhorn, R., 24, *84*
Schröder, U., 11, 26, 57, 58, 73, *84, 85*
Schroeder, T.J., 174, 237, 267, 271, 272, 273, 276, 284, *289, 292*
Schubert, D., 276, *293*
Schuck, P., 147, *222*
Schumacher, D., 142, *219*
Schwedt, G., 45, *85*
Scozzafava, M., 198, 203, *225*
Sefton, J., 142, 147, 163, *220–21*
Sekkat, Z., 147, 196, 198, 199, 203, *221*
Senyshyn, J., 272, 273, *292*
Shakesheff, K.M., 142, 147, 163, *220–21*
Sharp, T., 237, *289*
Shaw, S.J., 10, 11, 28, 30, 31, 32, 33, 74, 77, *83, 84*
Shen, H., 281, *293*
Shen, Y.R., 142, 196, 198, *218, 219, 225*
Shepherd, G.M., 229, 230, 231, *288*
Shibuki, K., 259, *291*
Shih, R.D., 63, *85*
Siegel, G.J., 286, *294*
Sigal, G.B., 142, 147, 173, *220, 222*
Silin, V.I., 147, *222*
Simon, H.J., 142, *219*
Sinclair, D.A., 106, *138*

Author Index

Singer, K.D., 198, 203, *225*
Sipe, J.E., 142, *220*
Siperko, L.M., *224*
Slepushkin, V.V., 2, 4, *82*
Small, R.D., *225*
Smirnova, T.P., 5, *82*
Smith, C.P., 196, 213, *224*
Smith, D.C., 130, *139*
Smith, E.L., *224*
Smith, L.M., 192, 193, *224*
Smolyaninov, I.I., 142, *221*
Sohn, J.E., 198, 203, *225*
Sohn, T.W., 249, *291*
Song, J.I., *292*
Songina, O.A., 7, *83*
Spinke, J., 142, 147, 163, 166, 167, *220*
Staehelin, M., 203, *225*
Stallcup, W.B., 279, *293*
Stamford, J.A., 228, 229, 237, 239, 243, 244, 251, 255, 260, 261, 262, 263, *287, 288, 289, 290, 292*
Stegeman, G.I., 142, *220*
Stelzle, M., 147, *222*
Stenberg, E., 142, 147, 148, *220, 223*
Stifatov, B.M., 4, *82*
Stößer, R., 52, *85*
Stock, J.T., 7, *83*
Stoecker, P.W., 249, *291*
Stojek, Z., 6, 77, 78, 79, *82, 86*
Stokes, R.H., 105, 106, *138*
Stole, S.M., 142, *218*
Stolz, L., 148, *223*
Stora, T., 147, *222*
Stranick, S.J., 168, *224*
Strein, T.G., 244, 255, 287, *290, 291, 294*

Strominger, J., 147, *222*
Stutts, K.J., 229, 239, 240, *288*
Suaud-Chagny, M.F., 243, *290*
Suetaka, W., 146, 149, *221*
Sulzer, D., 276, *293*
Sun, L., 166, *224*
Superfine, R., 142, *218*
Suzuki, S., 146, 149, *221*
Swalen, J.D., 142, 143, 144, 146, *219, 220, 221*
Symons, P.G., 10, 11, 30, 31, 32, 33, 74, 77, *83*
Szabo, A., 148, *223*
Szentirmay, M.N., 245, *290, 292*

Tadjeddine, A., 142, 143, 146, *219, 220*
Taha, Z., 286, *294*
Tainton, R.R., 286, *294*
Talapatra, G.B., 147, 199, *221*
Tan, W.T., 27, *84*
Tan, Z., 196, 208, *225*
Tao, Y.-T., 166, 168, *224*
Tarlov, M.J., 142, 146, 147, 156, 182, *220, 224*
Tascón, M.L., 7, *82*
Tascón Garcia, M.L., 7, *83*
Tauc, L., 238, 252, *289, 291*
Taylor, D.M., 163, 166, *223*
Tendler, S.J.B., 142, 147, 163, *220–21*
Terrettaz, S., 147, *222*
Texter, J., 8, *83*
Than, K.A., 147, *222*
Thiel, A.J., 192, 193, *224*
Thierfelder, C., 15, 35, 37, *85*
Tischler, A.S., 276, 279, *293*
Tomboulain, P., 286, *294*
Torri Tarelli, F., 230, 267, *288*

Trevin, S., 287, *294*
Tripathy, S.K., 163, *223*
Troughton, E.B., 166, *224*
Tsacopoulos, M., 259, *292*
Tsien, R.Y., 273, *292*
Turner, R.F.B., 248, *291*
Turyan, I., 79, *86*
Twomey, T., 142, *220*

Uhlen, M., 147, *222*
Ulman, A., 142, 213, *219*, *220*, *225*
Ulmer, R.W., 248, *291*
Ungerstadt, U., 237, *289*
Urankar, E.J., 198, 203, *225*
Urbaniczky, C., 148, *223*
Urquhart, R.S., 147, *222*
Ushioda, H., 248, *291*
Ushioda, S., 142, *219*

Valtorta, F., 230, 267, *288*
van Esch, J., 163, *223*
Van Steveninck, R.F.M., 12, 28, 29, 74, *84*
Vavrin, Z., 286, *294*
Vázques, M.D., 7, *82*
Vázques Barbado, M.D., 7, *83*
Verbiest, T., 203, *225*
Vidrevich, M.B., *82*
Vieth, M., 147, 148, *222*
Vigdergauz, V.E., 4, 8, *82*
Villa, A., 230, 267, *288*
Villarta-Snow, R., 251, *291*
Viveros, O.H., 174, 237, 267, 271, 272, 276, 284, *289*, *292*
Vlasov, Y.G., 147, *222*
Vogel, H., 147, *222*, *223*
Vohnsen, B., 142, *221*
Volksen, W., 203, *225*

von Rüden, L., 237, 267, 269, 271, *289*
Vytras, K., 6, *82*

Wagner, J.A., 276, *293*
Walla, R., 248, *291*
Wallingford, R.A., 252, 255, 257, *291*
Walsh, C.A., 203, *225*
Walter, J.N., 10, 12, 28, 29, 74, *83*, *84*
Wang, J., 6, *82*, 247, 248, *291*
Wappler, G., 70, *85*
Ward, M.D., 2, 27, *82*, *84*, 163, 166, *223*
Ward, R.N., 142, *218*
Watanabe, M., 8, *83*
Watson, J.G., 142, *219*
Way, D.M., 27, *84*
Weber, K., 196, *225*
Wegner, G., 147, 196, 198, 199, 203, *221*
Weiß, A., 45, *85*
Weiner, N., 276, *293*
Weisman, R.S., 63, *85*
Weiss, P.S., 168, *224*
Weissmuller, G., 147, *222*
Welling, G.W., 147, *222*
Wendt, H., 77, *86*
Westhead, E., 267, 272, *292*
White, H.S., 196, 213, *224*, *225*
Whiteley, L.D., 228, 244, 245, 287, *290*, *292*
Whitesides, G.M., 142, 147, 166, 168, 173, *220*, *222*, *224*
Widrig, C.A., 196, 213, *224*
Wieck, H.J., 249, *291*
Wiedemann, D.J., 237, 243, 265, *289*

Author Index

Wightman, R.M., 174, 228, 229, 230, 234, 237, 238, 239, 240, 241, 242, 243, 244, 245, 246, 247, 252, 260, 261, 262, 263, 264, 265, 266, 267, 269, 271, 272, 273, 275, 276, 281, 284, 286, 287, *287*, *288*, *289*, *290*, *291*, *292*, *293*, *294*
Wijekoon, W.M.K.P., 147, 199, *221*
Wilchek, M., 163, *223*
Wilde, W., 19, *84*
Wilkerson, J.E., 245, *290*
Wilkins, M.J., 142, 147, 163, *220-21*
Willand, C.S., 174, 198, 203, *224*, *225*
Williams, D.J., 174, *224*
Williams, P.M., 142, 147, 163, *220-21*
Williamson, A.T., 106, *138*
Willig, F., 77, *86*
Wilson, G.S., 287, *294*
Wilson, W.L., 198, *225*
Winkler, M., 267, 272, *292*
Wipf, D.O., 228, 241, *288*
Wittke, W., 142, *219*
Wojclk, J.F., 91, *137*
Wong, D.K.Y., 244, 247, 248, 252, 253, 255, *290*, *291*
Wong, P.K.Y., 238, 253, 254, 255, *289*
Wooster, T.T., 8, *83*
Wrighton, M.S., 128, *139*

Xu, L.F.Y., 244, *290*

Ya, 79, *86*
Yacynych, A.M., 238, 248, 249, *289*, *291*
Yakovlev, V.A., 147, *222*
Yamaguchi, I., 147, 199, 217, *222*
Yamauchi, S., 248, *291*
Yang, H.C., 171, 174, *224*
Yang, Z., 6, *82*
Yau, S.-L., 171, 174, *224*
Yeatman, E.M., 147, 199, 217, *222*
Yee, S.S., 146, 147, 156, 199, 217, *221*, *222*, *225*
Yoo, C.S., 128, *138*
Yukiashi, T., 248, *291*

Zakharchuk, N.F., 5, 6, *82*
Zammit, A.C.A., 142, *219*
Zamponi, S., 9, *83*
Zawodzinski, T.A., Jr., 248, *291*
Zayats, A.V., 142, *221*
Zeppenfeld, A.C., 174, *224*
Zerby, S.E., 275, 276, 278, 279, 280, 281, *293*
Zetterstrom, T., 237, *289*
Zhang, H., 8, *83*
Zhang, S., 11, 39, 40, 41, 51, 52, *84*, *85*
Zhao, S., 163, 169, *223*
Zhizhin, G.N., 147, *222*
Zhou, R., 174, *293*
Zhou, Z., 271, 285, *292*, *294*
Ziemer, B., 21, 22, 23, 77, *84*
Zimmerman, J.B., 237, 239, 240, 246, *289*, *290*
Zimmerman, R.M., 163, 167, *223*
Zysset, B., 203, *225*

SUBJECT INDEX

Acetylcholine, 231, 233
Acids:
   calculation of pH in
      concentrated acids, 91–93
   colorimetric estimates of pH in
      concentrated acids, 91
Air-gap capacitors, modulated
   SPR measurement on,
   199–203
Alkaline solutions, potentiometric
   measurement of pH in,
   93–101
γ-Aminobutyric acid (GABA),
   233
Analysis in highly concentrated
   solutions, 87–139
   conductometric analysis,
      113–121
      conventional conductometric
         titration, 113–115
      conventional conductometric
         titration of concentrated
         solutions, 115–117
      differential conductometric
         methodology, 118,
         119–121
   differential densometric
      analysis, 121–127
      benefits and disadvantages,
         127
      differential densometric
         analysis, 121–122
      example, 123–127
      methodology, 123

[Analysis in highly concentrated
   solutions]
   evanescent (solvent) activity
      analysis, 103–113
      determination of mean
         solution activity from
         solvent activity, 104–105
      evanescent methodology,
         106–113
      solvent activity, 103–104
      traditional analysis of
         activity, 105–106
   potentiometric analysis, 89–103
      calculation of pH in
         concentrated acids and
         bases, 91–93
      colorimetric estimates of pH
         in concentrated acids and
         bases, 91
      conventional pH
         measurement, 89–91
      potentiometric measurement
         of pH in concentrated
         alkaline solutions,
         93–101
      tools for concentrated
         solution potentiometric
         analysis, 101–103
   submicrometer path
      UVA/VIS/IR
         spectroscopy, 127–136
      conventional absorption
         spectroscopy limits,
         127–128

[Analysis in highly concentrated solutions]
  submicrometer path length absorption spectroscopy, 128–131
  submicrometer path length cells, 129
  submicrometer path length examples, 131–136

Basal plane pyrolytic graphite electrode, 10–11
Bases:
  calculation of pH in concentrated bases, 91–93
  colormetric estimates of pH in concentrated bases, 91
BIACORE SPR adsorption instrument, 147–148, 217
Biopolymer/protein adsorption, SPR thickness measurement of, 162–169
Biopolymer adsorption, SPR imaging of, 181–195
  ex situ SPR imaging, 181–186
  in situ SPR imaging, 186–195
Bovine adrenal chromaffin cells, 266–275
Brain studies, extracellular voltammetry in, 261–266

Carbon fiber vs. platinum microelectrodes, 238–239
Carbon-paste electrodes:
  with an electrolyte solution serving as binder, 6–7
  with an organic binder, 5–6

Colorimetric estimates of pH in concentrated acids and bases, 91
Compact electrodes, 4
Conductometric analysis in concentrated solutions, 113–121
  conventional conductometric titration, 113–115
  conventional conductometric titration of concentrated solutions, 115–117
  differential conductometric methodology, 118, 119–121

Differential densometric analysis in concentrated solutions, 121–127
  benefits and disadvantages, 127
  differential densometric analysis, 121–122
  example, 123–127
  methodology, 123
Direct path ultraviolet (UV) spectroscopy, 128
DNA, immobilized, multiple oligonucleotide hybridization onto, 192–195
Dopamine, 233
  monitoring intracellular dopamine, 251–257
  oxidation reaction for, 235

Electrochemical corrosion stability of metal alloys, 34–37

Subject Index

Electrochemically modulated surface plasmon resonance (EM-SPR), 143
EM–SPR measurements on multilayer films at electrode surfaces, 203–216
　electric field measurements, 203–210
　electric field profile measurements inside multilayers, 210–213
　interference effects within mixed HAPA/PY-AZO multilayers, 213–216
　theory, 196–198
Electrochemical reactions of immobilized solid particles, 14–34
　electrochemistry of molecular solids, 27–34
　oxidative dissolution of metals and alloys, 14–17
　reductive conversion of oxides and solids, 19–22
　reductive dissolution of oxides and oxide hydrates, 17–18
　solid compounds that possess redox sites, 22–27
Enzyme-modified microelectrodes, 248–251
Epinephrine, 233
　oxidation reaction for, 235
Evanescent (solvent) activity analysis in concentrated solids, 103–113
　determination of mean solution activity from solvent activity, 104–105
　evanescent methodology, 106–113
　solvent activity, 103–104
　traditional analysis of activity, 105–106
Exocytosis, 230–231
　single-cell system used in electrochemical studies of, 266
Extracellular voltammetry in the neuronal function, 259–286
　bovine adrenal chromaffin cells, 266–275
　brain studies, 262–266
　invertebrate and mammalian neurons, 282–286
　mast cells, pancreatic ß cells, and rat melanotrophs, 280–282
　rat pheochromocytoma (PC12) cells, 276–282
　single-cell systems used in electrochemical studies of exocytosis, 266

Glassy carbon electrodes, 11
Glucose, monitoring glucose in single cells, 257–258
Glutamate, 233
Glycine, 233
Gold surfaces, monolayer thickness on, SPR measurements of, 148–179
　examples of SPR thickness measurements, 162–179

[Gold surfaces]
  SPR experimental apparatus, 160–162
  SPR metal and dielectric film thickness parameters, 150–158

HAPA, 196, 198
  EM–SPR interference effects with mixed HAPA/PY-AZO multilayers, 213–216
Highly concentrated solutions, analysis in, 87–139
  conductometric analysis, 113–121
    conventional conductometric titration, 113–115
    conventional conductometric titration of concentrated solutions, 115–117
    differential conductometric methodology, 118, 119–121
  differential densometric analysis, 121–127
    benefits and disadvantages, 127
    differential densometric analysis, 121–122
    example, 123–127
    methodology, 123
  evanescent (solvent) activity analysis, 103–113
    determination of mean solution activity from solvent activity, 104–105
    evanescent methodology, 106–113

[Highly concentrated solutions]
    solvent activity, 103–104
    traditional analysis of activity, 105–106
  potentiometric analysis, 89–103
    calculation of pH in concentrated acids and bases, 91–93
    colorimetric estimates of pH in concentrated acids and bases, 91
    conventional pH measurement, 89–91
    potentiometric measurement of pH in concentrated alkaline solutions, 93–101
    tools for concentrated solution potentiometric analysis, 101–103
  submicrometer path UVA/VIS/IR spectroscopy, 127–136
    conventional absorption spectroscopy limits, 127–128
    submicrometer path length absorption spectroscopy, 128–131
    submicrometer path length cells, 129
    submicrometer path length examples, 131–136
Hydrogen and hydroxide ions in water, equivalent conductivity of, 114, 115
5-Hydroxytryptamine. *See* Serotonin

Subject Index

Imaging experiments using SPR, 179–195
  ex situ experiments of biopolymer adsorption, 181–186
  in situ experiments of biopolymer adsorption, 186–195
Immobilization of solid particles on electrode surfaces, 10–14
  cleaning of electrode surface after measurement, 13–14
  electrodes, 10–11
  measurement, 12–13
  transfer and immobilization of solid particles on electrode surface, 11–12
Immobilized solid particles, mechanisms of elctrochemical reactions of, 14–34
  electrochemistry of molecular solids, 27–34
  oxidative dissolution of metals and alloys, 14–17
  reductive conversion of oxides and solids, 19–22
  reductive dissolution of oxides and oxide hydrates, 17–18
  solid compounds that possess redox sites, 22–27
Infrared (IF) spectroscopy, 128
Intracellular voltammetry in the neuronal function, 252–259
  monitoring glucose in single cells, 257–258

[Intracellular voltammetry]
  monitoring intracellular and extracellular oxygen, 259
  monitoring intracellular dopamine, 252–257
Invertebrate neurons, 282–286

Mammalian neurons, 282–286
Mast cells, 280–282
Melanotrophs, 280–282
Merceptoundecanoic acid (MUA)/PL surfaces, avidin adsorption onto, 186–190
Metal alloys, quantitative analysis of, 34–37
Metal compounds, quantitative analysis of, 37–42
Metal electrodes, 11
Met-enkephalin, 233
Minerals, voltammograms of microparticles of, 47–50
Monolayer thickness on gold surfaces, SPR measurements of See Gold surfaces, monolayer thickness on, SPR measurements of Multiple oligonucleotide hybridization onto immobilized DNA, 192–195

Neuronal microenvironments, 227–294
  electrochemistry (methods and electrodes), 234–251
  carbon fiber vs. platinum microelectrodes, 238–239

[Neuronal microenvironments]
  potentiometric vs.
    voltammetric
    measurements, 234–238
  ultrasmall carbon electrodes,
    239–251
  extracellular voltammetry,
    259–286
    bovine adrenal chromaffin
      cells, 266–275
    brain studies, 262–266
    invertebrate and mammalian
      neurons, 282–286
    mast cells, pancreatic
      β cells, and rat
      melanotrophs, 280–282
    rat pheochromocytoma
      (PC12) cells, 276–280
    single-cell systems used in
      electrochemical studies of
      exocytosis, 266
  intracellular voltammetry,
    252–259
    monitoring glucose in single
      cells, 257–258
    monitoring intracellular and
      extracellular oxygen, 259
    monitoring intracellular
      dopamine, 252–257
Neurotransmitter systems,
    231–234, 286–287
Nitric oxide (NO) as a
    neutrotransmitter,
    286–287
Norepinephrine, 233
  oxidation reaction for, 235

Organic residues on solids,
    quantitative analysis of, 44

Oxygen, monitoring intracellular
  and extracellular oxygen,
  259

Pancreatic β cells, 280–282
Paraffin-impregnated graphite
  electrodes (PIGEs), 10
Pencil lead electrodes, 11
Permselective polymer-modified
  microelectrodes, 244–248
Phase analysis, 44–52
Pheochromocytoma (PC12) cells,
  276–280
pH measurement
  in concentrated acids and
    bases, 91–93
  in concentrated alkaline
    solution, 93–101
  conventional measurement,
    89–91
Platinum microelectrodes vs.
  carbon fiber, 238–239
Powder mixtures, quantitative
  analysis of, 42–43
Polarization modulation Fourier
  transform infrared
  reflection absorption
  spectroscopy (PM-
  FTIRRAS), 142
Polymer film on electrodes,
  voltammetry of solid
  particles using, 8
Poly-*t*-lysine adsorption,
  molecular weight
  dependence of,
  190–192
Potentiometric analysis in
  concentrated solutions,
  89–103

Subject Index

[Potentiometric analysis]
  calculation of pH in concentrated acids and bases, 91–93
  colorimetric estimates of pH in concentrated acids and bases, 91
  conventional pH measurement, 89–91
  potentiometric measurement of pH in concentrated alkaline solutions, 93–101
  tools for concentrated solution potentiometric analysis, 101–103
PY AZO, 196, 198–199
  EM–SPR interference effects within mixed HAPA/PY AZO multilayers, 213–216

Quantitative analysis of composition of solids, 34–44
  metal alloys and assessment of their electrochemical corrosion stability, 34–37
  metal compounds, 37–42
  organic residues on surfaces, 44
  powder mixtures, 42–43

Scanning electron microscopy equipped with an energy dispersive x-ray detector (SEM/EDX), 72–74
Serotonin (5–hydroxytryptamine), 232, 233
  oxidation reaction for, 235

Solid compounds, electrochemical studies of, 4–9
Solid state electrochemistry (SSE), 2–4
Submicrometer path UV/VIS/IR spectroscopy, 127–136
  conventional absorption spectroscopy limits, 127–128
  submicrometer path length absorption spectroscopy, 128–131
  submicrometer path length cells, 129
  submicrometer path length examples, 131–136
Surface plasmon polarizations (SPPs), 142–143
Surface plasmon resonance (SPR) measurements, 141–225
  background, 143–148
  future directions, 217
  of monolayer thickness on gold surfaces, 148–179
    examples of SPR thickness measurements, 162–179
    SPR experimental apparatus, 160–162
    SPR metal and dieletric film thickness parameters, 150–158
  SPR electric field measurements, 196–216
    EM-SPR theory, 196–198
    measurements on multilayer films at electrode surfaces, 203–216

[Surface plasmon resonance]
  modulated SPR
    measurements on air-gap
    capacitors, 199–203
  noncentrosymmetric
    zirconium phosphonate
    multilayer films, 198–199
  SPR imaging experiments,
    179–195
    ex situ experiments of
      biopolymer adsorption,
      181–186
    in situ experiments of
      biopolymer adsorption,
      186–195
Suspensions of solid particles,
  7–8

Theoretical description of
  voltammetry of
  microparticles,
  74–80
Thermodynamic data of solid
  compounds, 64–72

Ultrasmall carbon electrodes,
  239–251
  electrochemical treatment and
    preparation,
    242–246
  enzyme-modified
    microelectrodes,
    248–251
  fabrication, 240–241
  microelectrode geometrics,
    241–242
  permselective polymer-
    modified microelectrodes,
    244–248

Visible (VIS) spectroscopy, 128
Voltammetry of microparticles
  (VMP), 1–86
  methods used for
    electrochemical studies of
    solid materials, 4–9
    carbon paste electrodes with
      electrolyte solution
      serving as a binder, 6–7
    carbon paste electrodes with
      organic binder, 5–6
    compact electrodes made of
      solid material, 4
    suspensions of solid
      particles, 7–8
    voltammetry of solid
      compounds sandwiched
      between solid electrodes,
      8–9
    voltammetry of solid particles
      immobilized on electrodes
      in polymer film, 8
  of solid compounds
    immobilized on electrode
    surface, 9–80
    determination of
      thermodynamic data of
      solid compounds, 64–72
    immobilization of solid
      particles on electrode
      surface, 10–14
    in situ and ex situ
      combinations of
      voltammetry of solid
      microparticles, 72–74
    mechanisms of
      electrochemical reactions
      of immobilized solid
      particles, 14–34

Subject Index

[Voltammetry of microparticles]
   phase analysis, 44–52
   problems of theoretical description of voltammetry of microparticles, 74–80
   quantitative analysis of composition of solids, 34–44
   structure analysis, 52–64

Water, equivalent conductivity of hydrogen and hydroxide ions in, 114, 115

Zirconium phosphonate (ZP):
   on noncentrosymmetric multilayer films, 198–199
   SPR measurements of, 169–176